Wildflowers of the Eastern United States

WORMSLOE FOUNDATION PUBLICATIONS

NUMBER TWENTY

Wildflowers

of the
Eastern United States

by Wilbur H. Duncan
and Marion B. Duncan

The University of Georgia Press
Athens and London

© 1999 by the University of Georgia Press
Athens, Georgia 30602
All rights reserved
Set in Minion with Syntax by G&S Typesetters, Inc.
Printed and bound by C and C Offset Printing Co., Ltd.
The paper in this book meets the guidelines for permanence
and durability of the Committee on Production Guidelines
for Book Longevity of the Council on Library Resources.

Printed in Hong Kong
03   02   01   00   99   C   5   4   3   2   1

Library of Congress Cataloging in Publication Data

Duncan, Wilbur Howard, 1910–
Wildflowers of the eastern United States / by Wilbur H.
Duncan and Marion B. Duncan.
         p.   cm.
Includes index.
ISBN 0-8203-2107-9 (alk. paper)
1. Wild flowers—East (U.S.)—Identification.   2. Wild
flowers—East (U.S.)—Pictorial works.   I. Duncan,
Marion B.   II. Title.
QK115.D86   1999
582.13′0974—ddc21   98-43314

British Library Cataloging in Publication Data available

Wildflowers, as we knew them when we were children, bordered our paths through woods to school, floated in slow-moving streams, and filled our lives with pleasure. Our children enjoyed a somewhat similar background once we moved away from town. Now their children go to "special places" to daydream and lose themselves in the beauty of the outdoors.

We dedicate this book of wildflowers to our grandchildren:

Laramie, Amber, Laura, and Ross

knowing they will channel their energies toward preserving "special places" for coming generations.

Contents

Preface   *xi*

Acknowledgments   *xiii*

General Introduction   *xv*

Glossary   *xxix*

List of Abbreviations   *xxxv*

Species Descriptions   *1*

Color Plates   *181*

Index   *349*

Photo Credits   *379*

Preface

There were many reasons compelling us to write about wildflowers. Simple enjoyment would be reason enough. We were groping, though, for something more,
an intangible element that is felt, never fully defined. We wanted to promote
preservation of the natural vegetation but we wanted it motivated by appreciation.
Perhaps families could revive an old custom of walks in the woods or parks, on
mountain trails, or behind the beach and the dunes. A new generation of
preservationists-in-the-making might evolve. Let us hope.

Acknowledgments

There is no way we can describe adequately the support, encouragement, and confidence we have received from faculty, staff, and graduate students of the University of Georgia Botany Department. Dr. Melvin Fuller, department head at the time, initially provided research time and office support prior to Wilbur's retirement, and Dr. Fuller's successors have continued support in various ways. Dr. Michael Moore and Dr. Gary Kochert personally, with patience and good humor, at all hours and in all weather, supervised our passage into the computer era, which we accomplished with a minimum of hysteria because of their forbearance.

We also wish to recognize with gratitude members of the Georgia Botanical Society for their immediate and enthusiastic response to our request for photographs we lacked in our personal collection. Mr. Richard Ware was the leading force in pursuing our request, which resulted in excellent contributions from other Society members. Their names and all others contributing photographs are listed in the appendix together with the photograph(s) each contributed.

Mr. Bill E. Duyck, Asheville, NC, formerly with the US Forest Service, hearing of our need for photographs, became an immediate champion of the cause and not only supplied several sets of quality slides but checked regularly on our progress and other possible needs.

In the course of many decades of fieldwork throughout the eastern United States and elsewhere we encountered a host of other wonderful people who shared with us their knowledge and time. From them we have collected and stowed away a treasure of warm memories.

In some areas where we were less familiar with the flora or lacked access to specific localities, botanists, some known to us and many others not, took us to certain promising locations and directed us to others. We want them to know that their assistance meant more and deserves more than this mere mention of their names. Included are Mr. Marvin and Ms. Nancy Baker, The Panther Lick, Sugar Run, Pa; Dr. Lionel Eleutereus, Ocean Springs, Miss; Ms. Cheryl McCaffrey, then of The Nature Conservancy Barrier Island Preserve, Va; Dr. Max Medley, then at Univ. of Louisville, Louisville, Ky; Dr. Richard Mitchell, NY State Museum, Albany, NY; Dr. Douglas W. Ogle, Virginia Highlands Community College, Abingdon, Va; Dr. Peter Scott, Memorial Univ., St. Johns, Nfld, Canada; Dr. John Thomson, Univ. of Wisconsin, Madison; Dr. Sam Vander Kloet, Acadia Univ. Wolfville, NS, Canada. We only wish we could introduce all of them to each of you so that you, too, could be touched by their kindness and generosity.

To Dr. John Kartesz, Director of the Biota of North America Program of the North Carolina Botanical Garden, who volunteered to check all the botanical nomenclature for accuracy and current status, we are not only grateful but indebted. His interest in this book and his personal concern for us underscore our deep friendship for him. Any errors in the scientific names that may have slipped through are ours and not his.

General Introduction

This book is about wildflowers and specifically those of the eastern United States. It is written primarily to provide enjoyment for the user and in so doing provide a means for positive identification of each species included. Any concomitant values it may inspire or worthwhile causes it may stimulate are happily acknowledged and, indeed, hoped for. There is need for a more widespread concern for and understanding of this major component of our landscape for which we all share responsibility. This responsibility should be buttressed with knowledge, and acquiring it while pursuing an enjoyable pastime provides a painless beginning.

By wildflowers we mean herbaceous nonwoody plants. This includes forbs, grasses, rushes, and sedges, although subshrubs sometimes are mistaken for wildflowers. By eastern United States we mean that area east of the Mississippi River, but we have excluded the tropical and subtropical areas of peninsular Florida. Plants obviously are not constrained by prescribed boundaries, and many of them in this book extend west to the prairie or north into Canada. These extensions are included within species ranges.

Our treatment of the species included has been limited, for the most part, to a minimum of characteristics needed for accurate identification. Photographs of species augment the written descriptions. Priority in taking the photographs was given to exposing as many diagnostic features as possible, and in some instances beauty or artistic composition was sacrificed.

The naming of plants follows a binomial system by which all species are assigned a double name in Latin. The first word is the genus and represents an association of similar plants with closely related characteristics. The second word is the specific epithet and applies to all plants having the same unique identifying characteristics. This procedure is understood and accepted internationally.

The scientific name of a species is followed by the name, sometimes abbreviated, of the botanist(s) who first described that species. A person's name is sometimes placed in parentheses and followed by a second name, or even two, which indicates that the second person or persons changed the scientific name from a previous combination to the one shown. The names are sometimes necessary in finding the proper species in manuals or other publications.

Occasionally a species is given one scientific name by one botanist and a different one by another. This may be due to one's ignorance of the other's work or to a difference of opinion. More recently, sophisticated equipment and the resulting techniques for more detailed analyses are disclosing new bases for plant relationships, often revealing instances of prior misclassification.

No species may bear two scientific names; when this occurs the older name preempts the newer one. If evidence is sufficient to support a name change, it is given official recognition according to the International Rules of Botanical Nomenclature. We have tried to conform in using names that are currently recognized and have placed former names at the conclusion of descriptions as needed.

Genera bearing sufficient similar characteristics to relate them are grouped together as a family with a name ending in -*aceae,* as in Rosaceae, the Rose Family. Plants are arranged under the appropriate family headings. In many cases, plants of more than one family appear on a single page in the text section and/or in the photo section. Family names are not given in the photo section running heads to avoid confusing those readers who may not recognize where the families break on a page. As you become more familiar with families, we encourage you to write in the family names. We have introduced dicotyledonous plants first, then monocots, and the genera in them are roughly in phylogenetic order.

Common names, insofar as we were able to collect them, are given, but with reservations on our part because of their unreliability from one locality to another. We accepted those we thought had the most widespread usage.

The name for each plant is followed by a description that provides recognition characteristics of the illustrated species. Where a feature is either obvious in the photograph or too obscure for detection without magnification it generally is omitted, although there are occasional exceptions. Plants similar to the one described and with which it might be confused are included in a second paragraph in which points of contrast and often additional identifying characteristics are given. By carefully utilizing these tools and avoiding shortcuts, an amateur more than likely will arrive at a correct identification or at least come reasonably close. Some species are extremely complex, and a more advanced student may be required to interpret the critical clues and make a sure identification.

If more than one species of a genus is included, characteristics common to species of that genus are presented with the first one described. Characteristics common to all genera in the family occasionally are described under that family heading.

Choosing which species to include was a daunting task and the process of elimination an unhappy one. Common sense has to dictate the limits imposed by weight and size if a knapsack-compatible product is to result. Ferns and other nonflowering plants were not considered, nor were trees and shrubs. The inclusion of grasses, sedges, and a few rushes was mandated because they too often are overlooked and because the natural niche they occupy needs to be recognized.

There are over 100 genera and perhaps 600 species of grasses in the eastern US, which, despite their importance, is too large a number for more than a sampling here. The genera included give some idea of the differences grasses exhibit; a few additional species illustrate ornamental possibilities.

Rushes and sedges are around us in various habitats and, along with grasses, are present in essentially all wetlands, in pond and lake margins, and in saltwater marshes. The number of genera in this area, however, is relatively small, and we have tried to include representatives of as many of them as practical.

The cultivation of a vast number of grass species as a reliable source of food for the populace and their livestock is a matter of historical record. A more recent development has been their uses for ornamental purposes, and a smartly clipped front lawn no longer represents the number of cows, goats, or sheep the household owns. Today, graceful and colorful species and varieties of native and exotic grasses are being added to our borders and foundation plantings and even hedges.

Hardy and aggressive plants often carry the lowly title of "weed," but your weed may be the pride of my flower border. Many, perhaps most, are generally ignored but when examined closely and individually may surprise us with their beauty.

The choice of species included was influenced partly by our field experience in the eastern US. In the early 1930s Wilbur spent four summers as a park naturalist with the state of Indiana. The latter part of that decade he spent in mountainous areas of North Carolina and Tennessee.

In 1940 we settled in what would become our permanent home, Athens, Georgia, which is about equidistant from the seacoast and the spruce-fir forests to the north. At first our plant surveys and habitat studies were more intensive than extensive, but our range of exploration soon expanded and has continued.

In the late 1960s and 1970s we made extensive and somewhat exhaustive studies of the flora along and near the Atlantic coast north into Newfoundland and the Gulf of Mexico coast to the Mexican border, excluding the tropical and subtropical peninsula of Florida. We traveled in excess of 60,000 miles making these studies during all seasons except winter and regularly revisited representative sites.

A final evaluation of our data revealed to us that a clear majority of the genera and many of the species we encountered were pancoastal. Our field studies of the eastern interior were less extensive. We feel, however, that these data, with the addition of those derived from herbarium and library research, are sufficiently convincing to warrant a publication encompassing the entire area. It is our thought, too, that the inclusion of some endemics and a few of the less widely distributed species would enhance its appeal and usefulness.

The area we have defined has a flora surprisingly ubiquitous in many respects. In making such a general statement we are not ignoring the fact that many species cannot withstand the frigid northern winters while others scorch in southern summers.

This area has no deserts, no great plains, no active volcanoes. The mountains are old and rounded; the shorelines suffer little violent assault other than from occasional seasonal hurricanes. Overall changes from north to south and from east to west, for the most part, are gradual and subtle, the lines of integration mostly imperceptible. Plant migrations often go unnoticed unless some incident or special study brings them into focus. An example is Camphorweed, *Heterotheca subaxillaris*, which was recorded as having its northernmost limit at Chesapeake Bay in Virginia in the 1970s. It was later reported on Long Island, NY, and adjacent Connecticut.

Although large-scale vegetational changes appear to be gradual within the total area, local changes can interrupt an otherwise relatively uniform landscape due to an abrupt and clearly defined habitat alteration. This may be an ancient shoreline fostering a sandhill scrub oak climax, a rhododendron bald, wetlands with an aquatic flora, a limestone glade of cedar, and many others.

To identify the vegetation around us, you need a moderate knowledge of botanical terminology. Although we have tried to keep the language as simple as possible, there may be unfamiliar words that require defining. Even so, in a remarkably short time a working vocabulary of common botanical terms can be acquired. Word descriptions are provided in the glossary; photographs and line drawings illustrate graphically much of what the glossary describes. Both should be utilized routinely as some lines of differentiation can be disturbingly close.

In addition, there is a need for knowing how to look and how to interpret what is seen. This requires a rudimentary acquaintance with the elements that make up a plant, not only flowers and fruits but vegetative parts as well. This facilitates learning what constitutes a specimen, whether it is a simple leaf or a leaflet, a flower or a flower head.

Plants with Unusual Characteristics

Some groups of plants such as families or genera have characteristics sufficiently distinctive for use as identifying features or at least to alert one to that possibility. Exceptions obviously may occur within the group itself, or species outside the group may bear some of these same features. The list below provides shortcuts to reducing the choices, but the above caveat should be remembered.

1. Flowers usually small and packed tightly in heads: FABACEAE, ASTERACEAE, ERIOCAULACEAE, *Eryngium, Xyris.*
2. Flowers in simple umbels: LILIACEAE, PRIMULACEAE, ASCLEPIADACEAE.
3. Flowers in compound umbels: APIACEAE.
4. Sepals and petals 3 each and all similar: LILIACEAE, AMARYLLIDACEAE, JUNCACEAE, IRIDACEAE.
5. Stamens numerous: RANUNCULACEAE, ROSACEAE, HYPERICACEAE, MALVACEAE, ALISMATACEAE.
6. Plants lacking green color: ERICACEAE, OROBANCHACEAE, ORCHIDACEAE.
7. Plants with milky juice: ASCLEPIADACEAE, APOCYNACEAE, EUPHORBIACEAE, ASTERACEAE.
8. Petals united into a tube and with a distinct upper and lower lip: LAMIACEAE, SCROPHULARIACEAE.
9. Flowers with 4 separate sepals, 4 separate petals, and 6 stamens (2 short, 4 longer): BRASSICACEAE.
10. Leaves thick, succulent, smooth: PORTULACACEAE, CRASSULACEAE.
11. Inflorescences narrow and coiled backward at tips: BORAGINACEAE, HYDRO-PHYLLACEAE.
12. Inflorescences apparently fastened between nodes: SOLANACEAE, *Phytolacca.*

Guide to Species Groups

Although there are a few species that will be out of place in the scheme given below, time can be saved and identifications made easier by using it. If your plant has none of the characteristics under "Plants with Unusual Characteristics," then read both items headed by the letter A. Make a choice and then proceed to the pair of letters immediately below.

A. DICOTS. If the sepals and petals are 4 or 5 each, the leaves netted-veined, and the vascular strands in the stem arranged in one circular line, then check *B* below. (A few Dicots are often interpreted as having parallel veins. Species of *Plantago* are the most likely to be encountered.)

    B. If the petals are separate, look on pages 1–60.

    B. If the petals are united, perhaps only at their bases, then check *C* below.

        C. If flowers are in heads surrounded by bracts, look on pages 103–36.

        C. If flowers are not in heads surrounded by bracts, then check *D* below.

      D. If the petals are of equal size and shape, look on pages 61–77, 85, 97–102.

      D. If the petals are of unequal size and/or shape, look on pages 78–84, 86–96, 103.

A. MONOCOTS. If the sepals and petals, which may be similar, are 3 each, the leaves parallel-veined, and the vascular strands scattered (sometimes scattered in a ring such as in a bamboo stem), then check *E* below. (Some species of Monocots do not appear to have parallel-veined leaves. The most common species are in *Trillium, Sagittaria, Peltandra,* and *Arisaema.*)

      E. If the ovulary is superior, look on pages 137–70.

      E. If the ovulary is inferior, look on pages 171–80.

Structure of Flowering Plants

Vegetative structures consist of stems, leaves, and roots and include buds, which enclose miniature leaves and a stem. Roots generally are characterized as underground members from which stems emerge. They may be distinguished from underground stems by the absence of leaves (including scales) and/or their scars. Basal leaves of some plants may appear to arise directly from the root, but there is always a diminutive or vestigial stem tissue involved.

Stems commonly are considered aboveground parts as these are the most visible ones. They have leaves and the point of attachment is a node; the section between nodes is an internode.

Leaves have a blade and some also a stalk, the petiole; those without the petiole are sessile. They may be simple or compound. Simple leaves have a single blade that may be entire, lobed, or deeply lobed but never completely. Compound leaves have two to many discrete blades called leaflets. They may be arranged palmately, in which case each leaflet radiates from the same central point at the petiole summit, like a fan. If pinnately compound there is a linear axis (rachis) on which the leaflets are borne, as on a feather, and leaflets may be in pairs opposite each other or alternately, with or without a terminal one.

Stipules are another form of vegetative growth often present on the base of the petiole, or on the stem beside the petiole base, or partially on both. They are always paired and appear in varying forms such as a blade, a small point, or a spine. They are sometimes ephemeral and leave a scar upon falling.

Vegetative Structures

SHAPE OF BLADES (LEAF, PETAL, SEPAL, BRACT)

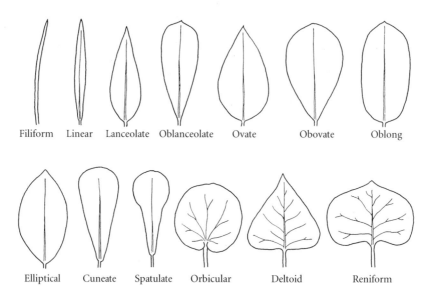

Filiform   Linear   Lanceolate   Oblanceolate   Ovate   Obovate   Oblong

Elliptical   Cuneate   Spatulate   Orbicular   Deltoid   Reniform

BLADE TIPS

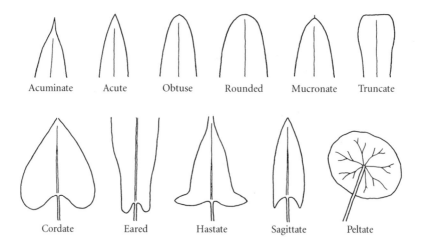

Acuminate   Acute   Obtuse   Rounded   Mucronate   Truncate

Cordate   Eared   Hastate   Sagittate   Peltate

Vegetative Structures (*continued*)

SIMPLE LEAVES

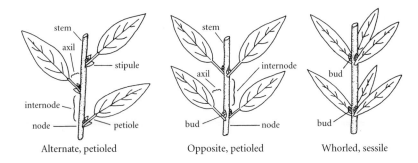

Alternate, petioled   Opposite, petioled   Whorled, sessile

ONCE-COMPOUND LEAVES

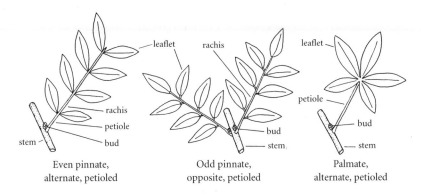

Even pinnate,   Odd pinnate,   Palmate,
alternate, petioled   opposite, petioled   alternate, petioled

TWICE-COMPOUND LEAF

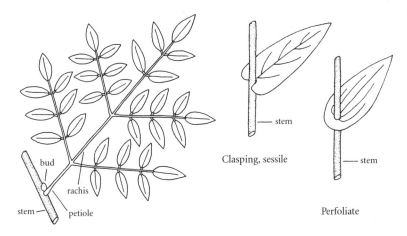

Clasping, sessile

Perfoliate

Vegetative Structures (*continued*)

BLADE MARGINS

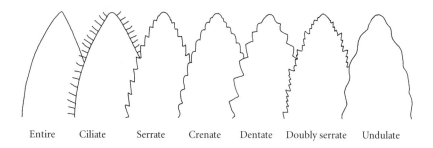

Entire   Ciliate   Serrate   Crenate   Dentate   Doubly serrate   Undulate

LOBING

Pinnately lobed   Palmately lobed

Floral Structures. A complete flower is made up of four sets of parts attached to a special stem called a *pedicel.* The part of the pedicel to which the four sets are attached is called the *receptacle.* Each of the four sets is in either one or more whorls or one or more spirals. Members of each set may be completely separate or partly or completely fused. The outermost of the sets is the *calyx,* consisting of *sepals,* which are usually green. Above these are the petals, which collectively are called the *corolla.* The corolla is usually the most colorful part of the flower. The third set of parts is the *stamens,* which bear the pollen. The final set is the *pistils.* They may be one to many, and contain *ovules.* The pistils grow into fruits and the ovules into the seeds in them.

The presence or absence of another flower part, the *hypanthium,* and variations in it are especially helpful in recognizing many species. The hypanthium is a cup-, saucer-, or disc-shaped structure found in some flowers. It supports the sepals, petals, and stamens, being between these parts and the receptacle. The diagrams below show positions and relationships of various flower parts and kinds of flower clusters.

Floral Structures

COMPLETE FLOWERS WITH SUPERIOR OVULARIES AND NO HYPANTHIUM

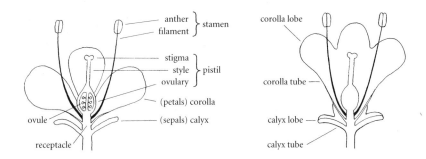

COMPLETE FLOWERS WITH A HYPANTHIUM (H)

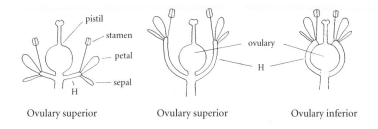

Ovulary superior   Ovulary superior   Ovulary inferior

FLOWER FORM

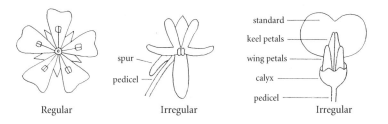

Regular   Irregular   Irregular

Floral Structures (*continued*)

INFLORESCENCES (FLOWER OR FRUIT CLUSTERS)

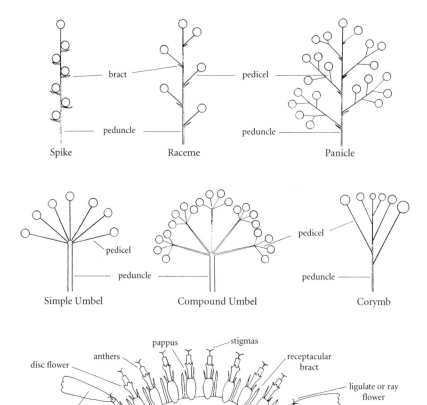

Head of Flowers (Asteraceae)

Fruiting Structures. Fruits develop from pistils, together with any other structures that sometimes may adhere to the matured pistil. A common added part is the hypanthium. For example, in the apple this structure forms the outer fleshy part that is eaten. In the banana the peels are developed from the hypanthium. In members of the sunflower family (ASTERACEAE) the hypanthium forms part of the husk covering the kernel inside. In a strawberry the receptacle is enlarged and is a major part of the fruit. A tomato, an English pea pod, and a milkweed pod, by contrast, develop from the pistil only. Diagrams of all these fruits except the banana follow:

Fruiting Structures

FRUIT TYPES (INTERIOR OR SURFACE VIEWS)

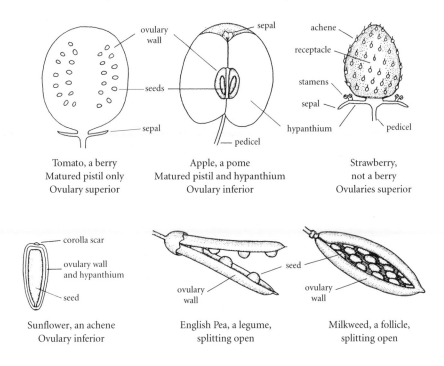

| Tomato, a berry
Matured pistil only
Ovulary superior | Apple, a pome
Matured pistil and hypanthium
Ovulary inferior | Strawberry,
not a berry
Ovularies superior |

| Sunflower, an achene
Ovulary inferior | English Pea, a legume,
splitting open | Milkweed, a follicle,
splitting open |

Comments on Texts

Measurement. The metric system is used in this book because of its utility. By no other practical method can comparative sizes of the small parts of flowers be demonstrated. For your convenience a metric scale 20 centimeters (cm) long has been placed inside each cover. Twenty cm equals 200 millimeters (mm). A meter is 5 times that long, i.e., 100 cm, 1000 mm, or about 39.4 inches. Ten mm equals 1 cm. These are the only metric units used in this book.

Value. The value of herbaceous plants too often is in the eye of the beholder. Scientists in several disciplines related to the environment (ecology, botany, anthropology, and others) recognize that all herbaceous plants in some way influence the environment. This value can be a negative one, as with exotics such as Japanese Honeysuckle or Kudzu that can overwhelm a native flora. Many other negative aspects often result from artificial disturbances, such as the introduction of nonnative plants and overcollecting and overcutting. Plants can be gathered for food, medicinal uses, and aesthetic purposes with little or no deleterious results if only foliage, flowers, or fruits are used. In instances where the root is taken (Bloodroot for dye, Sassafras for tea) or the entire plant removed, for whatever reason, that plant in that habitat is gone and in some few cases extinction of the species is threatened. One effective alternative to this is the gathering of seeds, fruits, or cuttings, leaving the plant for the future. Another is to visit sites designated for clearing preliminary to construction or areas to be inundated by dams and ask to collect there. Permission to do so is usually granted and transplanting in this manner is a sane means of sustaining some species.

Abundance. The local abundance of many species has been established, often by numerical tabulations. Such studies are of considerable interest to ecologists and others, but these data very rarely involve the abundance of a species over its entire range. Nevertheless, a general subjective evaluation of the abundance for each species can be made on the basis of field experience, comments of other specialists, and analyses of scientific studies and specimens preserved in herbaria.

Conclusions concerning abundance for each species in this book are reported as common, occasional, or rare. Because most methods are of necessity subjective, only common and rare should be considered as significantly different ratings. Any species listed as occasional is neither quite rare nor very common. Species that have limited distribution, although they may be abundant where found, are listed as rare, as are those that have wide distribution but are represented by few plants numerically. Species listed as common may also be occasional, rare, or even absent in a given locale.

Distribution. The natural occurrence of a species may be conveniently considered in two categories: (1) the kind of habitat in which it grows; and (2) its geographical distribution, also called range. Different habitats typically result from variations in light, moisture, soils, and kinds of associated species. Knowing that a species usually occurs in a certain habitat can be helpful in identifying a specimen or in finding plants of a given species. Geographical distributions are similarly useful, and for these reasons, as well as for general interest, the kinds of habitats and ranges are given for each species in this book.

Distributions for each species were determined from manuals, especially from published studies of families and genera, and by examining specimens in herbaria. Specific geographical sections of states have been indicated when helpful. Sections of states or provinces are indicated by uncapitalized abbreviations. The basic positions of the sections and their abbreviations are shown in the following diagram:

nw	cn	ne
cw	c	ce
sw	cs	se

Flowering Period. The period of flowering is given for each species over its entire distribution in the Southeast by months. "Apr–June" means that flowering occurs during part or all of April, the entire month of May, and part or all of June in all or some portion of its range. For some species, flowering during the earliest month of spring and the last month in the fall may occur only in Florida or the southern parts of Georgia, Alabama, Mississippi, and Louisiana.

Comments on Photographs

During our years of field research we traveled in excess of 80,000 miles, photographing extensively as we went. If we felt any uncertainty as to the identity of a species photographed, we collected a voucher specimen and later checked it at the University of Georgia Herbarium against collections there. Manuals and published treatments of groups relating to the voucher specimen were studied, a few unpublished theses and dissertations were examined, and an occasional specimen was sent to a specialist for verification. We believe these efforts were sufficient to assure a high degree of accuracy in identifying the species photographed.

Despite the extensive size of our photograph library, we needed color pictures of several species to provide a more balanced representation for the area covered by this book. Through our own contacts and through vigorous help from the Georgia Botanical Society, we obtained the loan of color slides of 89 species from 15 individuals. These slides, too, were critically examined and the names as labeled were corroborated to the best of our ability.

We also tried to achieve a high degree of accuracy in the size given for each photograph as compared to the plant's actual size. In our personal 542 color slides, most values (those to as small as $\frac{1}{10}$) were recorded directly from the camera. If the ratio was smaller, the size of a definite portion of the plant was recorded and compared with the image on the photograph under 10× magnification. In checking the 89 slides on loan we measured a given portion of the image, perhaps a flower or leaf, and compared it with the size of the same subject on an average herbarium specimen. This method was less accurate but still sufficient to convey a good impression of the size of the plant photographed. The relative size of the image and the part of the plant photographed is indicated by a number placed after the scientific name. For example, ×2 means the picture is twice that of the actual plant size, ×1 means the sizes are the same (natural size), ×$\frac{1}{3}$ means the picture is one-third the size of the actual subject. These values can be very helpful in getting from the picture a realistic conception of the natural size.

Glossary

achene A small dry 1-seeded indehiscent fruit smaller than a nut and thinner-walled than a nutlet.

acuminate Tapering to a point. (Compare with *acute.*)

acute Applied to tips and bases of structures ending in a point less than a right angle. (Compare *acuminate* and *obtuse.*)

alternate leaves A single leaf per node.

alternate stems A single stem per node.

annual Plant growing from seed to fruit in one year, then dying.

anther The pollen-bearing part of a stamen.

antrose Directed upward or forward.

appressed Lying flat against.

aristate With a bristle-shaped tip.

ascending Rising obliquely or curved upward.

awn A terminal slender bristle on an organ, as in grasses.

axil The space between any two organs, such as stem and leaf.

axile A placenta along the central axis of an ovulary with 2 or more cells.

axillary In an axil.

beaked Ending in a firm prolonged slender tip.

berry Any fruit with fleshy walls and with several to many seeds inside, such as a grape, tomato, or pepper.

biennial A plant growing for two years then dying naturally.

blade The flattened and expanded part of a leaf, or parts of a compound leaf.

bract Reduced leaf, particularly at bases of flower stalks and on the outer part of the heads of flowers in the Composite Family.

bristle A stiff, strong, slender hair or hairlike structure on the surface, edges, or top of an organ.

bulb A swollen structure composed of circular layers as in an onion, usually subterranean.

calyx The outer set of parts of a flower, composed of sepals.

capillary Hairlike in shape.

capsule A dry fruit with 2 or more rows of seeds that splits open at maturity.

carpel Each separate simple pistil, or section of a compound pistil.

cauline On or pertaining to a stem. Cauline leaves are attached to the stem distinctly aboveground, in contrast to basal leaves.

central placentation Ovules and seeds on a central axis in a 1-celled pistil or fruit; axis summit usually unattached.

ciliate Marginally fringed with spreading hairs. Margin may be entire or toothed.

circumboreal Around northern regions.

claw The narrowed parallel-sided base of some petals.

cleft Deeply cut.

column The united part of the stamens as in the Mallow Family; in Orchids, the fused style and stamens.

columnar Elongated solid, as is a rod.

compound Composed of 2 or more separate parts united into one whole.

compound leaf Leaf with 2 or more blades (leaflets).

compound pistil With 2 or more united carpels.

connective The extension of the filament between and sometimes beyond the 2 pollen sacs of the stamen.

cordate As the outline of a heart with the point at the terminal end.

coriaceous Resembling stiff leather in texture.

corm A short solid thickened portion of the underground base of a vertical stem, with poorly developed scale-leaves or more commonly leafless. (Compare *bulb.*) A swollen base of a stem, bulblike but solid.

corolla All petals of a flower, either separate or united, i.e., the inner series of the perianth.

corymb A flat-topped or rounded inflorescence with the outer flowers on the longest stalks and opening first.

crenate An edge with rounded teeth.

cultivar A kind of plant that has originated in cultivation.

cuneate Wedge shaped.

cyathia (cyathium) The ultimate inflorescence of *Euphorbia*. Consists of cuplike involucre bearing flowers from its base.

cylindric Elongated hollow object, as is a pipe.

cyme An inflorescence of broad flower clusters in which the terminal flower of each cluster blooms first.

decumbent With a prostrate or curved base and an erect or ascending tip.

decurrent With an adnate wing or margin extending down the stem or axis below the point of insertion.

deflexed Turned abruptly downward.

dehiscent Opening by natural splitting, as an anther in discharging pollen, or a fruit its seeds.

deltoid Triangular in shape.

dentate Toothed, the apex of each tooth sharp and outwardly directed. (Compare *serrate.*)

disc flower Any rayless flower in many members of the ASTERACEAE. See illustrations.

divergent Inclined away from each other.

divided Any blade cut into sections reaching three-fourths or more of the distance from the margin to the midvein or to the base.

drupe A fleshy fruit with a firm to hard inner part that permanently encloses the usually solitary seed (as in Peach) or with 2 or more portions of inner part each separately enclosing a seed (as in hollies).

elliptical Oblong with the ends about equally rounded.

entire A margin without teeth. Margins may be lobed and the lobes entire.

epiphyte A plant without connection to the soil, growing upon another plant, but not deriving its food or water from it. (Compare *parasite.*)

erose With an irregular margin, as if gnawed.

evergreen Holding live leaves over winter until new ones appear, or longer.

female flower With pistils but no pollen-bearing stamens.

fibrous Composed of or resembling fibers, as many thin elongated roots.

filament The part of the stamen below the anther, usually slender.

filiform Threadlike; long and very slender.

floret A little flower; an individual flower of a definite cluster, as in the head of a composite or the spikelet of a grass.

follicle A simple seedpod opening down one side.

forb An herbaceous plant not belonging to the grasses, rushes, and sedges.

fruit A mature ovulary, together with such parts as are regularly attached to it, i.e., the seed-bearing part of a plant and any attached parts.

fusiform Cylindrical except thick near the middle and tapering to both ends.

geniculate Abruptly bent or twisted, zigzag.

glabrous Lacking hairs or other protuberances.

glandular Bearing swollen structures. These may be sessile, on short stalks, or on the tips of hairs.

glaucous Surface with a very fine white substance that will rub off, as on many grapes and blueberries.

glume One of a pair of bracts, found at the base of a grass spikelet, that do not subtend flowers.

grain A fruit resembling an achene with the seed adhering to the inside of the thin ovulary wall and forming one body, as in a corn grain.

hastate Like an arrowhead but diverging at the base.

herb Any plant, annual, biennial, or perennial, with stems dying back to the ground at the end of the growing season. Any plant that has seasoning or medicinal use.

hypanthium A saucer-shaped, cup-shaped, or tubular organ below, around, or adhering to the sides of the ovulary. The sepals, petals, and stamens are attached at or near the outer or upper margin of the hypanthium.

indehiscent Not opening naturally at maturity.

inferior Descriptive of an ovulary surrounded by and sides fused to the hypanthium, the ovulary therefore appearing to be located below the sepals and petals.

inflorescence Any complete flower cluster, including any axis or bracts. Clusters separated by vegetative leaves are separate inflorescences.

internode A portion of the stem between one node and the next, i.e., between places where leaves are or were attached.

involucre A set of bracts below or around a head of flowers or a single flower.

irregular Flowers in which the members of one or more sets of parts are not the same size or shape.

lanceolate With the outline of a lance-head, much longer than wide and widest below the middle.

leaflet Any one of the blades of a compound leaf.

legume A one-carpelled fruit, such as in the FABACEAE generally, in which the fruit opens along both top and bottom edges.

lemma Bract attached at the base of each grass flower.

lenticular Shape of cross-section of a lens, i.e., convex on both surfaces.

ligule A flattened petallike terminal part of some corollas. Common in the ASTERACEAE, most often on the marginal flowers of each head as in the Sunflower, but sometimes on all the flowers as in the Dandelion.

linear Narrow and elongate with essentially parallel sides.

lip Either the upper or lower projecting parts of the corolla or calyx of mints, snapdragons, and other plants; the odd petal (usually the lowest) in the Orchids.

loment A legume composed of 1-seeded sections that separate transversely into joints, e.g., Desmodium.

male flower With stamens and no functional pistil.

membranaceous Thin, soft, and flexible.

mesic Moderately moist.

mucilaginous Slimy or mucilage-like.

mucro A short, slender tip of an organ, as the projected midrib of a leaf.

mucronate Having a short, sharp point at the apex (tip).

node That section of a stem from which leaves or branches arise.

nutlet A hard 1-seeded fruit, smaller than a nut, and thicker-walled than an achene.

oblanceolate Reverse of lanceolate, the terminal half the broader.

oblong Elongate and with parallel, or nearly so, sides.

obovate Reverse of ovate, the terminal half the broader.

obovoid Having the form of an egg with the broad end apical.

obtuse Blunt or rounded at the end, the angle at the end over 90°. (Compare *acute.*)

ocrea A tubular stipule surrounding the stem above the node.

opposite That arrangement of 2 leaves at a node with one attached opposing, or 180° from, the other. Also true of branches and buds.

ovate Having the outline of an egg with the broader half being basal.

ovulary That part of the pistil containing the ovules, the future seeds. Also sometimes called *ovary.*

palea Bract opposing lemma, attached at the base of a grass flower.

palmate Radiately lobed or arranged.

panicle A rebranching flower cluster of the raceme type.

papilionaceous Having a standard, wings, and keel petals, as in the corolla of many legumes.

papillate Bearing minute nipple-shaped projections.

pappus The outgrowth of hairs, scales, or bristles from the summit of achenes of many species of the ASTERACEAE. Generally considered to represent the calyx.

parasite A plant that derives its food and water chiefly from another plant to which it is attached. (Compare *epiphyte.*)

pedicel The stalk of each single flower.

peduncle The main flower stalk, supporting either a cluster of flowers or the only flower.

peltate Shield-shaped and attached to the support by the lower surface.

pendent Drooping or hanging loosely.

perfect Descriptive of flowers having both functional stamens and pistils.

perfoliate A leaf blade completely surrounding the stem, which appears to pass through the leaf, or two opposite leaves with bases fused to each other.

perianth The calyx and corolla collectively, or the calyx alone if the corolla is absent.

perigynium The special bract that encloses the pistil and achene of a *Carex.*

petal One of the sections, separate or united, of the corolla.

petiole The basal stalk of a leaf. Sometimes absent.

phyllaries Bracts surrounding a single head of flower in the ASTERACEAE, the Composite Family.

pinnate Blades of a leaf arranged along the sides of a common axis.

pistil The central organ of a flower, sometimes several separate ones. Contains the ovules.

pistillate Having pistils and no pollen-bearing stamens.

placenta Tissue of the ovulary to which ovules are attached.

plumose Structured like a feather, the lateral divisions being like fine fibers.

pod A dry dehiscent fruit.

prickle Sharp elevation of the epidermis of stem, leaves, bracts, and outer parts of some flowers.

prostrate Lying parallel with the ground.

raceme A type of flower cluster in which 1-flowered stalks are attached along the sides of a common axis.

rachis The main axis of a pinnately compound leaf, excluding the petiole.

ray See *Ligule*. Also a branch of an umbel or umbellike inflorescence.

receptacle The part of the pedicel to which the other flower parts are attached. Also the enlarged summit of the peduncle of a head to which the flowers are attached.

reflexed Abruptly bent downward. See *retrorse*.

regular Flowers in which the members of each set of parts are the same size and shape.

reniform Having the outline of a kidney, rounded but with a wide basal notch.

reticulate Veins in the form of a network.

retrorse Directed backward or downward. See *reflexed*.

rhizome A horizontal underground stem; distinguished from a root by the presence of nodes, buds, or scalelike leaves that are sometimes quite small.

rootstock The somewhat enlarged part of a plant at or under the ground surface and from which the roots and stem(s) grow.

sagittate Shaped like an arrowhead. Also see *hastate*.

saprophyte A plant that lives on dead organic matter, neither parasitic nor making its own food. Often used loosely to include nongreen plants that get their food from symbiotic fungi.

scabrous Rough or harsh to the touch due to the structure of the epidermis or to presence of short stiff hairs.

scale Applied to many kinds of small, thin, flat, usually dry, appressed leaves or bracts, often vestigial. Sometimes epidermal outgrowths, if disclike or flattened.

scape A leafless stem bearing flowers and rising from the ground or near it.

scurfy Surface with small scale or branlike particles.

sepal One of the parts of the calyx or outer set of parts of a flower, either separate or united.

serrate Having sharp teeth pointed terminally.

sessile Without any kind of stalk.

sheath A tubular structure surrounding an organ or part, such as the lower part of the leaf of grasses or an ocrea.

sinus Base of the space between two lobes of a leaf, calyx, corolla, or united bases of stamens.

spathe A large, usually solitary bract subtending and often enclosing an inflorescence; the term is used only in the monocotyledons.

spatulate Somewhat broadened toward a rounded end.

spike A type of flower cluster in which stalkless flowers are attached along the sides of a common axis.

spikelet Closely aggregated glume(s) and 1 or more flowers of grasses and sedges.

spine A sharp structure consisting of an entire leaf (as in barberries) or portion of a leaf (as in some hollies).

spur A saclike or tubular extension of some part of the flower.

stamen A pollen-producing organ of a flower, usually consisting of anther and filament.

staminate Having stamens and no functional pistil.

staminode (staminodia) A sterile stamen or any structure corresponding to a stamen and without an anther.

standard The upper dilated petal of a papilionaceous corolla.

stellate Star-shaped, with several similar parts spreading out from a common center. Usually applied to branched hairs.

stigma The pollen-receptive part of a pistil, usually terminal and often enlarged.

stipe; stipitate Basal stalklike support of an ovulary or fruit, not the pedicel; having a stipe.

stipules A pair of structures, usually small, on the base of the petiole of a leaf or on the stem near the petiole; sometimes fused together. Sometimes absent.

stolon A horizontal branch arising at or near the base of a plant, taking root and developing new plants.

style That portion of the pistil between the stigma and the ovulary, often elongate, sometimes apparently absent.

sub- As a prefix, usually signifying about or nearly.

subulate Awl-shaped; narrow and tapering evenly from base to tip.

succulent Juicy and fleshy. Either thin as in lettuce leaves, or thick as in cactus.

superior Descriptive of an ovulary that is free of other floral organs.

tendril A thread-shaped clasping or twining structure serving as a holdfast organ.

tepal Used for any sepal and petal of similar form.

terete Circular in cross-section.

thorn A hardened sharp-pointed stem.

tomentose With matted soft woollike hairiness.

truncate An apex or base nearly or quite straight across, as if cut off.

tubercle A small swelling or projection, usually distinct in color or texture from the organ on which it is borne, as the tubercle on the fruit of *Eleocharus*.

tuberculate Bearing small raised places.

umbel A flower cluster in which the flower stalks arise from the same point, as do the ribs of an umbrella. In a compound umbel this arrangement is repeated.

undulate With a wavy surface or margin.

unisexual One sex, bearing stamens or pistils only.

urticle A small, thin-walled, 1-seeded, more or less inflated fruit.

whorl Three or more structures (leaves, stems, etc.) in a circle, not spiraled.

zygomorphic See *Irregular*.

Abbreviations

Appal	Appalachian, Appalachians
BR	Blue Ridge Mountains
ca	circa, about
cm	centimeter, centimeters
CP	Coastal Plain Province
E, E.	east, indicating a general direction; East (as in E. Indies)
m	meter, meters
mm	millimeter, millimeters
mt	mountain
N, N.	north, indicating a general direction; North (as in N. Amer)
Pied	Piedmont Province
RV	Ridge and Valley Province
S, S.	south, indicating a general direction; South (as in S. Amer)
sp.	species (singular)
spp.	species (plural)
ssp.	subspecies
W, W.	west, indicating a general direction; West (as in W. Indies)

Dicotyledons

Leaves usually pinnately or palmately netveined; floral parts, when of definite number, typically in sets of 5, often 4, seldom 3 (carpels often fewer); embryo generally with 2 cotyledons; vascular bundles of stems in a ring that encloses a pith.

SAURURACEAE: Lizard's-tail Family

1 **Lizard's-tail**
Saururus cernuus L.
These perennials may reach almost 1 m and often form extensive colonies, spreading by rhizomes. Although the spike of flowers is drooping at the tip, the spike becomes erect as the seeds mature. Reestablishment of the beaver after near extinction in the sUS has increased the Lizard's-tail habitat. Common. Swamps, margins of streams and lakes, and low woodlands; Fla into eTex, seKan, sMich, swQue, and RI. Apr–July.

URTICACEAE: Nettle Family

2 **Stinging Nettle**
Urtica dioica L.
Erect rhizomatous perennial to 2 m with obvious stinging hairs. Leaves opposite, blades 5–15 cm, serrate, teeth averaging 2–3.5 mm deep, apex acute to acuminate, stinging hairs commonly confined to underside; stipules 5–15 mm, lance-linear. Inflorescences axillary, many-flowered; flowers minute, unisexual. Fruit an ovoid achene 1.5 mm. Common. Woods, waste ground; nearly throughout, scattered in sUS, more abundant northward. May–Aug.

3 **False-nettle**
Boehmeria cylindrica (L.) Sw.
Perennial without stinging hairs, to 1.3 m. Leaves opposite. Flowers on axillary spikes, without petals, unisexual and on same plant; male flowers tiny with a 4-parted calyx and 4 stamens; female flowers with an ovoid tubular calyx 2–4 toothed at summit and surrounding the ovulary. Fruits small achenes enclosed in the enlarged narrowly 2-winged calyx. Common. Moist to wet soils; under shrubs or trees or in open; Fla into NM, Neb, Minn, sOnt, sQue, Me, and NC. June–Sept.

ARISTOLOCHIACEAE: Birthwort Family

4   **Wild-ginger**
Asarum canadense L.
A creeping, hairy perennial. Leaves deciduous, cordate, mostly in pairs. Flowers single between the petioles of 2 leaves. Calyx with a cup and 3 pointed lobes. Corolla absent. The 12 stamens are closely associated with the style. The rootstock has an odor and taste suggestive of ginger and has been used as a seasoning. Residents of the sAppal have told us that this species has been used in treatment of pregnant women, but we have no other information supporting such a use. Occasional. Rich woods; cSC into eOkla, eND, and nNB. Apr–June. *A. c.* var *acuminatum* Ashe; *A. c.* var. *reflexum* (Bickn.) Robins.

5   **Heart-leaf**
Hexastylis arifolia (Michx.) Small
Members of this genus are aromatic perennials with thick evergreen leaves, no aboveground stems, and no petals. The 12 stamens are fastened to the style. The species are sometimes included in the genus *Asarum.*

This species has mostly triangular to ovate leaves with eared bases, sometimes in clusters to 40 cm across. Calyx flask-shaped and strongly constricted to a neck below the spreading lobes, which are 3–6 mm, the tube 15–30 mm. Common. Moist to dry woods; wFla into seLa, eTenn, and seVa. Mar–May.

H. naniflora H. L. Blomquist has circular-cordate leaves, the calyx tube is cylindrical and less than 10 cm and the calyx lobes 5–6 mm. Rare. Deciduous woods; nSC and adjNC; sVa. Apr–May.

6   **Large-flowered Heartleaf**
Hexastylis shuttleworthii (Britt. & Baker) Small
Rhizomes 1–2 cm, a very large calyx tube (3–5 cm), calyx lobes 10–15 mm, and hairs in throat of calyx under 1 mm identify this species. Occasional. Rich woods; cGa into c and nAla, eTenn, nWVa, cVa, eNC, and nwSC. May–July.

H. lewisii (Fern.) H. L. Blomquist & Oosting also has long calyx lobes (8–15 mm) but rhizomes 5–20 cm; calyx tube 1.4–2 cm, and hairs in throat of calyx 1–4 mm. Rare. Deciduous woods; lower Pied and inner CP of NC into cVa. Apr–May.

POLYGONACEAE: Buckwheat Family

7   **Dog-tongue; Wild-buckwheat**
Erigonum tomentosum Michx.
The genus consists of about 225 species, most in the wUS.

Seeds and other parts of many species are important as food for various kinds of wildlife.

This species is a perennial to 1 m, erect or often leaning. Stems hairy. Leaves light gray to tan, densely hairy beneath. Basal leaves prominent, longer than the whorled upper ones, sometimes dying from age or drought. Flowers in clusters of 10–20, surrounded at their bases by tan-colored bracts. Fruits about 6 mm,

3-ribbed, surrounded by enlarged sepals. Dry sandy pinelands and sandhills; Fla into upper CP of SC. July–Sept.

8   Wild Sorrel
Rumex hastatulus Baldw.
Annual or short-lived perennial to 1.2 m. Stems single or in large clumps, often forming extensive and colorful masses in open areas. Leaves usually with 2–4 widely divergent lobes. Male and female flowers on separate plants. Sepals at fruiting stage expanded into broad wings much wider than the achene. Various species are nibbled for the acid taste and to quench thirst. When eaten in large amounts poisoning may result from the oxalates. Common. Roadsides, fields, and thin woods; Fla into e and cnTex, seKan, sIll, Pied of Ga, seVa, and locally along coast into Mass. Mar–June.

   R. acetosella L., Sheep-sorrel, is similar but smaller, a perennial from slender running rootstocks, and sepals at fruiting stage just equal the achene and enclose it. Common. Fields, pastures, and roadsides; in most of US and Can. Apr–Jul.

9   Curly Dock
Rumex crispus L.
Plants to 1.5 m. Blade of basal leaves 15–25 × 2–6 cm, margin strongly rippled or wavy, larger blades commonly rounded to subcordate at base. Pedicels limber, 5–10 mm. Fruits with broadly ovate entire or nearly so wings 4–5 mm and as wide; grains 3, two-thirds as long as wide, ca half as long as wings. Native of Europe. Leaves commonly used as food, alone or in combination with other wild herbs. Common. Weed at roadsides; in fields, pastures, along railroads; waste places; nearly throughout US and sCan. Mar–June.

   R. obtusifolius L. To 1.5 m. Blade of basal leaves 15–25 × 6–15 cm, margin flat, base cordate. Flower clusters touching or nearly so; pedicels conspicuously longer than fruit. Fruits with 1 plump grain; wings triangular-ovate, 3.5–5 mm, each with 1–4 spiny teeth on margin. Native of Europe. Common. Similar habitats but more commonly in moist places; nGa into cLa, Ariz, BC, NS, NC, and cSC. Apr–June. *R. pulcher* L., another native of Europe, is similar but flower clusters well separated; pedicels about equal to or shorter than fruit. Occasional. Similar usually moist habitats; Fla into Tex, Okla, NY, eNC, and SC; Calif and Ore. May–June.

10   Arrow-vine; Tear-thumb
Polygonum sagittatum L.
Members of this genus have stipules that form cylindrical sheaths (ocreae) around the stem, the leaves are alternate and entire, the flowers are in spikelike racemes, and the fruits lenticular or triangular achenes.

   This species is a sprawling freely branched annual, inconspicuous except when seen as a mass of growth. Physical contact with the plant readily brings attention because of the sharp backwardly turned prickles on the stems and midveins of the undersides of the leaves. Leaves sagittate. Flowers in small tight rounded clusters. Common. Wet places, usually in the open; Fla into eTex, sSask, and Nfld. May–frost.

P. arifolium L. is also prominently prickly but the leaves are wider and hastate, the flowers in loose elongate racemes. Occasional. Wet open places, often in tidal marshes; seGa into s and eWVa, neO, Ind, Minn, and NB. July–frost.

11 Dock-leaved Smartweed
Polygonum lapthifolium L.

An annual to 2 m. Ocreae thin and hairless or with a few small hairs on the veins. The peduncles bear sessile glands or are glandless. The racemes are usually arching or somewhat drooping. Various species of *Polygonum* have been used in seasoning. The leaves and seeds of many species are peppery and should be utilized with caution. They are mild in other species and are reported good in salads. *Polygonum* seeds are important food for wildlife, being eaten by many ground-feeding song and game birds and seed-eating small mammals. Occasional. Low open places, especially in disturbed areas; nPied of Ga into Tex, Mex, BC, and Nfld. May–frost.

    P. densiflorum Meisn. is similar but is a perennial with racemes mostly erect to slightly arching. Occasional. Swampy woods, shallow water; Fla into CP of Tex, seMo, Ga, cNC, and sNS.

12 Lady's-thumb
Polygonum pensylvanicum L.

Erect to sprawling annual with stalked glands and/or hairs on the peduncles just below the flower clusters. The ocreae are thin and have no cilia. Common. Disturbed and often moist places; Fla into eTex, Minn, and wNS; scattered localities westward. Apr–frost. *P. bicorne* Raf.; *P. longistylum* Small.

    Other annual species with dense racemes include *P. persicaria* L., which is glabrous throughout and has cilia about 2 mm on the ocreae. The calyx is not glandular-punctate. Common. Disturbed, mostly damp places; throughout temperate N. Amer. Apr–frost.

13 Water-pepper
Polygonum hydropiperoides Michx.

Perennial to 1 m. Leaves usually less than 15 mm wide, 3.5 or more times longer than wide. Ocreae with cilia under 10 mm and sides with spreading or appressed hairs. The calyx lacks glands. Flowers are frequently infected with smut. Common. Swamps, water edges, in open or in woods; Fla into Tex, Minn, and NB; scattered to W. May–frost. *Persicaria h.* (Michx.) Small.

    P. setaceum Baldw. is similar but the leaves are mostly wider and the hairs on the side of the ocreae are spreading. Common. Similar places; Fla into eTex, cwMo, SC, and sNJ. *P. cespitosum* Blume is also similar but is an annual with leaves mostly under 3.5 times as long as wide. Occasional. Damp to wet, usually disturbed places; NC into nPied of Ga, sLa, ceMo, and Mass.

14 Japanese Knotweed
Polygonum cuspidatum Sieb. & Zucc.

Stout erect herbaceous rhizomatous perennials to 3 m, usually forming dense colonies; stems hollow. Leaf blades truncate to widely cuneate at base, those of main stem 10–15 × 6–11 cm. Flowers in axillary clusters; male and female on separate plants. Young underground rootstalks and young stems may be treated as a salad or cooked and used in same manner as asparagus. Native of Japan. Occasional. Waste ground; nGa into c and nwArk, eMo, Ill, sOnt, Nfld, Mass, Va, NC, and cSC. May–Sept.

P. sachalinene F. Schmidt is similar, to 4 m; leaf blade base cordate, those of main stem 15–25 × 12–20 cm. Native of eAsia. Rare. Waste ground; scattered localities in eUS. July–Aug.

15 **October-flower**
Polygonella polygama (Vent.) Engelm. & Gray
Erect to document perennial, sometimes shrubby at the base. Leaves narrowly club-shaped to widely spatulate, under 5 mm wide, attached to cylindrical sheaths (ocreae) that surround the stem. The ocreae are pointed on one side. The outer sepals vary from red and pink to white. Petals absent. The stigma and style together are less than 0.2 mm. Fruits triangular, 1-seeded. Common, Sandy places, especially sandhills and pine barrens; Fla into eTex, CP of Ga, and seNC; seVA. Aug–Oct. *P. croomii* Chapm.

P. gracilis Meisn. is taller, slimmer, more thinly branched, and is an annual. Common. Sandy places in open and thin woods and scrub; Fla into sMiss, CP of Ga, and seSC. Sept–Oct.

CHENOPODIACEAE: Goosefoot Family

16 **Goosefoot**
Chenopodium ambrosioides L.
Strongly aromatic, annual in cold winter areas to perennial where frost-free; to 1 m, rarely 2, unarmed. Leaf blades with numerous tiny yellow glands, nearly entire to coarsely dentate or even lobed. Flowers bisexual, sessile in small dense clusters in short to long spikes arranged in a panicle; calyx 5-lobed and not obviously glandular; petals absent. Seeds dark brown to almost black, shiny, thick-lenticular, 0.6–1 mm across. Source of an oil used to treat domestic stock for intestinal worms. Native of tropical Amer. Common. Waste grounds, fields, borders, about buildings; nearly throughout US and Can. Aug–frost.

17 **Lamb's-quarters; Pigweed**
Chenopodium album L.
Annual to 3 m, much-branched. Leaf blades lanceolate to broadly ovate or deltoid, not glandular but young ones with whitish-mealy. Flowers very small, in dense clusters, bisexual; sepals 4–5, fleshy, without wings or spines, persistent, tightly enclosing fruit; petals none. Seeds 1 per fruit, horizontal, black, shiny, 1.3–1.5 mm across; the thin fruit cover persists when fruit is rolled between hands. (We have used young shoots as a potherb; it has the flavor of spinach.) Common. Edge of fresh and brackish marshes; fields, waste places; widely distributed in the US and sCan. July–frost.

C. rubrum L., Alkali-blite, another annual, to 80 cm, has similar flower clusters. It differs in having main leaves glabrous, blade rhombic-ovate to oblong, and with a conspicuous tooth on each side near base, cuneate below the teeth, sometimes other teeth above. Young leaves not white-mealy. Seeds erect, dark brown, shiny, lenticular, 0.6–1 mm across. Rare. Saline and brackish soils; Nfld into Wash; S into NJ, wNY, Ind, Mo, and Ia; sCalif into Ariz and NM. Aug–frost.

AMARANTHACEAE: Amaranth Family

18   Cottonweed
Froelichia floridana (Nutt.) Moq.
Loosely hairy annual to 180 cm, usually with a few well-developed erect branches, mostly from the upper nodes. Leaves few and opposite. Upper internodes progressively longer. Spikes 10–12 mm across. Mature fruit including beak, 5 mm. Occasional. Sandy soil, pinelands, sandhills, fields; Fla into sMiss, cAla, CP of SC, and seDel; La into Tex, NM, SD, nwInd, and Ark. June–Oct.

    F. gracilis (Hook.) Moq. is similar but smaller, to 70 cm, mostly branched at the base, spikes 7–8 mm across and with fruits 3–4 mm. Occasional. Sandy soil and along railroads and highways; ceGa into cnSC, sVa, and seNY; Ark into Tex, Mex, Ariz, Colo, and eInd.

19   Iresine
Iresine rhizomatosa Standl.
Perennial to 1.5 m, stolon-bearing, mostly one inflorescence per stem. Leaves opposite, glabrous or nearly so, blade 5–15 × 1.5–8 cm. Distinctive when in flower; flowers very small, numerous, in spikes to ca 13 mm arranged in a terminal panicle much longer than wide; unisexual, sexes on separate plants; calyx of 5 sepals. Occasional. Damp woods, interdunes, swales, marsh edges. Md into Fla, eTex, Kan, sIll, and La. Aug–frost.

PHYTOLACCACEAE: Pokeweed Family

20   Pokeweed
Phytolacca rigida Small
Glabrous perennial to 3 m. Leaves pliable, with the feel of thin kid leather. Flowering and fruiting racemes erect and appearing to be attached laterally on the stems but fundamentally terminally attached. Fruits a dark purple 5–12-carpelled berry, the flesh of which is edible when ripe. Tender young leaves and shoots when *properly* prepared make a safe and tasty cooked vegetable and are eaten by many. Older parts and the roots are poisonous. Discard the first water in which the greens are cooked to rid them of any poison. Sufficiently abundant and weedy to be considered for food by all. Considered by some to be *P. americana* var. *rigida* (Small) Caulkins & Wyatt. Common. Usually in open disturbed habitats; Fla into seTex, sGa, and swSC. May–Oct.

    P. americana L. is very similar but has divergent to declined racemes. Common. Fla into Tex, sWisc, sQue, and sMe. May–Oct.

HOLLUGINACEAE: Carpetweed Family

21 Carpetweed; Indian-chickweed
Mollugo verticillata L.
Annual with mostly prostrate stems, spreading radially from a central root. Leaves in apparent whorls of 3–6, mostly unequal in size. Flowers 2–5 from each node; sepals 5, green with white margins, ca 2 mm; petals none; fruit a 3-carpelled many-seeded capsule; seeds dark reddish-brown with several parallel ridges. Weedy immigrant from tropical Amer. Common. Gardens, borders, fields, sandy soils; usually moist soils. Fla into Tex, Wash, sOnt, sQue, NS, and NC. Apr–frost.

22 Sea-purslane
Sesuvium portulacastrum L.
Perennial with decumbent stems, rooting at the nodes. Leaves opposite, fleshy. Flowers solitary in leaf axils, on pedicels 3 mm or longer. Sepals 5, green outside and pink within, 7–10 mm. Petals absent. Stamens numerous. Capsule 8–10 mm, opening along a line circling its base. Occasional. Coastal dunes and beaches, upper parts of saltmarshes; Fla into Tex and seNC. May–Nov, or all year.

    S. maritimum (Walt.) B.S.P. is similar but is an annual, is more erect, has sessile flowers with only 5 stamens. The sepals are only 3–5 mm. This species has been used as a potherb. Young shoots are the best. We have no record of *S. portulacastrum* being used thusly but it might be tried. Occasional. Wet coastal sands and salt flats; seTex into Fla and NY. May–Nov, or all year.

PORTULACACEAE: Purslane Family

23 Talinum; Rock-portulaca
Talinum teretifolium Pursh
Succulent perennial to 35 cm, the leaves terete. Flowers in long-peduncled inflorescences, opening only in the afternoon, or not at all on heavily overcast days. Petals 5–8 mm. Stamens 15–20. Ovulary superior. Seeds shiny black. Common. Thin soil on dry rocks, dry sand; cCP of Ga, nwAla, cTenn, Va, eWVa, sePa, and Pied of NC. May–Oct.

    In *T. mengesii* Wolf there are 50–90 stamens and the petals are 9–12 mm. Rare. Thin soil on dry sandstone or granitic rocks; ce and nwAla into cPied of Ga; cCP of Ga. *T. parviflorum* Nutt. has only 5 stamens. Rare. Thin soil pockets on granitic rocks; cAla only. *T. appalachianum* Wolf. In *T. calcaricum* Ware there are 25–45 stamens, the petals are 8–10 mm, and the seeds are dull gray. Rare. Thin soil on calcareous rocks in cedar glades; nwAla and cTenn. May–Sept.

24 Spring-beauty
Claytonia virginica L.
Perennial from a small globose corm. Stems 1–several from each corm. Leaf blades linear to linear oblanceolate, more than 8 times as long as broad, the basal ones none to many, stem leaves 2 and opposite. Flowers up to 15. Unusual in having instability of chromosome numbers, with about 50 different chromosomal

combinations. Occasional. Rich woods or open areas; swGa into eTex, Minn, sQue, and NB. Feb–Apr.

C. caroliniana Michx., which is similar, has leaf blades lanceolate, elliptic, to spatulate, and less than 8 times as broad. Mts of nGa into nwArk, eMinn, sQue, swNfld.

25 Hairy Portulaca
Portulaca pilosa L.

Annual to 20 cm, much-branched. Hairs in the leaf axis. Leaf blades linear to spatulate or oblanceolate. Petals to 6 mm. Stamens 15 or more. Fruit with a lid, opening near middle at even circular line. Seeds many, reddish-black. Occasional. Dry sandy soils in thin scrub or open; Fla into seLa, CP of Ga and of NC. June–Oct.

P. smallii P. Wilson is quite similar but has silvery-black seeds. Rare. Granitic outcrops; Pied of Ga; Pied of NC. June–Oct. *P. grandiflora* Hook. also has hairs in the leaf axils but the petals are 15–25 mm. Rare. Escaped from cultivation in scattered localities.

26 Common Purslane
Portulaca oleracea L.

A prostrate much-branched annual. Leaves alternate and opposite, fleshy, spatulate to obovate. Fruit with a lid opening just below the middle at an even circular line. Seeds dark red or black. When cooked and seasoned like spinach the tender young branches and leaves make a tasty potherb. Used as food in India for over 2000 years and in Europe for hundreds of years. Frequently offered for sale in Mexican markets. It may be eaten raw, being a tasty addition to salad dishes. Large amounts may cause oxalate poisoning. Occasional. Cultivated areas, waste places. Throughout most of the US. May–frost.

P. umbraticola ssp. *coronata* (Small) Mathews & Ketron is similar but the leaves are generally smaller, the seeds silvery gray, and the cap of the fruit with a rim around its base. Rare. In open, granitic rocks, sandy soil; cnSC; cGa; sMiss. June–Sept. *P. coronata* Small.

CARYOPHYLLACEAE: Pink Family
All members of this family have 4–5 persistent sepals, the same number of petals, and 1-celled capsular fruits with central placentation.

27 Giant Chickweed
Stellaria pubera Michx.

This species is a perennial with erect to spreading very finely hairy stems and opposite elliptic to ovate-lanceolate leaves. Flowers in loose terminal clusters. Petals white, apparently 10 but actually 5 deeply split. Sepals obtuse to acute, less than 6 mm. Common. Rich woods, sometimes among rocks; nwFla into Ala, Mo, Ind, neIll, and NJ. Mar–June.

S. corei Shinners is a similar species, but the sepals are acuminate and over 8 mm. Rare. Rich woods; mts of NC and Tenn into Ind and NY. Apr–June.

28 **Common Chickweed**
Stellaria media (L.) Vill.
A highly variable opposite-leaved annual with weak stems. Petals are white and appear to be 10 but actually are 5 deeply cleft. Stigmas 3. Sepals and some other parts are from glandular hairy to glabrous. Plants usually compact at first but later loosely branched and often forming dense masses. The young growing tips have been used as a cooked green vegetable although the plant has little taste. A complete list of birds that use chickweed for food would be very long. Common. Widely distributed, especially noticeable in yards and gardens. Dec–May in sUS, usually dying completely by June. Flowering later northward, sometimes until frost under favorable conditions.

Some species of *Cerastium* and *Arenaria* are somewhat similar. *Cerastium* can be recognized by its 5 stigmas and *Arenaria* by its entire or notched petals.

29 **Common Stitchwort**
Stellaria graminea L.
Perennial with weak angled stems, 30–50 cm. Leaves 1.5–5 cm × 1.5–7 mm, blade linear to lance-linear. Flowers abundant, in dichotomous cymes; sepals 4.5–5.5 mm, strongly 3-veined; petals white, surpassing the sepals. Fruits straw-colored, about length of sepals. Native of Europe. Occasional. Fields, grassy places, roadsides; Kan into Minn, Que, Nfld, NS, Mass, Va, se and nSC, and Tenn. May–July.

30 **Mouse-eared Chickweed**
Cerastium viscosum L.
Matted sticky-haired winter annual to 30 cm with many flowering stems. Margin of bracts below flowers green. Pedical equaling or shorter than fruit. Flowers in dense clusters; sepals with long forward-pointing glandless hairs protruding well beyond the tip. Fruit a capsule 5.5–9 mm, less than twice as long as calyx. Common. Cleared woods, fields, gardens, yards, roadsides, waste places; Fla into Tex, SD, BC, Ont, Nfld, and NC; Calif. Feb–May. *C. glomeratum* Thuill.

C. nutans Raf., an annual, is also sticky-hairy but 5–60 cm, capsule more than twice as long as calyx, and hairs do not surpass sepal tips. Occasional. Rich wooded slopes, calcareous areas, alluvial soils; neGa into Tenn, La, Ariz, Ore, Mack, swQue, NS, Va, cnNC, and nSC. April–May.

31 **Sandwort**
Minuartia caroliniana (Walt.) Mattf.
Perennial to 30 cm with a dense basal cushion of decumbent to prostrate stems from a single stout taproot. Leaves opposite, linear-subulate, rigid, overlapping the ones above. Flowering stems erect, with many small glands. Sepals and petals 5 and separate. Stamens 10, 5 of them short and between the petals. Styles 3. Occasional. Well-drained sands; Fla into CP of Ga, sCP of NC, and scattered into RI. Apr–June. *Arenaria c.* Walt. *Sabulina c.* (Walt.) Small.

M. uniflora Walt. is another conspicuous species of this genus. It is an erect glabrous annual to 9 cm, often occurring in dense masses. Leaves herbaceous, oblong to linear, 2–7 mm, acute, and entire. Sepals 2–3 mm. Petals white, 3–

4 mm. Occasional. Shallow soil of granite rocks of Pied of ceAla into csNC and of other rocks in the CP of Ga. Apr–May. *Sabulina brevifolia* (Nutt.) Small. *Arenaria uniflora* (Walt.) Muhl.

32   Fringed Campion

Silene polypetala (Walt.) Fern. & Schub.

Members of this genus have 3, or rarely 4, styles, and the sepals are partly united into a tube that has 10 main veins.

This species is a perennial to 25 cm. Some stems erect but most decumbent and rooting. Leaves mostly spatulate, occasionally to elliptic or oblong. Calyx finely hairy. Petals separate. Stamens 10. Rare. Rich deciduous woods, usually on hillsides; nFla into cwGa. Mar–May. *S. baldwinii* Nutt.

S. ovata Pursh is another species with fringed petals. The stems are erect, to 1.5 m, the petals white and with only 8 linear lobes. Leaves about 6 pairs. Rare. Rich deciduous woods; ceMiss into seKy, and csNC.

33   Fire Pink

Silene virginica L.

Perennial to 60 cm, with several basal leaves and 2–4 pairs of cauline leaves, about 5 times as long as wide or longer. The basal stalk of the pistil is 2–4 mm. Sepals glandular hairy. Common. Thin woods and slopes; Ala into seOkla, seMinn, seNY, and seSC. Apr–July.

Two other pinks have similar deep red to scarlet flowers. *S. rotundifolia* Nutt. has 5–8 pairs of cauline leaves, nearly as broad as long. Rare. Cliff ledges and slopes beneath; cnAla into sO, eWVa, and nwGa. June–July. *S. regia* Sims grows to 1.6 m and has 10–20 pairs of broader cauline leaves, mostly 2–3 times as long as wide. Rare. Thin woods and prairies; seMo into swO, eKy. June–Aug.

34   Wild Pink

Silene caroliniana Walt.

Tufted perennial to 25 cm, from a thin deep taproot. In var. *caroliniana,* which is shown in the photograph, the basal leaf blades are obtuse, broadly oblanceolate, and finely hairy. Occasional. Sandy soils, usually open woods, often on slopes; seGa into eTenn, seO, sNH, and eNC. Mar–June. *S. pensylvanica* Michx.

In var. *wherryi* (Small) Clausen the basal leaf blades are acute, mostly narrowly oblanceolate, and glabrous. Occasional. Rocky upland woods, usually calcareous areas; cAla into cMo and csO. Apr–May.

35   Forking Catchfly

Silene dichotoma Ehrh.

Densely hairy annual or biennial to 45 cm, often branched at base. Leaves 4–9 × 0.4–3 cm, those at midstem opposite and lanceolate to oblanceolate. Inflorescence composed of 1 flowering branch per node. Sepals united, 10–15 mm in fruit; cup with hairy ridges and not inflated, lobes 2–3 mm; petals white, 15–22 mm, cleft. Capsule ovoid; seeds reddish. Native of Eurasia. Occasional. Fields, waste places; wNC into eTenn, ceMo, WVa, Mont, Que, and n and csVa.

S. noctiflora L. (Sticky-cockle; Catchfly), an annual or winter-annual, also has a hairy calyx and cleft petals but calyx lobes are 5–10 mm, petals white to pink, 5–10 mm. Flowers fragrant, opening in evening. Capsule opening by 6 teeth. Native of Europe. Easily confused with the much more common *Lycnis alba* Mill. (White Campion), but it has the calyx with 20 longitudinal veins of which 10 are prominent and 10 faint, whereas *S. noctiflora* has only 10 veins. Rare. Cultivated areas, waste places; cNC into Tenn, Mo, Utah, Wash, sCan, cVa. May–Sept.

36   Soapwort; Bouncing-bet
Saponaria officinalis L.
Glabrous perennial, spreading from seed and rhizomes. Stems to 150 cm, decumbent to erect. Leaves opposite. Sepals united into a cylindrical, or nearly so, tube and obscurely 5-nerved, the lobes about 2 mm. Petals separate, white to light pink. Styles 2. The scientific name comes from *sapo*, soap. The mucilaginous juice forms a lather with water and was used as a soaplike material in ancient Greece. The plant is poisonous when eaten. Occasional. Waste places, roadsides, fencerows, fields; nearly throughout the US. May–Oct.

NELUMBONACEAE: Lotus-lily Family

37   Yellow Nelumbo; Lotus-lily
Nelumbo lutea Willd.
Perennial from large tuber-bearing rhizomes. Leaf blades peltate, usually raised above the water, orbicular, to 70 cm wide. Flowers erect, above the water. Perianth parts numerous, the sepals grading into petals, stamens numerous. Pistils several, embedded in the top of an otherwise continuous and inverted fleshy cone. The tubers, which are starchy, are tasty when baked or boiled and seasoned. The interior of immature fruits is reported to be good either raw or cooked. In ripe fruits the shell is quite thick but can be removed by parching or cracking and the kernel inside eaten dry, baked, or boiled. Rare but common locally. In ponds, lakes, quiet streams, pools; Fla into eTex, eNeb, Minn, and Mass. June–Sept.

   N. nucifera Gaertn. is quite similar but the petals are pink. In scattered localities, Fla into La, Mo, and NC.

NYMPHACEAE: Water-lily Family

38   Water-shield
Brasenia schreberi J. F. Gmel.
Perennial with slim rhizomes creeping in mud; submerged parts heavily coated with mucilage. Leaf blades all floating, peltate, elliptic, not notched, 3.5–11 × 2–6.5 cm. Petals slightly longer and narrower than sepals. A food for ducks. Sometimes mistaken for a water-lily because of its leaves, but the flower is unlike those of native water-lilies. Occasional. In and at edge of freshwater ponds, lakes, sloughs, sluggish streams. CP in US, rarely inland; Fla to Tex, Neb, Man, BC, and Ore; Ont into NS and NC; Mex into C. Amer; W. Indies; Asia; Afr; Australia. June–freezing.

39   **Water-lily; Water-nymph**
Nymphaea odorata Ait.
Perennial from a large rhizome. Leaf blades floating or lying on mud, purple beneath, nearly orbicular, notched at the base. Sepals 4, nearly separate. Stamens fastened all over the ovulary. Seeds mature under water. Flowers fragrant. Petals white or rarely pinkish. Flower buds have been boiled, seasoned, and then eaten. Seeds of foreign species are known to be edible; those of our species are probably edible also. Common. In or at edges of fresh water; Fla into seTex, sMan, and Nfld. Mar–Sept. *Castalia o.* (Ait.) Wood.
    N. mexicana Zucc. is similar but petals are yellow. Rare. Fla into CP of NC and Ga, and e and sTex. *Castalia flava* (Leitner) Greene.

40   **Yellow Pond-lily**
Nuphar lutea (L.) S. Sm.
Perennial from a large rhizome. Leaves submersed, floating, or emersed. Leaf blades deeply cut at base. Stamens many, fastened under the ovulary. Seeds have been used as food in the same manner as those of *Nymphaea*. Common. In fresh water; Fla into csTex, seKan, seWisc, and sMe. Apr–Oct.
    Subspecies *sagittifolia* (Walt.) Beal is similar but leaf blades are more than twice as long as wide. Rare. eSC into ceVa. In ssp. *orbiculata* (Small) Beal leaf blades are orbicular. Rare. nFla into swGa; neGa.

RANUNCULACEAE: Crowfoot Family

41   **Golden-seal**
Hydrastis canadensis L.
A perennial to 50 cm from a yellowish rhizome. From this develops 1 leaf and/or stem with 2 leaves near the top. The leaves enlarge after flowering to as wide as 30 cm. The 3 sepals fall early. There are no petals, the flower being conspicuous because of the many stamens. Ovularies becoming a head of 1–2-seeded berries in fruit. Rare. Rich woods; nGa into Ark, eNeb, seMinn, cwVt, and swNC.

42   **Marsh-marigold; Cowslip**
Caltha palustris L.
Plants 20–60 cm; stems hollow, branched above. Basal leaves long-petioled, blade broadly cordate-rounded and usually with a deep and narrow sinus; main stem leaves with upwardly progressively shorter petioles and wider sinuses. Flowers 1.5–4 cm across; sepals 5–9, bright yellow to orange, 1–2.5 × 0.5–2 cm, broadly oval to narrowly obovate; corolla none; stamens 50–120, anthers 2 mm; pistils 4–14. Fruits follicles, 10–15 mm. Occasional. Wet woods and meadows, swamps, bogs, shallow water; circumboreal; S into Ia, Ill, Ind, WVa, se to n and swVa, and in the mts to Tenn and NC; Neb. Apr–June.
    C. natans Pallas has slender floating stems. Leaf blade kidney-shaped, 1–5 cm. Flowers 1 cm across, sepals pink to white; corolla none; stamens 12–25; pistils 20–40. Follicles 4–5 mm, in a dense head. Occasional. Ponds, lakes, and slowly flowing streams; circumboreal; S into nMinn and nMe. June–Sept.

43 **White Baneberry**
Actaea pachypoda Ell.
Perennial to 80 cm with 2 large compound leaves. The flowers are white in a single short compact raceme; sepals 4–5, falling quite early; petals 3–7, equal-sized, 3–5 mm, soon dropping; stamens numerous, the filaments thicker upward; stigma wider than the ovulary. Ovules and seeds several to many; fruiting pedicels very stout; fruit a white berry. The plants, especially the berries, may be poisonous. Occasional. Rich woods; nFla into seLa, eOkla, Mo, eMinn, and NS. Apr–May.
A. alba (L.) Mill.
    A. rubra (Ait.) Willd., Red Baneberry, has red, rarely white, berries; the fruiting pedicels are slender, and the ovulary is wider than the stigma. Occasional. Rich woods; circumboreal, in Amer. Alas into Lab and Nfld; S into Ia, nInd, nNJ, and Conn. May–June.

44 **Black Snakeroot**
Cimicifuga racemosa (L.) Nutt.
Perennial to 2.5 m, erect or the top bending. Leaves large with 3 pinnately compound divisions. Flowers in a large terminal raceme with a few to several smaller lateral ones. Sepals inconspicuous, falling as the flower opens. Petals absent. Stamens many. Pistil 1 and sessile. Fruit a several-seeded follicle. Common. Rich woods; neCP of Ga into nArk, sMo, and nwNC. May–July.
    C. americana Michx. is quite similar except there are 3–8 stipitate pistils, and later fruits, for each flower. Occasional. Rich woods; mts of nGa into BR of NC and Tenn, seKy, cPa, and mts of Md. July–Sept.

45 **Wild Columbine**
Aquilegia canadensis L.
The red and yellow flowers hang from slender nodding stems, causing the 5 long hollow spurs, 1 from the base of each petal, to point upward. The sepals are petal-like. The leaves are unusual in that they are divided 2–3 times, each time into 3 parts. Plants usually have leaves at the base and on the erect stems. The upper part of the plant is branched but not densely so. Occasional to common, especially in calcareous soils. In the open or in rocky woods, usually on slopes, and rarely in bogs; wFla into Tex, Man, NS, and nSC. Mar–Aug.

46 **Dwarf Larkspur**
Delphinium tricorne Michx.
All species of Larkspur are poisonous when eaten, but not equally so. In the wUS livestock losses have been heavy.
    This species is not always small but averaging smaller than our other species. A perennial rarely over 60 cm from tuberous roots. Stems with fine hairs. Flowers white to blue or violet, each with a single spur. Seed pods usually 3 and spreading, hence the name tricorne (three-horned). Occasional. In rich woods and in open rocky areas, especially calcareous; nwGa into eOkla, eNeb, wPa, nVa, and csNC. Mar–May.
    D. carolinianum Walt. is similar but is usually taller, has more flowers, and the seed pods are erect and not spreading. Occasional. Dry places in thin woods and

in open, often in rocky or sandy soils; neTex into swMo, cw and cIll, La, nw and Pied of Ga, and cwSC. Apr–July.

47   Windflower
Anemone lancifolia Pursh
Perennial to about 30 cm, from an elongate rhizome about 2 mm across, with basal and cauline leaves, the latter whorled and with 3 leaflets, the lateral ones not deeply cut. Flowers single on erect pedicels. Sepals usually 5–7, white. Petals absent. Fruits are achenes and numerous. Occasional. Rich woods; cSC into neAla, swPa, and cNC. Mar–May.

   A. quinquefolia L. is similar but has leaves with 5 leaflets, or with 3 leaflets and the lateral ones deeply cut. Common. Rich woods; Pied of Ga into nMiss, sQue, and NJ.

48   Liverleaf; Hepatica
Hepatica nobolis P. Mill. var. *obtusa* (Pursh) Steyerm.
Perennial from a short rhizome. Aboveground stems absent. Leaves evergreen, the blades with blunt lobes, purplish beneath. Flowers appear before the leaves of the year. Sepals petallike, bluish or less often pink or nearly white. The sepallike struc-tures beneath these are persistent bracts. Occasional. Rich woods, usually on slopes; nwFla into e and nArk, sMan, NS, and Va. Feb–Apr. *H. americana* (DC.) Ker.

   H. nobolis var. *acuta* (Pursh) Steyerm. is very similar, being generally larger and the leaf blades having acute lobes. Occasional. Similar habitats, nwSC into nGa, nArk, cMinn, sQue, and sMe. Mar–Apr. *H. acutiloba* DC.

49   Bulbous Buttercup
Ranunculus bulbosus L.
   Members of this genus have 3–5 green sepals, 5 yellow petals, many yellow stamens attached directly beneath the several to many greenish pistils. The fruits are achenes. There are perhaps 30 species of *Ranunculus* in the eUS. Flowers vary in width from 3 to 25 mm. Leaves are entire to lobed or palmately or pinnately divided. Identification to species is mostly dependent on the fruit characters and is often difficult.

   This species is a perennial from a firm swollen base, which is not a true bulb, as the name might imply. Plants may reach 60 cm and are conspicuous because of the many bright yellow flowers. This species and others have been reported occa-sionally to cause severe irritation when eaten. Common locally. Generally in low open areas; La into Ill, Ont, Nfld, NC, and cGa. Apr–June.

50   Early Buttercup
Ranunculus fascicularis Muhl. ex Bigelow
Hairy perennial usually with short tuberous roots. The achenes are smooth, flat-tened, and with a distinct margin. Beak of achene straight, more than half as long as the body. Common. Open areas or woods, wet to dry; Fla into eTex, Minn, and NH. Mar–Apr.

   R. hispidus Michx. var. *nitidus* (Chapm.) T. Duncan is similar but is glabrous or nearly so and the roots thick and fibrous but not tuberous. Occasional. Wet habitats, in open or woods; Ga into eTex, eNeb, sMan, and Me. Apr–July. *R. carolinianus* DC.

51   Kidney-leaf Buttercup

Ranunculus abortivus L.
An erect glabrous annual, conspicuous vegetatively. Often several stems originate from the plant's base. Flowers many, inconspicuous. Achenes plump, with a beak shorter than 3 mm. This is the most abundant of the species having crenate and unlobed blades on the basal leaves. Common. Usually in moist places, fields, gardens, borders of yards, low woods; Fla into eTex, Colo, Wash, Alas, and Nfld. Mar–June.

R. *allegheniensis* Britt. is similar but the achene beak is 0.6–1 mm and curved. Rare. Rich woods; wNC into eTenn, seO, sVt, and Mass. *R. micranthus* Nutt. ex T. & G. is also similar but the stem and leaves are hairy. Rare. Rich woods; nAla into neOkla, cInd, cMd, and cnNC.

52   Rue-anemone

Thalictrum thalictroides (L.) Eames & Boivin
Perennial to 25 cm from small tuberous roots. Somewhat similar to *Anemone lancifolia* but leaflet tips are rounded. Leaves at top of stem, with several stalked leaflets. Sepals 5–10, white to pinkish, 5–18 mm. The fruits prominently 8–10 ribbed, 1-seeded, and indehiscent. In some localities the tuberous starchy roots are cooked and eaten. Common. Rich woods, nwFla into eOkla, seMinn, and sNH. Mar–May. *Syndesmon thalictroides* (L.) Hoffmg. ex Britt. *Anemonella thalictroides* (L.) Spach.

Isopyrum biternatum (Raf.) Torr. & Gray is similar but leaves usually arise from more than one level on the stem and the fruits are 2–3-seeded and split open at maturity. Occasional. Rich woods; Fla into neTex, seMinn, sOnt, and cNC.

53   Tall Meadow-rue

Thalictrum pubescens Pursh
Perennial to 2.5 m. Leaves compound, the upper ones sessile. Leaflets glabrous, usually over 15 mm. Male flowers are conspicuous and on one plant, as seen in picture, and the less conspicuous female flowers are on another. Fruits sessile. Occasional. Wet places in thin woods or open; NC into nMiss, sOnt, and NS. May–Aug. *T. polygamum* Muhl.

There are several similar species. In *T. dioicum* L. the upper leaves have petioles. Occasional; cGa into nArk, sMan, and cMe. Mar–May. In *T. revolutum* DC. the leaflets and fruits have very small stalked glands. Occasional. Drier places in woods or open; nwFla into nArk, sOnt, and Mass. In *T. macrostylum* Small & Heller the leaves and fruits are glabrous and the leaflets less than 15 mm. Rare. Rich woods and moist meadows; SC into neAla and seVa.

54   Lady-rue

Thalictrum clavatum DC.
Inconspicuous but easily recognized by its stalked, curved (scimitar-shaped) fruits. Plants are perennial, to 60 cm. The leaves are twice-divided into 3 divisions.

The flowers have both stamens and pistils but lack petals. Occasional. Moist places, often in deep shade but occasionally in the open; nSC into nAla, cWVa, and mts of NC. May–July.

One other species, *T. coriaceum* (Britt.) Small, has stalked fruits but they are obliquely ovoid and not scimitar-shaped. Plants may grow over 1 m. Occasional. Rich woods; nGa into swPa, and nVa.

BERBERIDACEAE: Barberry Family

55   Mayapple; Mandrake
Podophyllum peltatum L.
Glabrous perennial to 50 cm from long rhizomes. Flowering plants with a single flower from the junction of the petioles of the 2 similar leaves. Nonflowering plants have a single leaf. Leaves peltate and deeply lobed. Sepals fall early. Petals 6–9. Stamens twice as many as the petals. Ripe fruit a yellow to reddish 1-celled berry with a peculiarly flavored pulp that is sparingly eaten raw and used to make marmalade. Seeds and vegetative parts poisonous. Long considered a medicinal plant. Indians and early settlers used it for a variety of ailments. Research has been conducted on at least 16 physiologically active compounds. Because of the poisonous nature of the plant, extracts should be used only when obtained under a doctor's prescription. Common. Rich woods or moist meadows; Fla into eTex, Minn, wQue, and Va; escaped eastward. Mar–May.

56   Twinleaf
Jeffersonia diphylla (L.) Pers.
Stemless glabrous perennial from a short rootstock. Petioles at first shorter than the flower stalk, later elongating to as much as 50 cm. The 2 leaf blades are entire or toothed. They enlarge after the plant flowers, sometimes to 15 cm. Sepals usually 4, petallike, and dropping early. Petals usually 8, to 2 cm. Stamens 8. Fruit erect, to 3 cm, opening by a halfway around horizontal cleft, the upper part making a lid that is hinged at the back. Named in honor of the third president of the US, Thomas Jefferson, for his knowledge of natural history. The only other species of this genus occurs in Asia. Occasional. Rich woods, usually calcareous soils; nwGa into nAla, seMinn, sOnt, nNY, and nVa; swNC. Mar–May.

57   Umbrella-leaf
Diphylleia cymosa Michx.
Flowering plants have a single cluster of white flowers and 2, rarely 3, leaves, which are at different levels and are dentate, peltate, and 2-cleft. Nonflowering plants have a single large leaf. Sepals fall early. Unlike most other dicots the sepals, petals, and stamens number 6 each. The flowers are similar to those of Mayapple but smaller. The blue fleshy fruits are conspicuous late in the growing season. The genus is represented by this species and one in eAsia. Occasional. Cool moist woods in mts; Ga into Va. May–June.

58   Blue Cohosh
Caulophyllum thalictroides (L.) Michx.
Glabrous perennial to 80 cm from a knotty rootstock. A leaf with 3 long stalks and many leaflets is on the upper part of the stem. Above this is another (rarely 2) similar but smaller leaf. At or near the stem tip are 1–3 clusters of inconspicuous flowers. Sepals 6, petallike, yellowish-green to greenish-purple, with 3–4 sepallike bracts beneath them. The 6 petals are merely small glandlike bodies. The seeds are blue when mature and are exposed, growing faster than the ovulary and splitting out of it. Considered poisonous, especially the seeds. Occasional. Rich deciduous woods; nGa into nAla, Mo, seMan, NB, and cnSC. Apr–May.

PAPAVERACEAE: Poppy Family

59   Bloodroot
Sanguinaria canadensis L.
Perennial from a thick rhizome with abundant red juice. Flower single on a leafless stem. The 2 sepals drop early. Petals usually 8 but vary to 16. The single leaf continues to enlarge after the petals drop and may reach 25 cm wide. Stamens about 24. Fruit an ellipsoid 1-celled capsule with 2 rows of seeds on the inside surface of the wall. The juice has been used as a dye. Baskets and cloth colored with the juice are still occasionally available locally. The plants are likely poisonous when eaten. Common. Rich woods; Fla into eTex, Man, and NS. Feb–Apr.

60   Celadine-poppy; Wood-poppy
Stylophorum diphyllum (Michx.) Nutt.
A 30–50 cm perennial. Leaves thin, mostly basal, long-petioled; blade broadly oblong to ovate in outline, pinnately divided almost or to the midvein into 5–7 oblong to obovate segments. The single pair of main stem leaves smaller than the basal ones. Sepals hairy; petals 2–3 cm; stamens 3–4. Fruit an ovoid capsule. Occasional. Rich woods and bluffs; nArk into Wisc, sMich, wPa, swVa, and nwGa. Mar–May.

FUMARIACEAE: Fumitory Family

61   Dutchman's-breeches
Dicentra cucullaria (L.) Bernh.
There is some evidence that members of this genus are poisonous when eaten.
Flowers not fragrant. Stems and leaves from a fleshy, loosely scaly, pink to light-colored bulb, the scales pointed. The 2 diverging spurs of the corolla base provide the legs of the "breeches," which are turned up because the flowers are nodding. Occasional. Rich woods; nGa into Ark, neOkla, ND, Que, NS, and cNC; eOre and eWash. Mar–Apr.

62   **Squirrel-corn**
Dicentra canadensis (Goldie) Walp.
Flowers fragrant. Stems and leaves arising from a rhizome bearing yellow rounded
pea- or cornlike structures from which this species gets its common name. The
corolla base is merely cordate. Occasional. Rich woods; nGa into Tenn, Mo,
seMinn, seQue, NS, and eNC. Apr–May.

63   **Bleeding-heart**
Dicentra eximia (Ker) Torr.
Flowers similar to those of squirrel-corn but flesh-colored to dark pink and in
panicles instead of a simple raceme. Robust. Rare. Rich, rocky woods; wNC and
eTenn into eVa, WVa, wNY, and NJ. Apr–Aug.

64   **Pale Corydalis**
Corydalis sempervirens (L.) Pers.
Erect biennial to 70 cm. Flowers in a terminal raceme or panicle; pedicels erect;
petals pink with yellow tips, the 1 spurred petal 10–15 mm. Fruit an erect or
nearly so many-seeded columnar capsule 25–35 mm. Occasional. In thin woods
or open; rocky places; nGa into Tenn, Ill, Mont, sBC, Alas, Nfld, n and swVa,
nwNC, and nwSC. May–Sept.
   C. flavula (Raf.) DC. has similar flowers and leaves but petals uniformly yellow,
the spurred petal 7–9 mm, and fruit drooping. Occasional. Rich woods on slopes
or in bottomlands; northward on rocky or sandy slopes, thin woods, shores; cnFla
into sw and nwGa, Tenn, Miss, La, eOkla, Minn, sOnt, Conn, Va, cSC. Mar–May.

BRASSICACEAE: Mustard Family
Members of this family have 4 separate sepals, 4 separate petals, 6 stamens (2 long
and 4 short) or rarely 2 or 4 of the same length, and a 2-carpelled fruit that splits
open at maturity and retains the central partition. Identification to genus and
species often is not easy, being based chiefly on fruit characters.

65   **Warea**
Warea cuneifolia (Muhl.) Nutt.
Erect annual to 1.2 m, branching in upper part. Leaves all simple, the blades
cuneate. Flowers clustered in tight terminal racemes. Petals 4, clawed. Stamens 6,
4 long and 2 shorter. This is the only genus in this family in which the pistil has a
stipe that elongates into a prominent stalk as the fruit matures. Occasional. Sand-
hills; cnFla into seAla, CP of Ga, and cwCP of NC. July–Nov.
   W. amplexifolia (Nutt.) Nutt. (*W. sessilifolia* Nash) has rounded leaf bases.
Rare. Sandy soils, pinelands or in open; cn and nwFla. *Cleome hassleriana* Chodat
(CAPPARACEAE) has similar but larger flowers, fruits, and upper leaves. The lower
leaves are palmately compound, leaflets 5–7, and the 6 stamens are the same size.
Rare. Waste places, cultivated areas; Fla into eTex, ceMo, and Conn. June–Frost.
C. houttenea Raf.; *C. spinosa* misapplied.

66 **Peppergrass; Pepperwort**
Lepidium virginicum L.
Glabrous or minutely hairy annual or biennial to 90 cm, the plant illustrated is
unusually small. Basal leaves not clasping stem; blade incised to pinnate. Petals 4,
equal, white; stamens 2. Fruits 3 – 3.5 mm, orbicular, smooth, faintly notched at
tip, 2-celled, winged and flattened perpendicular to partition. Seeds 1 – 2 per fruit.
Common. Dry or moist soil in open, fields, gardens, yards, roadsides, waste places;
nearly throughout US and sCan into Nfld. Apr – June, sporadically until frost.
    L. campestre (L.) Ait. f. (Field Cress) is similar but densely short-hairy, leaves
clasping stem, stamens 6, fruits 5 – 6 mm and conspicuously notched at tip. Native
of Europe. Common. Similar habitats and distribution but uncommon S of NC,
Tenn, and nArk. Mar – June.

67 **Wart-cress; Carpet-cress**
Coronopus didymus (L.) Sm.
Prostrate to ascending annual, often forming mats; distinctive because of its fetid
odor and fruits notched at both ends, with each of the 2 1-seeded halves nearly
globose. Leaf blades oblong, to 3 cm, deeply pinnately lobed. Fruits ca 1.5 × 2.7
mm, the surface deeply wrinkled. Native of S. Amer. Common. Yards, fields,
disturbed habitats, waste places, roadsides, cultivated areas. Tex into Mo, O, Que,
Nfld, e and cVa, ne and csNC, and Fla. Feb – Oct.

68 **Sea-rocket**
Cakile edentula (Bigel.) Hook.
Cakile species are succulent annuals with flowers in terminal racemes. Sepals
green. Petals 4, white to pink or purple. The unusual fruit is distinctly divided into
2 differently shaped segments. The terminal segment gradually tapers to a short
beak, is usually 1-seeded, is dry and corky when mature, eventually breaks off, and
can float great distances. The basal segment is persistent and 1-seeded or seedless,
usually falls later and remains nearby.
    In this species the inflorescence is linear; pedicel on fruits not as broad as axis
of infrutescence, fruit 4 – 7 mm across, and its upper segment 4- or 8-ribbed when
dry. Common. Sea beaches, low active dunes, overwash areas, edges of salt
marshes; eFla into Nfld. Mar – Oct.
    C. constricta Rodman is similar but mature fruits are only 3 – 4 mm across.
Common. Sandy shores, Fla into eTex. Mar – Aug. In *C. geniculata* (Robins.)
Millsp. the inflorescence is zigzag and the fruiting pedicels as broad as the axis.
Common. Sandy drift areas; La into Tex.

69 **Winter-cress; Yellow-rocket**
Barbarea vulgaris R. Br.
This species is a biennial to 80 cm, usually glabrous. Basal leaves with 1 – 4 pairs of
lateral lobes. Cauline leaves auricled. Fruit slender, 2 – 3 cm, the beak 1.8 – 3 mm.
Seeds in 1 row in each carpel of the fruit. Sometimes used as a salad or potherb
but somewhat bitter. The bitterness can be reduced or eliminated by changing the
cooking water 2 – 3 times. Common. Damp places, meadows, fields, thin woods;
NC into eOkla, Minn, sQue, and Nfld. Mar – June. *Campe barbarea* (L.) Wight.

70   **Watercress**
Rorippa nasturtium-aquaticum (L.) Hayek
Glabrous succulent perennial to 75 cm. Stem submersed or partly floating, creeping to decumbent, the ends erect. Leaves with 3–9 leaflets, the tip one the largest. Flowers in racemes. Fruit 15–20 × 1.5–2.5 mm. In the genus *Rorippa* the fruits are curved, pedicelled, and not stalked; the leaves are pinnately compound. Watercress is the only species of the genus in the eUS with white flowers. Frequently used as a green salad, but now often found in contaminated water. An important food for wildlife. Occasional, yet abundant locally. In clear running water or rooted in wet soil; nearly throughout the US. Apr–Aug. *Nasturtium officinale* R. Br.

   Cardamine pensylvanica Muhl. ex Willd. is often similar in general appearance but the fruits are straight. Also good as a green salad. Common. Aquatic and drier habitats. Throughout US and sCan.

71   **Bitter Cress**
Cardamine parviflora L.
Annual or biennial to 20 cm, glabrous or nearly so. Leaves pinnately compound and with 5–8 pairs of distinct entire leaflets. Fruits slender, erect. Seeds wingless. Common. In a variety of open and wooded habitats; Fla into Tex, Minn, and NS; Calif into BC. Mar–May.

   In the similar *C. hirsuta* L. the lower petioles are ciliate. Common. Fields, lawns, waste places; Fla into neTex, Ill, and seNY. In *Sibaria virginica* (L.) Rollins the seeds are winged and the leaves are only deeply divided and not compound. Common. In a variety of habitats, often a weed; Fla into Tex, seKan, sOhio, and eVa. *Arabis v.* (L.) Poir. The top parts of plants of *Arabidopsis thaliana* (L.) Heynh. are also similar to those of the above species, but the leaves are mostly basal, entire to indistinctly serrate, and not compound. Common. Throughout most of the US.

72   **Toothwort; Pepper-root**
Cardamine diphylla (Michx.) Wood
Perennial to 40 cm from an elongated rhizome of nearly uniform diameter. These should be uncovered only with care. Segments of basal leaves broad, similar to the cauline leaves. The rhizomes of this species and *C. angustata* have been used for food. With salt they taste somewhat like radishes. An interesting condiment when grated and mixed with vinegar. Occasional. Rich woods; upper Pied of Ga into nwAla, upper Mich, NB, NS, WVa, and nwSC; cNC. Apr–May. *Dentaria d.* Michx.

   C. angustata O. E. Schule has similar leaves but the rhizome is segmented, the segments 2–4 cm. Occasional. Rich woods; cnSC into nwMiss, cnO, and NJ. Mar–Apr. *Dentaria heterophylla* Nutt.

SARRACENIACEAE: Pitcher-plant Family

73   **White-trumpet; Pitcher-plant**
Sarracenia leucophylla Raf.
In *Sarracenia* species the leaves are hollow and liquid-containing to varying degrees. Insects caught in the hollow leaves are digested, thus providing some of the

plant's food and mineral needs. The flowers are also unusual in that they all are nodding and the end of the style is expanded into a large, persistent, 5-lobed umbrellalike structure.

This species is a perennial to 120 cm. The leaves are erect, their upper parts white and veined with reddish-purple. Occasional. Bogs, low savannas; sMiss into swGa and nwFla. Mar–Apr. *S. drummondii* Croom.

In *S. rubra* Walt. the petals are also red and the leaves erect, but the leaves are green, narrow, and shorter, to 50 cm, and have a small opening. Hoods are only 1–3 cm wide. Rare. Similar places; nwFla into seMiss, cAla, CP of Ga, and seNC; nwSC and adjNC. Apr–May.

74   Parrot Pitcher-plant
Sarracenia psittacina Michx.
Perennial with decumbent leaves in a basal rosette usually mostly hidden by other vegetation. The plant in the photograph was on the edge of a depression and easily seen. The hood almost covers the opening in the leaves. Occasional. Sphagnum bogs, wet savannas, open pinelands; seLa into nwFla, sAla, and CP of Ga; neFla. Mar–July.

S. purpurea L. also has decumbent leaves but the hood is turned away from the opening. The inner surface of the hood bears many stiff hairs that point toward the open part of the leaf. These hairs may help trap insects in the hollow leaves. Flowers are red and up to 40 cm above the ground. Rare. Scattered distribution; nwFla into seLa and swGa; seGa; neGa into swNC; se and cSC into eVa, eMd, Pa, neIll, cwMinn, sOnt, sLab, and Nfld.

75   Hooded Pitcher-plant
Sarracenia minor Walt.
Perennial, with erect leaves to 40 cm, rarely to 60 cm. Opening near top of leaves covered by a hood. Base of hood and upper part of leaf spotted with large white or translucent areas. Sepals yellowish, petals yellow. Common. Bogs, wet savannas, and open pinelands; cFla into seGa and seNC. Mar–May.

76   Trumpet Pitcher-plant
Sarracenia alata Wood
Perennial, the leaves erect, to 85 cm, yellow to greenish yellow, the veins reddish, the margins on the base of the hood only slightly turned backward. Petals yellow. Occasional, but abundant locally. Bogs, wet savannas, low open pinelands; swAla into seLa; cwLa into eTex. Mar–Apr. *S. sledgei* MacFarlane.

In *S. flava* L. the leaves may reach 125 cm but usually are a little over half that. They are yellowish, sometimes veined with maroon, or occasionally entirely maroon. The lower margins of the hood are strongly turned backward. Occasional, once common but now rapidly disappearing. nwFla into sAla, CP of Ga, and seVa; cwNC. Mar–Apr. *S. oreophila* (Kearney) Wherry is similar but has several shorter sword-shaped leaves with no cavities in addition to the erect hollow ones. Petals greenish-yellow. Rare. Wet places; c and neAla.

DROSERACEAE: Sundew Family

77   Sundew
Drosera brevifolia Pursh
Members of this genus have stalked glands on the leaves. These glands exude a clear sticky secretion that aids in insect catching.

In this species the leaves are basal, the blades much wider than the petioles, and the scape very finely glandular hairy. Occasional. Moist places, in thin pinelands or open; Fla into seLa, neAla, and adjTenn, CP of SC, and c and ceNC. Mar–May. *D. leucantha* Shinners.

In *D. capillaris* Poir. the scape is glabrous and the leaf blades longer than wide. Common. Similar places; Fla into eTex, seTenn, CP of Ga, and seVa. Mar–Aug. In *D. rotundifolia* L. the leaf blades are wider than long. Rare. Sphagnum bogs and seepage places; nwSC into neGa, neIll, Mont, Calif, Alas, and Greenl. June–Aug.

CRASSULACEAE: Orpine Family

78   Woods Stonecrop; Sedum
Sedum ternatum Michx.
Perennial to 15 cm, forming mats, the flowerless branches prostrate or spreading, and with about 6 broad flat crowded leaves at their tips. Lower leaves whorled, not as wide. Leaves below flower clusters spatulate. Petals white. Stamens dark red. Occasional. On rocks, often calcareous, rich woods; nSC into nAla, cwArk, neMo, sMich, and wMass. Apr–June.

S. nevii Gray is similar but all leaves are alternate and glaucous. Leaves below flowers are slender or absent. Rare. On limestone or shale rocks; swNC into nAla, cnTenn, neWVa, and cNC. May–June. *S. glaucophyllum* Clausen.

79   Lime Stonecrop
Sedum pulchellum Michx.
Winter annual to 45 cm. Plant greens, succulent. Leaves on flowering stems numerous, alternate, sessile, narrow, often almost as thick as wide, auricled at base. Petals rose-colored to white. Anthers dark red. Carpels separate, loosely ascending. Leaves of certain European species have been used in salads and as a potherb; some of our species may also be usable. Occasional. Shallow soil on and in crevices of limestone rocks; nwGa into eTex, seKan, cwMo, and wVa. Apr–May.

S. pusillum Michx. is also annual. Leaves green, succulent, thick, spatulate to obovate or elliptic, not auricled at base. Rare. Granitic rock outcrops; csNC into ceSC, and in Pied of Ga W to vicinity of Atlanta. Mar–Apr.

80   Roseroot
Sedum rosea (L.) Scop.
Succulent perennial from a thick scaly rootstock; stems erect, 10–40 cm, arising from scale axils. Leaves sessile, thick, oblanceolate to obovate, 2–4 cm. Flowers mostly or all unisexual, sexes on separate plants. Inflorescence compact, repeatedly branched; flower parts in (3)4(5)'s; petals 2–4.5 mm, oblanceolate to elliptic-oblong. Occasional. Cliffs, ledges; circumboreal, in N. Amer S into Calif and NM;

S along the Atlantic coast into NS and Me; separately inland in Pa into NY, NH, NC, and Tenn. May–July.

81 **Diamorpha; Elf-orpine**
Diamorpha smallii Britt. ex Small
Winter annual to 10 cm. Plants reddish, succulent. Leaves alternate, fleshy, the blades obtuse and nearly circular. Petals 4–5, white. Stamens 8 or 10. Carpels 4–5, united at base for one-third to one-half of their length. Carpels slow to lose their seeds. Occasional. Forms spectacular dense colonies in and around depressions on granitic outcrops, sometimes on exposed sandstone rocks; Pied and outer part of BR Mts of NC, SC, and Ga; neAla, nwGa, and the adj part of Tenn. Mar–May. *D. cymosa* (Nutt.) Britt. *Sedum smallii* (Britt.) Ahles.

SAXIFRAGACEAE: Saxifrage Family

82 **False Goatsbeard**
Astilbe biternata (Vent.) Britt.
Coarse perennial herb to 2 m. Leaves petioled, 2–3 × compound, terminal leaflet usually 3-lobed. Flowers in a terminal glandular-hairy panicle; sepals 5, persistent, 0.7–1.5 mm; petals 1.5–4 mm, withering and persisting, or absent in male flowers; stamens 8–10. Fruit a 2-celled follicle, partly fused or separate as 2 follicles. Closely resembling *Aruncus* of the Rosaceae. Occasional. Woods in mts; nGa into Ky, WVa, and wVa. May–July.

83 **Mountain Saxifrage**
Saxifraga michauxii Britt. ex Small
Perennial with basal leaves only. Sometimes with bracts similar to small leaves in the lower parts of the inflorescence. Blades oblanceolate to spatulate-oblong. Sepals reflexed in fruit. Petals unequal in size, 3 large and cordate and with a pair of yellow spots, the other 2 smaller and lanceolate. Fruit with distinct longitudinal veins. Tender young leaves of this and other species have been used in salads. Common. Moist rocks and slopes; nGa into cnTenn, nwWVa, and wVa. Mar–Nov.

   S. micranthidifolia (Haw.) Steud. also has relatively long leaves, the blades oblong to lanceolate or oblanceolate. Sepals reflexed in fruit. Filaments thickened above. Rare. Seepage slopes and moist rocks; nwSC into neGa, eTenn, sePa, and cVa. Mar–June.

84 **Early Saxifrage**
Saxifraga virginiensis Michx.
Perennial with basal leaves only. Leaf blades ovate to oblong, crenate or shallowly dentate. Sepals erect to spreading when flower is open. Petals equal in size. Stamens 10. Hypanthium partly fastened to the ovulary, which is 2-celled. Common. Rock exposures, hillsides, thin woods; ceGa into neArk, seMan, NB, and Va. Feb–May.

   Two species have relatively wider leaves and the hypanthium free from the ovulary. *S. careyana* Gray has filiform filaments and a fruit body under 3.5 mm. Rare. Moist rocky places and slopes; nwGa into eTenn, wNC, and wVa. May–

June. *S. caroliniana* Gray has filaments slightly broadened above and fruit bodies over 3.5 mm. Rare. Rocks and wet slopes; neTenn into nwNC, eVa, and adjKy. May–June.

85   Yellow Alpine Saxifrage

Saxifraga aizoides L.

Matted from a branching perennial rootstalk. Flowering stems 5–20 cm. Leaves alternate; blade thick, firm, linear to oblong, 10–20 × 2–4 mm, entire, often ciliate, apex with a narrow projection, bearing a minute pore just below. Stem leaves immediately below the inflorescence few, main basal ones circular to kidney-shaped. Flowers 5–8 mm across; corolla yellow usually dotted with orange. Rare. Circumboreal; S into BC, sAlta, nMich, c and wNY, and Vt. June–Sept.

86   Foamflower; False-miterwort

Tiarella cordifolia L.

A perennial with most, if not all, leaves basal. The flowers are in racemes and white or pink-tinged, except for the 10 yellow anthers. Common. Rich woods; swGa into Miss, Mich, and NS. Mar–June. *T. wherryi* Lakela.

The leaves of *Mitella diphylla* L., Miterwort, and some species of *Heuchera*, Alumroot, are similar. The Miterwort (nGa into Miss, eMo, Minn, and Me) has only 5 stamens, a very slender raceme, and beneath it 2 opposite leaves rather than a single leaf or none. Alumroots, which are widely distributed, have only 5 stamens and have panicles instead of racemes.

87   Alumroot

Heuchera villosa Michx.

Perennial with all leaves from a rhizome, the blades simple, sharply lobed, the terminal lobe triangular. Calyx with fine soft hairs, the tube at flowering under 3 mm. Occasional. Shaded cliffs and ledges; cAla into seMo, csInd, cWVa, cVa, and nwSc. June–Oct.

H. parviflora Bartl. also has a calyx with fine soft hairs but the leaf blades are nearly circular in outline. Occasional. Similar habitats; ePied of Ga into neAla, sIll, cw and ceWVa, and nwSC. July–Oct. In *H. americana* L. the calyx is very short glandular hairy, and the teeth and lobes of the leaves are obtuse. Common. Rich or rocky woods; swGa into neTex, ceMo, seMich, sOnt, Conn, and NC. Apr–June. In *H. longiflora* Rydb. the calyx is also glandular hairy but the tube is 3 mm or longer. Rare. Rocky woods; cwNC into neTenn, eKy, and sWVa; nAla. Apr–June.

88   Grass-of-Parnassus

Parnassia asarifolia Vent.

Instead of being a grass, as the name might indicate, it is related to Foamflower. Plants to 20 cm, each flowering stem with 1 nearly circular leaf, the other leaves basal, petioled, and with kidney-shaped blades. The petals have short stalks. The 5 fertile stamens are separated by shorter sterile ones. Occasional. Wet or moist places, in shade or open; neAla into eTex, eKy, ceWVa, and nwSC. Aug–Nov.

Two other species are similar but have the sterile stamens longer than the fertile: *P. grandifolia* DC. has pointed sterile stamens. Rare. Tex into seMo; nFla;

nwSC into eTenn, sWVa, and cwVa. Aug–Nov. *P. caroliniana* Michx. has sterile stamens with a rounded summit. Rare. Fla into sMiss; also in seNC. Aug–Nov.

ROSACEAE: Rose Family

Members of this family usually have stipules and many stamens, a hypanthium, and superior or inferior ovularies.

89 Goat's-beard

Aruncus dioicus (Walt.) Fern.
This species has male and female flowers on separate plants. Both kinds of flowers are almost white, the staminate flowers being the more conspicuous. Plants are erect, to 2 m, and bear a few large compound leaves and a large pyramid-shaped panicle of flowers. Occasional in rich woods; nGa into Okla, Ia, seNY, and nSC. May–June.

This species closely resembles *Astilbe biternata* (Vent.) Britt. (False Goat's-beard), a member of the SAXIFRAGACEAE, which may be identified by the 3-lobed terminal leaflet and by the 2 carpels and 10 stamens. The terminal leaflets of *Aruncus* have no lobes, and it has 3–4 carpels and 15 or more stamens.

90 Indian-physic

Porteranthus trifoliatus (L.) Britt.
A much-branched herbaceous perennial sometimes to 1 m. The alternate leaves have 3 leaflets and 2 narrow stipules 6–8 mm with the largest lobe about 1 mm across. The narrow white usually unequal petals are conspicuous. Fruits are small follicles. Common. Rich woods, frequently on slopes; Pied of Ga into Miss, seKy, sOnt, Mass, and nwSC; swMo. Apr–June. *Gillenia trifoliata* (L.) Moench.

P. stipulatus (Muhl. ex Willd.) Britt. has much longer stipules, the largest lobe to 10 mm or more across, and the lower leaves more dissected than the upper. Occasional. Rich woods; nGa into eTex, seKan, swNY, eVa, and cNC. May–June. *Gillenia stipulata* (Muhl.) Baill.

91 Cloudberry; Baked-apple-berry

Rubus chamaemorus L.
Stems 10–30 cm from creeping rhizomes, erect, unbranched, unarmed. Leaves simple, commonly 2–3, long-petioled; blade circular to kidney-shaped, to 9 cm across, with broadly rounded lobes. A solitary flower, 2–3 cm across, terminal on a long peduncle, unisexual. Fruit orange to red, 15–30 mm across, edible and much prized. Occasional. Bogs and other wet places; circumboreal; S into NB, Me, and NH. June–July.

92 Wild Strawberry

Fragaria virginica Duchn.
Perennial with long stolons. Leaves relatively thick, all basal, compound, leaflets toothed along margin. Flowers in peduncled clusters; sepals alternating with entire bracts; petals white. Fruit an enlarged red receptacle with many small achenes sunken in shallow pits in the surface. Wild strawberries are often preferred over cultivated ones. Common northward but rare southward. Barren fields, thin

woods on hillsides, cleared areas, prairies, roadsides; swGa into nLa, eOkla, Alta, seLab, Mass, NC, cSC, and neGa. Mar–July.

In *F. vesca* L. leaves are moderately thin and achenes are essentially borne on surface of hypanthium. Occasional. Rich often N-facing slopes, wooded bluffs, rocky woods, openings; Tenn into neMo, Neb, Man, Nfld, ce and swVa, and nwNC. May–Aug.

93   Cinquefoil: Fivefingers

Potentilla canadensis L.

Potentilla is a variable genus: leaves are palmately or pinnately compound; petals are white, yellow, or missing; stamens are 5–many. The species included here have palmately compound leaves, the 5 sepals are interspersed with bracts, and each flower produces several achenes.

This species is a perennial from a short rhizome to 20 × 8 mm. Stems soon become prostrate. Leaflets 5. First flower usually from node above the first well-developed internode. Common. Open well-drained places; Ga into Ark, Mo, Ky, neO, swOnt, and wNS. Mar–May.

P. simplex Michx. is similar but coarser, the stems more erect at first then arching and often rooting at tip, and the first flower usually from the node above the second well-developed internode. Common. Open, dry to moist places; Ga into eTex, Minn, and swNB. Apr–June.

94   Rough-fruited Cinquefoil

Potentilla erecta L.

Plants with several stems from a perennial base. Leaves with 7–9, rarely 5, leaflets, the lower leaves abundant and on long hairy petioles. Flowers abundant, 15–25 mm wide, on erect stalks in a broad terminal cluster. Petals showy, usually pale yellow, broadly notched at apex. Seedlike fruits with low curved ridges. Common in a variety of open habitats; Fla into Tex, Minn, and Nfld. Apr–July, occasionally to Sept.

P. intermedia L. is similar but has larger leaves with only 5 leaflets, flowers 8–10 mm wide, and petals about as long as the sepals. Rare. Mich to Nfld and S in mts to NC. *P. argentea* L., of a similar distribution but more frequent, is smaller, and leaflets are densely white tomentose beneath.

95   Silver-weed

Potentilla anserina L.

Plants bear a single rosette of basal leaves initially, soon developing long slender stolons and forming smaller rosettes of leaves at the nodes; blade densely hairy-matted overlaid with long straight hairs. Flowers 15–25 mm across, solitary on slender pedicels from the rosettes; petals yellow. Achenes ca 2.5 mm, about as thick, deeply furrowed on summit and back. Common. Circumboreal; S into NM, Ia, nInd, NY. May–Sept.

96   Indian-strawberry; Mock-strawberry

Duchesnia indica (Andr.) Focke

Plants with stolons, leaves, and fruits similar to those of Strawberry but fruits are tasteless and bracts that alternate with sepals are conspicuous and toothed. Native of Asia. Common. Usually in moist shady places; nFla into Okla, Ia, O, Nfld, and NC. Feb–frost.

97   **Barren-strawberry**
Waldsteinia fragarioides (Michx.) Tratt ssp. *doniana* (Tratt.) Teppner
A perennial much like a strawberry except the flowers have yellow petals and the fruits are dry. Leaves and flower clusters from rootstocks at the ends of creeping rhizomes. Leaflets 3 per leaf, each usually longer than wide. The petals are about as long as or shorter than the sepals and are 1–1.5 mm wide. Attractive as a wildflower in woodland gardens. Occasional. Rich woods, usually deciduous and on slopes; cwSC into nwAla, seKy, csVa, and cNC. Apr–June. *W. parviflora* Small.

   In *W. lobata* (Baldw.) T. & G. the leaves are simple, though shallowly lobed. Rare. Pied and adj mts of Ga. Apr–May. In *W. fragarioides* ssp. *fragarioides* the leaflets are about as wide as long and the petals are 2.5–6 mm wide and prominently longer than the sepals. Occasional. Moist or dry woods; cnNC into Ky, csMo, csInd, O, neMinn, sOnt, and sNB. Mar–June.

98   **Alchemilla**
Alchemilla microcarpa Boiss. & Reut.
Freely branched winter annual. Leaves palmately dissected, stipules ca ½ as large as leaves. Flowers in axillary clusters of 3–7, surrounded by the stipules and often unnoticed; sepals 4, ca 0.3 mm; petals absent; stamen 1. Fruit an achene ca 0.6 mm. Easily overlooked, and probably more common than records indicate. Native of Europe. Common. Weed of lawns, fields, and pastures; mostly Pied, rare in CP and mts; Del into Fla and Miss. Apr–May.

FABACEAE: Bean Family

Those members of this family represented here have a 1-carpelled pistil. Fruit commonly a 1-celled pod, most splitting along 2 sides but some are indehiscent and 1-seeded and others transversely divided into 1-seeded joints.

99   **Sensitive-brier**
Mimosa microphylla Dry.
This species is a perennial with prostrate to weakly arching stems. Leaves, stems, and fruits have numerous prickles. The many leaflets "go to sleep" when disturbed and at night; veins on underside obscure. Flowers bisexual or unisexual, in heads. Fruits with prickles on the ribs. This and next species are considered by some to be a single species, *M. quadrivalvis* L. Common. Open drier places; Fla into eTex, wIa, SD, Ky, WVa, and Va. May–Sept. *Schrankia microphylla* (Dry.) Michx.

   M. nuttallii (DC.) B. L. Turner is very similar, having raised lateral veins on the underside of the leaflets. Occasional. Ala into e and neTex, SD, and Ill. *Schrankia n.* (DC.) Standl. May–Sept.

   Desmanthus illinoensis (Michx.) MacM. ex B. L. Robins. and Fern., Bundleflower, has similar leaves but the plants are erect and the flowers are white or greenish and in smaller heads. Occasional. Pied of SC into nwGa, NM, ND, and O. All summer.

100   Wild Senna

Senna marilandica (L.) Link

Species of this genus have pinnately compound leaves with an even number of leaflets. Stamens 5 or 10. Petals 5, slightly unequal.

This species is an erect little-branched perennial to 2 m. Leaflets 5–10 pairs, over 25 mm, oblong to elliptic, mucronate. Stipules falling early. Upper 3 stamens lacking normal anthers. Fruit straight or curved, 50–100 × 8–11 mm, about 3 mm thick, with small appressed hairs or none, the segments of fruit longer than wide. Occasional. Usually moist places, thin woods or open; Fla into eTex, cnKan, Ind, swPa, and ceVa. July–Aug. *Cassia medsgeri* Shafer; *C. marilandica* L.

S. herbicarpa (Fern.) Irwin & Barneby is similar but the fruit has spreading hairs and the segments are as wide as long. Rare. Pied of Ga into sWisc and NH. July–Aug. *Cassia hebecarpa* Fern.

101   Coffee-weed; Sickle-pod

Senna obtusifolia (L.) Irwin & Barneby

Erect annual to 1.5 m. Leaflets 2–3 pairs, mostly 3–6 × 2–4 cm, broadly obtuse, an elongate gland between or just below the 2 lower leaflets. Flowers on pedicels over 10 mm. Upper 3–4 stamens lacking normal anthers, the 3 lowest with very large anthers. Fruit 15–45 mm, very slender, usually strongly curved, 4-angled or nearly so. Important as the alternate host of the tobacco etch virus disease. Toxic when eaten in considerable amounts. Common. Waste places, cultivated land, pastures; Fla into eTex, cwMo, Ill, sMich, and sePa. June–frost. *Cassia o.* L.

Some books incorrectly refer to this species as *Senna tora* (L.) Roxb. (*Cassia t.* L.), which is an Old World species with flowering pedicels under 10 mm, fruiting pedicels under 15 mm, and differences in the seeds. Rare if present in the eUS.

102   Coffee Senna

Senna occidentalis (L.) Link

Erect ill-scented annual to 2 m. Leaflets 3–6 pairs, 3–9 × 1.5–3 cm, chiefly acuminate. Stipules fall early. A gland near base of petiole. Upper 3 stamens much reduced. Fruit linear, 70–140 × 5–10 mm, straight or slightly curved, the margins thickened. Reported to be weakly toxic. Roasted and ground seeds have been used as a coffee substitute. Seed extracts are reported to have an antibiotic activity. Occasional. Waste places and cultivated land; Fla into sTex, sInd, Tenn, NC, and eVa. July–frost. *Ditremexa o.* (L.) Britt. & Rose. *Cassia o.* L.

103   Partridge-pea

Chamaecrista fasciculata (Michx.) Greene var. *fasciculata*

Annual to 1.2 m, erect to nearly prostrate. Leaflets 6–26 pairs, to 18 mm. Flowers 10–30 mm across. Fruit to 9 mm. Common. Open places, usually disturbed; Fla into s and cTex, eSD, sPa, and Mass. June–Sept. *Cassia f.* Michx.

Chamaecrista deeringiana Small & Penn., which also has large flowers, is a perennial from long horizontal roots. Rare. Rocky and sandy soils, open pinelands and scrub oak; Fla into sAla, and wCP of Ga. June–Sept.

104   Small-flowered Partridge-pea
Chamaecrista nictitans (L.) Moench
Stems glabrous or with small incurved hairs. Leaflets 6–15 mm. Flowers less than 10 mm across. Eastern US plants are var. *nictitans.* Common. Dry, especially sandy, soils, thin upland woods, dunes, often in disturbed places; Ark into Kan, cwMo, Ill, sMich, O, sVt, Mass, NC, cFla, and eTex. June–frost. *Cassia n.* L.

    Chamaecrista aspera (Muhl.) Greene also has small flowers but stems with longer and spreading hairs. Common. Sandy soils; sSC into coastal Ga and Fla. June–Oct. Perhaps best considered as *C. nictitans* ssp. *nictitans* var. *aspera* (Muhl. ex Ell.) Irwin & Barneby.

105   Thermopsis
Thermopsis villosa (Walt.) Fern. & Schub.
Perennial to 1.6 m, unbranched or with a few short vegetative branches. Leaves palmately compound with 3 leaflets. Stamens 10, separate. Calyx persistent. Pods hairy, compressed, not inflated, sessile or nearly so, erect. Sometimes used as an ornamental. Occasional. Thin woods or in open; mts of Ga and Ala, BR of Tenn and NC; cTenn. May–July. *T. caroliniana* M. A. Curtis.

    Members of this genus are often mistaken for species of *Baptisia*, which may be recognized by their stipitate ovulary and fruit, and inflated fruit.

106   Bush-pea
Thermopsis fraxinifolia Nutt. ex M. A. Curtis
Perennial to 90 cm, with several to numerous zigzag branches. Bracts usually shorter than pedicels. Calyx persistent, the tube glabrous or nearly so. Stamens 10, separate. Ovulary and pods sessile or nearly so. Occasional. Roadsides, thin woods, or in open, usually dry; mts of NC, SC, and Ga. Apr–July.

    T. mollis (Michx.) M. A. Curtis ex Gray is similar but has bracts longer than pedicels and the calyx tube with small fine hairs. Thin woods and open, drier places; mts Ga into neAla, eTenn, cVa, and Pied of NC. *T. hugeri* Small. Apr–June.

107   Yellow False-indigo
Baptisia tinctoria (L.) R. Br. ex Ait. f.
Members of this genus are perennials, have 10 separate stamens, stipe on the ovulary and fruit, and inflated fruit. Some species have been reported poisonous when eaten but information on this subject is lacking for most.

    This species is bushy-branched, to 1 m. Leaflets mostly shorter than 4 cm, black when dried. Flowers about 10 mm, in terminal racemes. Pod almost black at maturity. Plants used locally as a dye. Common. Thin woods, dry places; nFla into La, seOnt, and swMe; seMinn. Apr–Aug.

    B. lecontei Torr. & Gray is similar but the leaflets do not dry black and the lower flowers of each raceme have a pair of small bracts on their stalks. Occasional. Dry sandy soils, thin pinelands and scrub; csGa into n pen Fla. Apr–June.

108   Cream Wild-indigo
Baptisia bracteata Muhl. ex Ell.
Plants to 60 cm, loosely branched, with soft spreading grayish hairs. Flowers on pedicels mostly less than 1.5 cm in drooping, 1-sided racemes with persistent

bracts 5 mm or wider. Petals cream-colored. Occasional. Dry thin woods; neAla into Pied of Ga and SC; cwNC. Apr–May.

  B. bracteata Muhl. ex Ell. var. *leucophaea* (Nutt.) Kartesz & Gandhi is similar but with stiffer hairs and the flowers on pedicels mostly over 1.5 cm. Common. Sandy soils; eLa into e half of Tex, seMinn, swMich, nwInd, and wKy. In *B. cinera* (Raf.) Fern & Schub. the racemes are usually erect and have bracts less than 3 mm wide that drop off soon after the petals. Occasional. Sandy soils, scrub, and thin woods; cwSC into seVa and CP of NC; cNC. *B. leucophaea* Nutt.

109   White Wild-indigo
Baptisia alba (L.) Vent
Plants to 1.2 m, with few to many divergent branches. Leaflets 20–35 mm. Flowers 12–18 mm, in 1–several terminal racemes. Pods cylindrical, brown, not over 11 mm thick, divergent to erect. Common. Dry thin woods, low pinelands; e half of nFla into Pied of Ga, and cs and eVa. Apr–July. *B. albescens* Small.

  B. alba var. *macrophylla* (Larisey) Isley has flowers 2–3 cm and black pendent fruits 8–25 mm across. Thin woods, prairies, dry places; csAla into eTenn, seOnt, and O; cnFla into cPied of Ga, and cn and ceNC. *B. leucophaea* Nutt.; *B. minor* Lehm.

110   Blue Wild-indigo
Baptisia australis (L.) R. Br. ex Ait. f.
Glabrous glaucous plants to 160 cm, bushy branched, the upper branches ascending and the lower ones spreading. Racemes 1–several, erect or nearly so, with a few to 35 flowers. Petals blue. Fruits thin-walled, plump, 3–6 cm, and with a long persistent claw-shaped style. Attractive and easily grown. Occasional. Thin woods and borders, limestone glades; nGa into neAla, ceInd, swNY, cMD, and cnNC; also cnTex into seNeb, cwIll, and neArk; escaped elsewhere. Apr–May. *B. minor* Lehm. Some divide this species into two varieties. *B. australis* var. *australis* has leaflets 4–8 cm and the stipe on the fruit about twice as long as the calyx. In var. *minor* (Lehm.) Fern. the leaflets are 1.5–4 cm and the stipe about as long as the calyx. A few persons treat these as species.

111   Pineland Wild-indigo
Baptisia lanceolata Walt. var. *lanceolata*
Plants to 90 cm, widely and bushy-branched. Leaflets spatulate to oblanceolate, mostly 4 cm or longer. Flowers single in axils of outer leaves or also 2–4 in short terminal racemes, standard 15 mm or longer. Pod subglobose to ovoid, 15–22 mm. Common. Dry sandy hills and open sandy pinelands; swAla into cwSC and nFla. Mar–May.

  Two other species are similar. *B. nuttalliana* Small has obovate-cuneate leaflets and pods 5–13 mm. Common. Woodlands, sandy soils; csLa into e and cTex, seArk, and cwMiss. Apr–May. *B. l.* var. *tomentosa* (Larisey) Isley has broadly elliptic leaflets. Occasional. Dry sandy pinelands; swGa into nwFla and swAla. Apr–May. *B. elliptica* Small.

112   **Hairy Wild-indigo**
Baptisia arachnifera W. H. Duncan
The dense tight cobwebby hairs on stems, leaves, ovulary, and fruits identify this species. Rare. Thin sandy pinelands and along railroads. Repeated fires favorable for this species; Wayne and Brantly Cos., Ga. Late June–Aug.

113   **Showy Crotalaria**
Crotalaria spectabilis Roth
Members of this genus have 10 united stamens, the anthers alternately of 2 sizes, and sessile inflated fruits.
   This species is an annual, to 1.5 m. Leaves simple, glabrous, mostly 6–20 cm. Stipules and bracts at base of flower stalks ovate, 5 mm or longer. Any part of the plant, especially the seeds, is poisonous when eaten. Growing plants are reported to reduce or eliminate nematodes from the soil. Occasional. Fields, roadsides, orchards; Fla into e half of Tex, cMo, Tenn, NC, and seVa; absent in BR. Mar–Oct.
   C. retusa L. is similar but the stipules and bracts are narrow and under 2 mm. Rare. Fla; eTex; seNc. July–Sept.

114   **Rabbit-bells**
Crotalaria rotundifolia Walt. ex J. F. Gmel.
Perennial to 40 cm from a prominent taproot. Stems erect, several to many, with appressed hairs; or decumbent and with appressed or spreading hairs. Leaves simple, upper surface hairy. Plants poisonous. Sandy areas, pinelands, thin woods; Fla into seLa, cAla, cGa, and seVa; also neGa and cNC. Mar–Nov.
   Other similar species are: *C. sagittalis* L., which is erect to strongly ascending and has spreading hairs on the stems. Upper surface of leaf hairy. Dry places in thin woods or open; Ala into eTex and Kan, cMinn, swMich, eKy, Tenn, NC, csVa, eWVa, sVt, and Mass. May–Oct. *C. purshii* DC. with glabrous upper surface of leaf. Sandy areas, pinelands, thin upland woods; nFla into seLa, sGa, and seVa. Apr–Aug.

115   **Lady Lupine**
Lupinus villosus Willd.
Our species of this genus are perennials with 10 united stamens, the anthers alternately of 2 sizes, the fruits sessile and flattened, and possibly poisonous.
   This species to 50 cm, from a deep woody taproot. Stems mostly decumbent, a few to many in a dense clump. Leaves evergreen and simple, which is unusual for lupines since almost all have deciduous palmately compound leaves. Standard purple to reddish with a deep reddish-purple spot. Pods shaggy with hairs 4–5 mm. Common. Sandhills and thin woods on dry sandy places; coastal Miss into nFla, lower CP of Ga, and seNC. Mar–May.
   L. diffusus Nutt. is similar but the standard is blue with a white spot and the fruits have appressed hairs about 2 mm. Occasional. Similar habitats; coastal Miss into Fla, eCP of Ga into sCP of NC.

116   Sundial Lupine

Lupinus perennis L. ssp. *perennis*

Plants to 70 cm from creeping underground stems. Leaves palmately compound, with 7–11 leaflets. Petals rarely white. Pods with short to long hairs. Thin woods and in open sandy soils; seTex into cGa, nFla, neNC, cVa, and sMe, and W into neWVa, nO, nIll, and ceWisc. Mar–June, sometimes Aug–Sept.

The more slender plants with the upper calyx-lip deltoid and with 2 lanceolate lobes, instead of being half-orbicular with 2 deltoid lobes, have been separated as *L. perennis* ssp. *gracilis* (Nutt.) D. B. Dunn. It is reported from seTex into nFla, CP of Ga, and ceSC.

117   Sour Clover

Melilotus indicus (L.) All.

Spreading to ascending annual with alternate compound leaves; leaflets 3, finely serrate, terminal leaflet stalked. Flowers 2–3 mm, densely crowded in slender racemes 1–5 cm, on pedicels less than 1 mm; corolla yellow, standard and wings about equal. Fruits 1–2 seeded, ca 2 mm. This and other *Melilotus* species valued as bee plants for honey production. Occasional. Weedy; roadsides, waste places, around buildings, back beaches, stable dune areas. Common. In seUS to Pacific area, less common northward; sSC. Mar–Oct.

M. officinalis (L.) Lam. (Yellow Sweet Clover) also has yellow petals but flowers are 5–7 mm, loosely arrayed, pedicels 1.5–2 mm. Fruits cross-ribbed and honeycombed. Used as hay but if moldy can cause internal hemorrhaging and even death. Probably native of Asia. Common. Fields, grasslands, along roadsides and railroads, various soils. Nfld into NC, cwSC, W to Pacific; in Can from BC into Que; Europe; Australia. Apr–Oct. *M. albus* Medic. is similar but the petals are white and the standard somewhat longer than the wings. Probably native of Eurasia. Occasional. Fields, along roadsides and railroads, waste places; especially calcareous soils; throughout most of US and much of Can; Mex; W. Indies. Apr–Oct.

118   Rabbit-foot Clover

Trifolium arvense L.

Members of this genus have compound leaves with 3 finely serrate or dentate leaflets and united uniform-sized stamens. The corolla withers and persists concealing the ripened pod.

This species is an annual, when in flower unlike any other clover. The clusters of flowers are soft and simulate a rabbit's foot. The stems branch freely and are topped by numerous heads in vigorous plants. Ingestion of mature heads may cause intestinal irritation in livestock. Common. Dry fields and roadsides, waste places; natzd., Fla into eTex, Minn, Que, and Me; also scattered to W. Apr–Aug.

119   Low Hop Clover

Trifolium campestre Schreb.

Flowers usually 20–40 per head, each flower stalked. Terminal leaflet on a stalk much longer than those of the other 2 leaflets. Petioles usually shorter than the leaflets. Although weedy, plants are valuable as soil builders. Common. Natzd. In lawns, fields, roadsides, and waste places; Fla into eTex, seND, and NS; also in scattered localities to W, especially on Pacific slopes. Apr–Oct. *T. procumbens* L.

120   Yellow or Hop Clover
Trifolium aureum Pollich
Much-branched annual with ascending stems to 45 cm, glabrous to variously hairy, leaflet stalks very short including that of terminal leaflet. Heads short-columnar, 1–2 × 1–1.2 cm, many-flowered; pedicels 0.3–0.6 mm, reflexed as fruit matures. Flowers 5–7 mm, calyx strongly 2-lipped; corolla yellow, standard more than 2 mm across and conspicuously striated when old. Seeds globose. Planted for forage and soil improvement. Native of Eurasia. Common. Roadsides, dry fields, swales, stable dunes, waste places; SC into Ark, BC, Nfld, and n and wNC. May–Sept. *T. agaricum* L.

   T. dubium Sibth. (Hop Clover; Shamrock) also has yellow flowers but terminal leaflet obviously is stalked; heads 3–18 flowered; standard not striated, less than 2 mm across. Stems erect to decumbent, to 35 cm. Planted for forage and soil improvement. Native of nEurope. Common. Lawns, fields, roadsides, borders of wet areas, thin rocky woods, waste places; Fla into Tex, Wisc into BC, sOnt, NS, NC, Fla, and Tex. Apr–Oct.

121   White Clover
Trifolium repens L.
Glabrous or nearly so, perennial with creeping stems that root at the nodes, often forming large masses. Stipules at base of petiole usually less than 1 cm, pale and thin. Flower heads on peduncles 10–25 cm that arise from the creeping stems. Individual flowers 7–11 mm, distinctly pedicelled; calyx teeth equal to or shorter than tube. Clover honey is derived largely from this species. A number of cultivars have been developed. Native of Eurasia. Common. Roadsides, yards, fields, grass-lands, thin woods, around buildings, waste land; cnFla into E half of Tex, Kan into sCan, E into Nfld, and NC. Apr–Oct.

   T. stoloniferum Eat. also has creeping stems and forms masses but calyx teeth 1.5–3 times as long as tube. Stipules at base of petiole 1–2 cm, green, and leaflike. Occasional. Thin woods and prairies; WVa into Mo, eKan, and SD, May–Aug.

122   Alsike Clover
Trifolium hybridum L.
Erect to ascending perennial to 80 cm with glabrous stems. Flowers 30–40 in each head, individual flower 8–11 mm, with a distinct pedicel; corolla nearly white, washed with rose. Extensively planted for forage and ground improvement. Common. Fields, pastures, roadsides, around buildings, shores, swales between stable dunes. Essentially throughout N. Amer. Apr–Oct.

123   Red Clover
Trifolium pratense L.
Perennial with several stems from a strongly developed taproot. Stipule tips triangular and abruptly awned. Flower heads globose, sessile. Flowers over 1 cm and sessile. Common. In fields, roadsides, other open places, extensively cultivated; nearly throughout temperate N. Amer. Apr–Sept.

T. reflexum L., Buffalo Clover, also has large globose heads but the flowers are stalked, the standard brighter red, and the wing and keel petals almost white to pink. Occasional. Thin woods; Fla into eTex, eKan, Ill, and Va. Apr–Aug. In *T. incarnatum* L., Crimson Clover, the petals are scarlet or deep red (rarely white) and the flower heads are peduncled and, except at first, cylindrical. Overripe Crimson Clover can be dangerous to horses; its short stiff hairs may become impacted in the digestive tract. Occasional. Frequently cultivated and escaped, especially in the South. Natzd. from Eurasia. Apr–June.

124   Birdsfoot-trefoil; Cat-clover

Lotus corniculatus L.

Prostrate to nearly erect perennials with main stems to ca 60 cm. Leaves with 5 elliptic or oblanceolate to obovate leaflets, 5–15(25) cm × 2–8 mm; stipules represented by dark glands. Peduncle ca 2 times as long as leaves; pedicels 1–3 mm. Inflorescence subtended by 3 leaflike bracts. Flowers mostly 4–8 in umbels; calix tube and lobes about equal; petals yellow, becoming orange with brick-red markings, 8–16 mm. Stamens alternately long and short, the long ones swollen just below the anthers. Fruit a dehiscent columnar legume 15–40 mm. Native of Europe. Common. Meadows, fields, roadsides, disturbed habitats; sArk into Mo, Minn, O, Nfld, Va, neNC; Asia; Afr; Australia; New Zeal. June–Sept.

125   Hairy Indigo

Indigofera hirsuta Harv.

Mostly erect plants. Stems, leaves, and fruits with long spreading reddish-brown hairs. Leaves compound; leaflets an odd number, entire. Flowers in racemes; sepals united, lobes ca equal; petals 5, wing petals with a spur or pocket in the base of the blade. Rare. Floodplains; cnFla into sGa. Sept–Nov.

126   Summer-farewell

Dalea pinnata (J. F. Gmel.) Barneby

Members of this genus have pinnately compound leaves with an odd number of leaflets, and 5 stamens, an unusual number for the Bean Family. The keel and wing petals are attached to the united bases of the stamens. The fruits are 1–2-seeded pods that usually do not split open.

   This species is a perennial to 1.2 m. Leaflets 3–11. Flowers in heads with bracts at their bases, the lower bracts brownish and broadly ovate to orbicular. Petals white. Common. Dry sandy places, pinelands, and scrub; Fla into seMiss, ceGa, and seNC. Aug–Oct. *Kuhnistera p.* (J. F. Gmel.) Kuntze; *Petalostemon pinnatus* (J. F. Gmel.) Blake; *P. caroliniensis* (Lam.) Sprague.

127   Purple-tassels

Dalea gattingeri (Heller) Barneby

Perennial to 50 cm, with several to many decumbent to erect stems. Leaflets 5–7. Bracts at base of flower heads long pointed from broad bases. Calyx loosely hairy. Petals rose-purple. Rare. Rocky calcareous open areas of cedar glades; nwGa into nAla and cTenn. May–June. *Petalostemon g.* Heller.

   D. purpurea Vent. is similar but the bracts at the base of the flower heads are acuminate from narrower bases and the calyx is appressed-hairy. Petals violet to rose-purple. Common. Prairies, dry hills; cwAla into cLa, nNM, sAlba, sMan, and eInd. June–July. *Petalostemon purpureus* (Vent.) Rydb.

128 **Goat's-rue**
Tephrosia virginiana (L.) Pers.
Members of this genus are perennials with deep roots. The leaves are pinnately compound with an odd number of leaflets. Lateral veins of the leaflets parallel and run from midrib to margin. Stamens 10 and all united.

This species is stiffly ascending to erect, 20–70 cm. Variable in amount and character of hairs. Leaflets 15–25. The inflorescence is terminal. Fruit 35–55 × 4–5.5 mm; straight or curved slightly downward. The roots have been shown to contain rotenone, which is an insecticide and fish poison but not poisonous to mammals. Common. Dry places, in thin pine or hardwood stands, or in open; Fla into e half of Tex, cnKan, cwWisc, and sNH. Apr–July. *Cracca v.* L. *C. mohrii* Rydb.

129 **Bladder-pod**
Glottidium vesicarium (Jacq.) Harper
Annual to 4 m. Leaves evenly pinnate. Flowers 8–15 mm in racemes shorter than the leaves. Petals yellow, tinged with red, the wings sometimes mostly dark pink. Fruit tapered at both ends; the body flattened; seeds 2, rarely 1 or 3; at maturity the firm outer layer separates from a thin soft layer enclosing the seeds. Seeds are poisonous, with consumption as low as 0.05% of body weight causing death. Cattle sometimes eat the fruits; hundreds from a single herd have been killed under special circumstances. Occasional. Low areas in open places; Fla into e third of Tex, seOkla, ceGa, and e half of NC. July–Sept. *Sesbania vesicaria* (Jacq.) Ell.

130 **Sesbania**
Sesbania exaltata (Raf.) Rydb. ex A. W. Hill
Annual to 2(4) m, thinly branched. Leaves pinnately compound; leaflets an even number, mostly 20–70. Flowers in peduncled axillary clusters. Corollas shaped like those of Sweet Pea; stamens 10, 1 filament separate from the 9 united into a tube. Fruit a glabrous legume 10–23 cm × 3–4 mm, linear with cross partitions. Occasional. Thin low pinelands, ditches, margin of marshes, swales, fields, alluvial soils; sNC into ce and swGa, Fla, eTex, eOkla, Ark, and sIll. June–Oct. *S. macrocarpa* Muhl. ex Raf.

131 **Crown-vetch**
Coronilla varia L.
Perennial herb with trailing to ascending stems to 1.2 m. Leaves pinnately compound; leaflets 9–25 with 1 terminal, obovate to oblong, 1–2 cm. Flowers in umbels on peduncles 5–15 cm; umbels 5–20 flowered, pedicels 3–7 mm; stamens 10, 9 united basally, 1 free. Fruit a 4-angled linear legume 15–55 mm, breaking into 3–7 sections, each with 1 seed. Occasional. Fields, roadsides, old home sites, waste places; Tenn into c and nwArk, Mo, SD, Ky, WVa, Va, wNC, and Pied of Ga. June–Aug.

132   **Joint-vetch**
Aeschynomene indica L.
Bushy annual to 2.5 m. Leaves even-pinnate, sometimes appearing odd-pinnate; leaflets 20–70, not gland-dotted, the 1 major vein slightly off center. Flowers 8–10 mm; corolla yellow; filaments united into 2 groups of 5 each. Fruit with 5–12 joints, breaking into 1-seeded segments at maturity; seeds food for birds. Occasional. Moist to wet places; thin woods, marsh edges, wet meadows, in swales between stable dunes; ceNC into nwFla, Tex, and swArk; S into Braz. and Arg. July–Oct.
   A. virginica (L.) B.S.P. is similar but flowers are mostly over 10 mm. Occasional. Tidal shores of rivers and marshes, thin woods in swamps; sNJ into CP of SC, and nwFla. July–Oct. *A. viscidula* Michx. also has similar yellow corollas and similar stamens but is sticky-haired and fruits have only 2–5 joints. Fruits can be confused with those of some *Desmodium* species but the latter have only 3 leaflets and corollas are not yellow. Occasional. Sandy pinelands, scrub; sGa into Miss and Tex. June–Sept.

133   **Pencil-flower**
Stylosanthes biflora (L.) B.S.P.
Perennial with 1–several prostrate to erect stems from a stout rootstock. Fruit in 2 sections, the lower one usually aborting and becoming stipelike. Common. Dry soil, thin woods and barrens; cFla into eTex, seKan, cIll, Ky, sPa, and ceNJ. May–Sept. *S. riparia* Kearn.

134   **Beggar-ticks**
Desmodium paniculatum (L.) DC.
Our species of this genus have 3 leaflets, stipulelike structures just below each leaflet, and loments as fruits, i.e., they break apart into joints that have hooked hairs and readily cling to passing objects. Young seeds are tasty and are not reported to be poisonous. Loments abundantly eaten by wildlife. There are ca 30 species in the seUS and most are difficult to identify with certainty.
   This species is an erect perennial to 1.2 m. Stipules 3–6 mm and persistent. Terminal leaflet 3–6 times as long as wide. Petals 6–8 mm. Loments usually 2–6, triangular, and 5–7.5 × 3.5–4.5 mm. Common. Fields, meadows, clearings, thin woods, between stable dunes, borders of ponds and swamps; Ark into Neb, sOnt, sMe, NC, Fla, and Tex. June–Sept.

135   **Beggar-ticks**
Desmodium canescens (L.) DC.
A stiff erect to strongly ascending perennial to 1.4 m. Petiole much longer than stalk of terminal leaflet; terminal leaflet 6–14 cm. Stipules 5–15 mm. Pedicels 6–14 mm. Loments 3–6, 6–13 × 4–8 mm. Occasional. Fields, edge of woods, in thin woods, roadsides; Neb, swWisc, sOnt, Me, NC, Fla, and Tex. June–Aug.

136   **Hairy Lespedeza**
Lespedeza hirta (L.) Hornem.
Members of this genus have compound leaves; leaflets 3, entire. Fruits indehiscent, 1-seeded, not adherent. All species provide fruit for birds and nutritious forage. Although the genus is easy to recognize, most species are not.

This species grows to 1.5 m. Flower clusters numerous, axillary, varying from mostly sessile to mostly distinctly peduncled but always some peduncled; calyx 4 – 8 mm; petals yellowish-white with a purple spot, standard 6 – 8.5 mm. Occasional. Disturbed areas, thin woods; Ark into Wisc, sOnt, Me, NC, Fla, eTex, and Okla. July–Oct.

L. capitata Michx. is also hairy and has similar flowers but peduncles are usually shorter than the subtending leaves; calyx 7–12 mm; standard 8–12 mm. Common. Thin dry woods, fields, thickets, sand dunes, prairies; se and nwArk into SD, Minn, sQue, Me, NC, nwFla, and eTex.

137   Creeping Lespedeza
Lespedeza repens (L.) Bart.
Prostrate perennial with several to many well-spaced thin branches, the ends sometimes ascending. Stems with sparse and very short appressed hairs. Keel petals about equal wings. Common. Dry open places, roadsides, thin woods, sandy pinelands; nFla into eTex, seKan, swWisc, sInd, and ceNY. Apr–Oct.

Two species are similar: *L. violacea* (L.) Pers. has weakly ascending stems and keel petals usually longer than the wings. Common. Dry open places, roadsides, thin woods, prairies; cGa into neTex, sMo, sWisc, and Mass. July–Sept. *L. procumbens* Michx. bears spreading hairs. Common. Similar places; nFla into e and cnTex, sWisc, O, and sNH. May–Sept.

138   Narrow-leaved Vetch
Vicia sativa L. ssp. *nigra* (L.) Ehrh.
Vetches have pinnately compound leaves with a tendril on the end, leaflets an even number; style bearded around the top like a bottle-brush. Vetches are easily confused with species of *Lathyrus* but in the latter the style is bearded along one side of the tip like a toothbrush. Vetches are an important source of food for animals. Seeds of some species may be poisonous.

This species is a glabrous or nearly so annual to 60 cm. Leaves with 4 – 10 linear to narrowly elliptic leaflets. Flowers 10 – 18 mm, commonly in pairs, nearly sessile in upper leaf axils; calyx tube 4 – 6 mm, lobes 3 – 6 mm. Mature fruits blackish. Seeds subglobose. Native of Europe. Common. Weedy; fields, fencerows, roadsides, grassy areas; eTex into Ark, Mich, sQue, NS, Mass, NC, and Fla. Mar–Oct. *V. angustifolia* L.

139   Smooth Vetch
Vicia villosa Roth ssp. *nigra* (Host) Corb.
Annual or rarely perennial with tendrils on ends of leaves. Flowers short-stalked, 10 – 30 in peduncled elongate axillary racemes. Pedicels attached at side of base of calyx. Raceme glabrous or with short appressed hairs. Longest calyx lobe 1 – 2 mm. Common. Fields, roadsides, waste places; nearly throughout the contiguous 48 US states. Apr–Sept. *V. dasycarpa* Ten.

V. cracca L. is similar but the pedicels are attached near the center of the base of the calyx. Rare. In similar habitats; NC into eTenn, Minn, BC, sQue, and Nfld. May–July.

140   **Woods Vetch**
Vicia caroliniana Walt.
Slender trailing or climbing perennial to 70 cm, with tendrils on ends of leaves.
Leaflets 8–20, linear, pointed. Flowers 5–12 mm, 7–20 in each raceme. Pods have
1 seed. Occasional. Open woods; ceAla into nwSC. Apr–June. *V. hugeri* Small.
    Other species with racemose flowers and white or blue-tinged petals include
V. acutifolia Ell., which has 4–6 linear leaflets, flowers 7–8 mm and 4–10 per
raceme, and pods with 4–8 seeds. Common. Sandy moist soils in open, seSC into
ne and pen Fla. Feb–June.

141   **Everlasting Pea**
Lathyrus latifolius L.
Perennial with decumbent to high-climbing broadly winged stems. Leaves with a
pair of blades and a branched well-developed terminal tendril. Flowers 15 mm or
longer, in axillary racemes. Petals commonly purple, or red, pink, or white. Style
strongly flattened. Ovulary and fruit glabrous. Seeds are poisonous. Occasional.
Roadsides and waste places, open situations; Pied of Ga into eTex, Mo, csMe, and
NC. May–Sept.
    Two other species also have 2 blades on each leaf, with the tendrils and flowers
smaller and the flowers fewer. *L. hirsutus* L. has a hairy ovulary and fruit. Rare.
Central CP of Ga into e and cnTex, ceMo, and Va. Apr–July. *L. pusillus* Ell. is an
annual and the smallest of the species. The flowers are 12 mm or shorter and the
ovulary and fruit glabrous. Rare. Similar distribution. Apr–July.

142   **Butterfly-pea**
Clitoria mariana L.
Perennial from a deep narrow taproot. Stems to 1 m, little-branched, spreading or
sometimes twining. Leaves compound, with 3 entire leaflets. 1–several flowers
open at a time, persisting longer than a day. Flowering even during drought when
few other attractive plants are in flower. Calyx tubes much longer than the lobes.
The standard petal is 25 mm or longer, about twice as long as the wing and keel
petals, and different from most legumes in that the standard petal is below the
others. Fruit with a stipe, dehiscent, several-seeded, the seeds sticky. Common.
Dry places, thin woods or open; Fla into cTex, sMo, sO, and sNY; Ariz. May–Sept.
Martiusia m. (L.) Small.

143   **Climbing Butterfly-pea**
Centrosema virginianum (L.) Benth.
Twining perennial from a tough elongated root. Leaflets 3. Calyx tube shorter
than the lobes and hidden by conspicuous bracts. Petals last 1 day. Standard petal
25–35 mm, about twice as long as the wing and keel petals. Fruit sessile, flattened,
many-seeded, 7–14 cm, with a long persistent style. Common. Drier places, thin
woods or open; Fla into cTex, Ark, and sNS. Mar–Sept. *Bradburya virginiana* (L.)
Kuntze.

144   Cardinal-spear; Coral Bean
Erythrina herbacea L.
Perennial to 1.2 m. Leaves alternate with 3 leaflets, which are hastate to widely
deltoid and occasionally prickly beneath. Stipules are curved spines. Inflorescence
of 1 or more terminal spikelike racemes. Calyx red, tubular. Corolla scarlet, the
standard to 53 mm, folded so that the entire flower appears long and narrow.
Wing and keel petals 13 mm or less. Fruit to 21 cm, constricted between the few
to many brilliantly scarlet seeds, which hang on after the pod splits open. Occasional. Open woods, sandy soils; Fla into s tip of and neTex, seSC, and seNC.
Apr–July.
    In the warmer parts of the range of the species, aboveground stems often live
over a number of winters. This has led some persons to believe that there is another species, *E. arborea* (Chapm.) Small, but this is not supported by more thorough studies.

145   Groundnut; Indian-potato
Apios americana Medik.
Perennial twining vine with tuberous enlargements on the roots. These may be
6 cm across and are edible raw, boiled, fried, or roasted. When eaten raw they
leave an unpleasant rubberlike coating in the mouth, a quality lost in roasting or
frying. When well prepared the tubers can be one of our best wild foods. Seeds
have also been used as food. The leaves have 5 or 7 ovate to lanceolate pinnately
arranged leaflets. The flowers are as many as 30 in a tight cluster but do not open
all at once. The petals are brownish-purple to reddish-purple. The fruits are 6–
12 cm long and about 60–120 × 6 mm thick when mature and usually have 8–
10 seeds. Common. Moist places in woods or open; Fla into cTex, seND, and NS.
June–Sept. *A. tuberosa* Moench, *Glycine apios* L.

146   Galactia
Galactia elliottii Nutt.
Our members of this genus are perennials and vines with prostrate or trailing
stems, one species having erect stems only. Leaves once-pinnate, the leaflets entire
and odd numbered. Each flower has a pair of very small bracts at or near the top
of the pedicel. Pod few- to many-seeded.
    This species is a twining, climbing perennial vine often with a long horizontal
rootstock and some tuberous roots. Leaflets 7 or 9. Flower clusters are long-
peduncled and axillary. Petals sometimes tinged with red. Fruits 3–5 × 6–13 cm,
densely covered with short appressed hairs. Common. Open sandy areas, usually
low; Fla into coastal Ga and sSC. May–Sept.

147   Climbing Milk-pea
Galactia volubilis (L.) Britt.
Twining and climbing unless there is no support available. Hairs on stems usually
retrorse-appressed to uncommonly retrorse-spreading. Inflorescences 5–65 mm,
all flower-bearing nodes well separated; longest pedicels 3–4 mm at flowering.
Flowers 10–14 mm; longest calyx at pollen-shedding 6–10 mm; maximum number of ovules in pistils and later seeds in pods (counting aborted ovules) on any
given plant 10–12(13). Common. Thin woods, thickets, clearings; n three-fourths

of Fla into csLa, CP of Ga, and e edge of Va. July–Sept. *G. regularis* (L.) B.S.P. as used in many manuals.

The most abundant and widespread Climbing Milk-pea is *G. regularis* (L.). B.S.P. Stem hairs inconspicuous, usually retrorse but frequently antrorse and sometimes spreading. Inflorescences 3–15 cm; longest pedicels at flowering 2–2.5 mm. Longest flowers 7–9 mm, longest calyx 4–5 mm, and the most ovules per pistil (and seeds per pod counting aborted ovules) (4)5–7(8) on any given plant. Common. Thin woods, thickets, or in open; barrens; nFla into e half Tex, Kan, sw and ceMo, sO, ePa, sNY, NJ, and Va. July–Aug. *G. mollis* Nutt. also twines and climbs but hairs on stem are dense, divergent, and prominently visible macroscopically. Inflorescences 3–20 cm; calyx brown to reddish-brown on innerside when dried; corolla dark when dry, persisting after withering, sometimes partly present when fruit is mature. Occasional. Sandy soils; thin woods or in open; cn and neFla into ceGa, and seNC. July–Sept.

G. glabella Michx. is the most widespread of the Trailing Milk-peas. It may be recognized by several to most internodes much longer than the largest leaflet of adjacent nodes, longest inflorescences 3–15 cm, nodes near the tip more closely spaced, upper flower clusters congested, longest flowers 12–18 mm, maximum number of ovules per pistil and seeds per pod (counting aborted ovules) 6–8(9) on any given plant. As explained above, plants of this species for many years improperly have been called *G. regularis*. Common. Sandy soils; sandhills, thin woods, in open; cn and neFla into CP of SC and NC, and seVa. *G. floridana* T. & G. is also trailing and may be recognized by fine close spreading hairs, the longest at least 0.7 mm; calyx greenish-yellow to tan on innerside when dried, and corollas falling as they wither or soon thereafter. Occasional. Sandy soils in thin woods or open; sAla into n half of Fla and seGa. July–Sept. *G. minor* W. H. Duncan is quite similar to *G. floridana* but the hairs on stems 0.05–0.25 mm and antrorse, internodes only a little longer to usually shorter than largest leaflet at adjacent nodes. Occasional. Similar habitats; sMiss into Fla panhandle, CP of Ga, and Fall Line of SC and NC.

148 Erect Rhynchosia
Rhynchosia tomentosa (L.) Hook. & Arn.
Our members of this genus have resinous dots on the foliate, blades entire; 9 filaments united, 1 separate; and 1–2-seeded fruits.

This species is an erect little-branched perennial 15–85 cm; leaflets usually 3. Corolla equal to or shorter than calyx. Common. Dry places; thin woods, pinelands, roadsides, clearings; cFla into cLa, seKy, sMd, sDel, and NC but scanty in mts. Aug–Oct. *R. erecta* (Walt.) DC.

Other species with 3-foliate leaves have trailing or twining stems. Species with similar, but generally broader, ovate to ovate-rhombic leaflets: *R. difformis* (Ell.) DC. with flower clusters on peduncles 1–35 mm. Occasional. Fla into eTex, ceGa, and seVa. May–Aug. *R. latifolia* Nutt. ex T. & G. with flower clusters on peduncles 3–30 cm. Common. Woodlands, prairies, alluvial woods, rocky places; e third of Tex into cOkla, sMo, wKy, and cMiss. May–Aug.

Leaves of 2 species have only 1 blade, rarely 2. *R. reniformis* DC., an erect perennial to 22 cm. Common. Sandy soils; pinelands, thin woods, open places; cFla into seTex, cAla, ceGa, and cCP of NC. *R. michauxii* Vail. has similar unifoliate leaves but stems are elongate and trailing or twining. Occasional. Similar habitats; Fla.

149   Wild-bean
Strophostyles umbellata (Muhl. ex Willd.) Britt.
Seeds of this genus are probably edible; we know of no report of their being poisonous.

This species is a trailing or sometimes twining perennial vine. Leaves with 3 leaflets. Flowers 9–15 mm, nearly sessile at top of axillary peduncles. Bract at base of calyx half as long, or less, as the glabrous (or nearly so) calyx-tube. Keel petals strongly curved and beaked. Occasional. Fields and thin woods, often sandy soils; nFla into eTex, Mo, seNY. June–Sept.

S. helvula (L.) Ell. is similar but is an annual. The bract at the base of the calyx equals or exceeds the calyx tube. Fields, roadsides, thin woods, low places between dunes; Fla into e and cnTex, eSD, swQue, and sMe. In *S. leiosperma* (Torr. & Gray) Piper flowers are 5–8 mm long and the calyx-tube is quite hairy. Ala into se and cnTex, neColo, sWisc, and cwInd. July–Oct.

150   Vigna
Vigna luteola (Jacq.) Benth.
Perennial, much like *Strophostyles* vegetatively but the corollas are yellow and the keel petals less strongly curved. Flowers in a tight cluster in axillary racemes on peduncles several times longer than the subtending leaf. Stamens with 9 filaments united and 1 separate. Legume 3–7 cm. Occasional. Thin woods, roadsides, lagoon shores, dunes; sNC into Fla and seTex; tropical Amer. Mar–Nov.

GERANIACEAE: Geranium Family

151   Wild Geranium; Cranesbill
Geranium maculatum L.
Perennial to 60 cm from a dark thick rhizome. Basal leaves with long petioles, the upper ones sessile or short-petioled. Petals 15–25 mm, rose-purple, rarely white. Stamens 10. Ovulary 5-carpelled. Fruit with a thick base and a beak 19–25 mm, hence the name "Cranesbill." Base of fruit splits apart and coils upward at maturity. Common. Rich woods, meadows; SC into swGa, eOkla, sMan, and sMe. Apr–June.

152   Carolina Cranesbill
Geranium carolinianum L.
Annual to 55 cm. Petals 4–6 mm, pale pink. Pedicels about as long as the calyx. Sepals with subulate tips 1 mm or longer. Stamens 10. Carpels with hairs over 0.6 mm. Common. Dry places, thin woods, fields, waste places; nearly throughout the contiguous 48 states. Mar–June.

Similar species include: *G. dissectum* L., which has dark purple petals and carpel hairs under 0.6 mm. Apr–July. Rare. Similar places; ceNC into Miss, neTex, sMich, and Mass. *G. columbinum* L., which has pedicels over 2 cm. Occasional.

Similar places; nGa into seInd, and NY. May–Aug. *G. pusillum* L. with awnless sepals and finely hairy carpels. Unusual in having only 5 stamens. Rare. Similar places; mts of NC into eKy, WVa, and NJ. May–Sept. *G. molle* L. also with awnless sepals but with glabrous carpels. Rare. Similar places; nSC into cTenn, eMo, NY, and cnNC. Apr–Aug.

153   Herb-Robert

Geranium robertianum L.

Weak annual or biennial, strong-scented; stems branching, spreading, hairy, to 60 cm. Leaves 3–5, cleft to near base, compound as the terminal segment is stalked. Flowers (1)2 on peduncles from upper nodes; sepals 7–9 mm; petals 9–13 mm, long-clawed, entire, bright pink to red-purple. Fruits unusual in that carpels separate at the base and remain attached at beak. Native of Eurasia. Common. Damp rich woods, ravines, gravelly shores; Man into Nfld; s into Ill, O, WVa, Md, seNY, and NS. May–Oct.

154   Heron's-bill; Redstem-filaree

Erodium cicutarium (L.) L'Her. ex Ait.

Procumbent to semierect winter annual with numerous branches radiating from a taproot. Leaves deeply pinnately dissected, 5–20 cm. Flowers in groups of 6–9 on peduncles 5–15 cm. Sepals 3–6 mm; petals pinkish-purple, 6–8 mm, slightly unequal, the 2 shorter with a white spot and base speckled with purple or black; stamens 10, bases united. Fruits bristly hairy, similar to those of *Geranium* in shape. Native of Mediterranean region. Common. Weed of fields, lawns, borders, waste places; throughout most of US. Mar–June.

OXALIDACEAE: Wood-sorrel Family

155   Yellow Wood-sorrel

Oxalis stricta L.

Perennial to 50 cm from long slender rhizomes. Leaflets 1–2 cm broad. Stems decumbent to erect, with small whitish appressed hairs. Flowers 7–11 mm, in 1–4 flowered umbels. Common. Woods, waste places; Fla into eTex, ND, several states in the wUS, sQue, and sMe. May–Frost.

In *O. grandis* Small the flowers are larger, over 12 mm. Leaflets are 2–5 cm across and with narrow, reddish or purplish margins. Occasional. Rich woods; nGa into nAla, seIll, neO, cMd; *O. priceae* Small ssp. *colorea* (Small) Eiten also has large flowers (13–18 mm) but leaflets are only 0.5–1.4 cm across. Occasional. Thin woods, dry sandy or rocky places; Fla into eTex, Ky, sWVa, and cVa. *O. recurva* Ell.

156   Violet Wood-sorrel

Oxalis violacea L.

All species of *Oxalis* have sour watery juice and if eaten in excess may cause poisoning. Small amounts are probably safe.

This species is a stemless perennial, the leaves and scapes all from a bulbous base. Sepals and pedicels glabrous, with calloused orange tips. Occasional. Rich woods, rocky places, pinelands, prairies; Fla into eTex, eND, and Conn. Mar–May and Aug–Oct.

O. debilis var. corymbosa (DC.) Lourteig is similar but sepals and pedicels are hairy. Rare. Commonly cultivated and escaped locally. Fla into seTex, cwSC, and seVa. O. corymbosa DC.; O. martiana Zucc. O. montana Raf. is also stemless but lacks a bulbous base and is creeping by slender rhizomes. Flowers often 1 per peduncle, petals mostly white, veined with pink. Rich woods; nwGa into high mts of Tenn and NC, N in the mts to eO, sOnt, neMinn, and sNfld. June–Sept. O. acetosella L.

157   Northern Wood-sorrel; Wood-shamrock

Oxalis montana Raf.

Stemless plants with a slender rhizome and no bulb. Leaflets 8–20 mm. Sepals 15–20 mm, not callus-tipped; petals white with a reddish to pink band near base, 10–15 mm. Common. Rich moist forests, often under hemlocks, spruce, or fir at high alts; Wisc into Sask, Mich, eQue, NY, and S in mts into nwGa. *O. acetosella auct. non* L.

ZYGOPHYLLACEAE: Caltrope Family

158   Puncture-weed

Tribulus terrestris L.

Prostrate herb with stems to 1 m, any branches alternate. Leaves opposite, pinnate, leaflets of even numbers to 18, 4–15 × 2–5 mm, base oblique. Flowers solitary from leaf axils. Fruit in 2–5 sections; bearing 2 lateral spines about midway; a nuisance to man and animals, injurious to bare feet. Once established difficult to eradicate. Rare. Roadsides, lawns, waste ground; seNC into sGa, nwFla, and Tex. June–frost.

POLYGALACEAE: Milkwort Family

159   Bachelor's-button

Polygala nana (Michx.) DC.

Members of this genus have flowers in racemes or spikes. Flowers are perfect and irregular with 3 small sepals and 2 larger petallike ones called wings. The corolla resembles those of some legumes, as do the stamens, which are united by their filaments and to the petals. Polygalas are easily separated from the legumes by the 2-carpelled ovulary and fruit.

This species is a biennial to 15 cm. Leaves succulent. Pedicels winged. Largest sepals with cusps at least 1 mm. Flowers turn a dark bluish-green when dried. Common. Wet pinelands, or low open areas; seTex into nwGa, nwFla, cSC, and pen Fla; Rhea Co., Tenn. Mar–June. *Pylostachya n.* (Michx.) Raf.

160   Candyweed
Polygala lutea L.
Glabrous biennial with succulent leaves. Stems decumbent to erect, to 50 cm.
Cusp on largest sepals less than 1 mm. Racemes light to dark orange, turning pale
yellow upon drying. Pedicels winged. Common. Wet pinelands and savannas,
bogs; Fla into sLa, CP of Ga, and Long Isl., NY. Apr–Oct. *Pylostachya l.* (L.)
Small.

161   Large-flowered Polygala
Polygala grandiflora Walt.
Perennial to 50 cm. Stems usually unbranched, 1–several, with appressed or
spreading hairs. Leaves alternate, oblanceolate to linear-oblanceolate, 1.5–5 cm.
Flowers 6–7 mm, the largest sepals as long as wide. Common. Sandy soils, pinelands, fields; Fla into seLa, CP of Ga and SC, and swCP of NC. Apr–Sept.
    P. polygama Walt. is similar but the flowers are a little smaller, about 4 mm, and
the largest sepals about twice as long as wide. Small flowers that never open are
found on horizontal branches at the base of the plant. Common. Usually in dry
sandy or rocky soils, thin woods, pinelands; Fla into eTex, seTenn, NC, Va, sMan,
and Me. Apr–July.

162   Tall Milkwort
Polygala cymosa Walt.
Glabrous biennial to 1.2 m. Leaves in a basal rosette, linear, to 8 cm. Stems usually
1. Racemes numerous in terminal cymes. Flowers turn greenish-yellow to dark
green on drying. Seeds glabrous. Common. In shallow water of cypress ponds,
swamps, depressions; Fla into seLa, CP of Ga, seNC, and locally to Del. May–Aug.
Pylostachya c. (Walt.) Small.
    P. balduinii Nutt. also has numerous racemes in a terminal cyme but the flow-
ers are white or nearly so, the plants only to 60 cm, and the lower leaf blades
spatulate to obovate. Rare. Low pinelands and swamps; Fla into seLa, swGa. June–
Aug. *Pylostachya b.* (Nutt.) Small.

163   Short Milkwort
Polygala ramosa Ell.
Glabrous biennial to 50 cm. Leaves elliptic to spatulate. Stems usually 1, rarely to
4 as seen in the picture. Racemes several to numerous in terminal cymes. Flowers
yellow, turning dark green on drying. Seeds hairy. Common. Wet pinelands and
savannas, swamps, pond margins; Fla into seTex, CP of Ga, ceNC, and scattered
locally north to sNJ. May–Aug. *Pylostachya r.* (Ell.) Small.

164   Drum-heads
Polygala cruciata L.
Erect annual to 35 cm. Stems freely branched above. Leaves mostly whorled.
Flowers in dense racemes, 10 mm or more across, sessile or sometimes with peduncles to 35 mm. Bracts remain on the raceme after the fruits fall. Wings acuminate. Common. Wet places, pinelands, savannas, bogs; Fla into eTex, cTenn,
cWVa, nO, eNeb, Minn, and sMe. June–Oct. *P. ramosior* (Nash) Small.
    P. brevifolia Nutt. is similar but most of the leaves are opposite or whorled and

the wings are acute. Rare. Wet places, pinelands, and savannas; Fla into seMiss; CP of NC; NJ. June–Oct. In *P. sanguinea* L. the leaves are alternate and the wings are 3 mm or more wide, ovate to oval, and obtuse to mucronate. Occasional. Open places, woods, prairies, meadows; neSC into Tenn, neTex, Minn, and NS. June–Aug.

165   Polygala
Polygala curtissii Gray
Annual to 40 cm, usually freely branched above. Leaves alternate, linear. Flowers in dense racemes about 10 mm across, the bracts persistent as subulate hooks on the racemes after the fruits fall. Wings elliptic over 3 mm. Common. Open places, often sandy or rocky, thin to normal woods; upper CP of Ga, ceAla, cTenn, sWVa, and Del. June–Nov.

   P. mariana Mill. is similar but the racemes are 9 mm or less across and the bracts fall from the racemes. Common. Open places, usually moist, pinelands, savannas; nFla into seTex, cwAla, seGa, neNC, and sNJ. June–Oct.

166   Slender Polygala; Pink Milkwort
Polygala incarnata L.
Very slender glaucous annual to 60 cm. Leaves linear, 5–15 mm, alternate and scattered, frequently falling early. Stems simple or with a few erect branches. Flowers nearly sessile, 7–10 mm. Sepals 2.5–3 mm, pink; corolla reddish-pink, twice as long as longest sepal; in all other species the corolla is less. Common. Thin pinelands, meadows, bogs, swales, roadsides; Fla into Tex, Kan, Ia, Wisc, Mich, seNY, and NC; rare to absent in mts. June–Nov.

167   Flowering-wintergreen; Fringed Polygala
Polygala paucifolia Willd.
This species is easily separated from the others by the 1–4 large flowers (15–23 mm). A perennial with a slender underground rhizome and slender stolons. Stems 8–15 cm bearing several scattered scale leaves 2–8 mm and near the summit 3–6 elliptic to oval leaves 1.5–4 cm. During summer and autumn bearing minute permanently closed flowers and suborbicular fruits on underground branches. Occasional. Rich woods; nIll into Minn, Sask, Que, NB, Conn, and S in mts to nGa. May–early July.

EUPHORBIACEAE: Spurge Family
Members of this family usually have stipules (often in form of glands), flowers are unisexual or rarely bisexual, and fruits are 3-seeded capsules except for one genus with a 1-seeded urticle.

168   Tooth-leaved Croton
Croton glandulosus L.
Annual to 60 cm with watery sap, bearing fine stellate hairs. Roots have a spicy odor. Leaf blade crenate-serrate, oblong to lanceolate, 2–9 cm. Male and female flowers on same plant; male flowers with 4-parted calyx and 4 white petals. Fruit a capsule with 1 seed in each of the 3 carpels. Common. Dry sand and loam soils,

thin woods, roadsides, meadows, between stable dunes; Ark into Neb, eKan, Ind, Del, Va, Fla, and Tex. May–Oct.

169 Short-stalk Copperleaf; Three-seeded Mercury

Acalypha gracilens Gray
An erect annual to 80 cm with watery sap, simple or branched; stems with small incurved ascending hairs. Petiole of main leaves 4–15 mm; blade 2–6 × 0.5–2 cm, margin entire or finely crenate. Flowers axillary, both sexes on same short axillary spike, female flowers at base. Bract at base of spike with 9–15 ovate to deltoid lobes to 2 mm. Seeds reddish to black. Common. Woods, fields, meadows, margin of marshes, waste places; eTex into Wisc, wOnt, swQue, sMe, NC, and nFla. June–frost.

There are 2 similar species, perhaps varieties of a single species. In *A. rhomboidea* Raf. the bracts have only 5–7 lobes, the longest 3 mm or longer. Occasional. Similar places; cMo into ND, Que, sNS, SC, ceGa, cnFla, and Tex. June–frost. *A. virginica* var. *r.* (Raf.) Cooperr. In *A. virginica* L. stems have straight spreading hairs, especially on upper part. Rare. Thin woods, fallow fields, disturbed areas, roadsides; seSD into Ill, Me, Mass, Va, cNC, cSC, nGa, and Tex. Aug–frost. *A. v.* var. *v.*

170 Stinging-nettle

Tragia urticifolia Michx.
Plants to 60 cm; main stem simple or sparsely branched, sometimes decumbent. Leaves triangular-ovate 2–6 × 0.7–4 cm, simply or doubly serrate; petioles 5–15 mm. Flowers in racemes 1–4 cm. Common. Sandy fields, thin woods, thickets, rocky areas in thin woods or open; Ariz into Colo, Kan, Mo, Ga, csVa, middle third of NC; Mex. May–Oct.

171 Tread-softly; Spurge-nettle

Cnidoscolus stimulosus (Michx.) Engelm. & Gray
Erect perennial to 1 m from a long root. Sepals white, petals absent. Fruits have 3 hard seeds. Stems, leaves, and female flowers bear long stiff sharp hairs with a caustic irritant that on contact produces a painful irritation and in some persons a severe reaction. Common. Sandy areas, including thin woods; Fla into seVa, Pied of Ga, and eLa. Mar–Sept. *Jatropa stimulosa* Michx.; *Bivonea stimulosa* (Michx.) Raf.

172 Queen's-delight

Stillingia sylvatica Garden ex L.
Perennial, usually with several stems from the enlarged rootstock. Male flowers in a yellowish terminal spike, female flowers few and at its base. Fruits 3-lobed, 3-carpelled, and 3-seeded. A western species of the genus has been shown to be poisonous and this species may be. Common. Sandy soils in open or thin woods; CP—seVa into Fla and Tex; then W into NM and N into Kan. Apr–July.

173 **Flowering Spurge**
Euphorbia corollata L.
Our members of this genus have milky juice. All have single-sex flowers. Both sexes are borne in cups (cyathia) usually with 4 –5 petallike lobes, the whole resembling a flower. The male flowers consist of a single stamen and the female of a single, 3-carpelled pistil. The fruits contain 3 seeds, 1 in each cavity, and split open. Plants of all species are probably poisonous. The juices may cause skin irritation, and plants when eaten may cause severe poisoning. Spurges often contaminate hay.
   This species is an erect perennial with symmetrical leaves, cyathia with white petallike lobes over 1 mm wide. Common. Thin dry woods, fields, roadsides, along railroads; occasionally in moist places; Fla into se and cTex, Minn, and sMe. May–Sept.

174 **Painted-leaf; Wild Poinsettia**
Euphorbia cyathophora Murr.
Annual to 60 cm, usually glabrous, with alternate midstem leaves, and small to large red patches on the upper leaves. The cyathia lack petallike lobes and usually have only 1 gland, about 1 mm, the depression in its top oblong. Seed not angular in cross-section. Occasional. Various natural habitats in the open, waste places; Fla into Ariz and eSC; adventive into eSD, Minn, and Va. Apr–Oct. *E. heterophylla* L. of most books; *Poinsettia h.* (L.) Small.
   E. heterophylla L. is similar but never has red patched leaves, the depression in the top of the gland is circular, and the seeds have 2 lateral angles. Rare. Open areas; Fla into sTex. June–Oct. *E. dentata* Michx. resembles the above two species but the leaves are mostly opposite and none has red patches. Rare. Dry open places, thin woods; SC into Ariz, Wyo, Minn, and NY. July–Oct.

175 **Prostrate Spurge**
Chamaesyce maculata (L.) Small
Prostrate annual. Stems finely hairy on all sides. Leaves with very small teeth near tip. Peduncles under 6 mm. Common. Cultivated areas, waste places, crevices of roads and sidewalks; Fla into Tex, ND, sQue, and Me. May–freezing. *Euphorbia supina* Raf.
   C. prostata (Ait.) Small (*Euphorbia p.* Ait.) is similar but the stem is hairy in a line on one side only. Rare. Similar habitats; Fla into Tex, Mo, and Va. Three prostrate species have entire leaves and glabrous stems: *C. cordifolia* (Ell.) Small (*E. cordifolia* Ell.) has leaves twice as long as wide or shorter. Occasional. Open pine-oak woods of sandhills; cCP of NC into Fla and s tip of Tex. June–Oct. In *C. bombensis* (Jacq.) Dugand (*E. ammannioides* H.B.K.) the leaves are more than twice as wide as long and the fruits are about 2 mm. Occasional. Sand dunes along the coast; seVa into Fla and Tex. In *C. polygonifolia* (L.) Small (*Euphorbia p.* L.) the fruits are 3 –3.5 mm. Occasional. Dunes and beaches; Ala into Fla and NB.

176 **Hairy Spurge**
Chamaesyce hirta (L.) Millsp.
Erect to decumbent taprooted annual 2 – 60 cm, densely hairy. Leaves strongly asymmetric at base, 15 – 45 × 5 –17 mm, a purple spot in center, serrate, hairy on

both surfaces. Flowers unisexual, both in the same small cup (cyanthia); cyanthia borne in dense conspicuous clusters on peduncles 1 cm or longer. Fruits finely hairy, sharply 3-angled, 1–1.2 mm. Occasional. Fields, pastures, gardens, roadsides, dock areas, around buildings; seSC into ce and sGa, nFla, and sTex; widespread in warmer parts of the world. June–Oct.

BALSAMINACEAE: Touch-me-not Family

177   Jewelweed; Touch-me-not
Impatiens capensis Meerb.
A succulent annual to 2 m, heavily glaucous, thus repelling water. Water drops usually roll off the leaves but sometimes stand on horizontal areas and appear like jewels in reflected light. Some flowers conspicuous, orange-yellow spotted with brown, on slim usually drooping pedicels. Sepals 3, colorful; forms a prominent sac with a curled spur at the base. Other flowers very small, these primarily producing the fruits. Fruits usually drooping, 5-carpelled, green and coiling elastically into 5 sections when mature, often projecting the seeds considerable distances, thus the second common name. Common. Wet places, usually in woods; SC into eTex, Alba, Alas, and Nfld. May–frost. *I. biflora* Walt.

In *I. pallida* Nutt. the flowers are light yellow to cream-colored. Occasional. Wet places, usually in woods; nGa into ceOkla, sSask, and NS. June–Sept.

MALVACEAE: Mallow Family

178   Bristly Mallow
Modiola caroliniana (L.) G. Don.
Procumbent perennial rooting at nodes; stems with stellate hairs. Leaves 20–70 × 15–25 mm, palmately 6–7 lobed or incised, serrate. Calyx with 3 bracts at base; petals orange. Fruit a ring of 15–25 beaked carpels that separate at maturity. Occasional. Lawns; eVa, into c and seNC, s two-thirds of Ga, nFla, and Tex; S into Arg. Mar–May.

179   Sida
Sida rhombifolia L.
Annual or southward a biennial to 1.2 m, the stems very tough. Flowers solitary in the leaf axil, on peduncles several times longer than the petioles. Common. Waste places; natzd., Fla into se and eTex, eTenn, NC, and seVa. Annuals and biennials in June–Oct; biennials also in Apr–May.

S. spinosa L. has sharp projection at the base of most leaf petioles, often with 2 or more flowers in each axil. Common. Open places; natzd., Fla into eTex, eNeb, sWisc, sPa, and Mass. June–frost. *S. elliottii* Torr. & Gray has narrow leaves, about 3–6 mm wide, and longer petals. Rare. Open areas, usually sandy soil; Fla into Miss, seMo, Tenn, and Va. July–Oct. *S. acuta* Burm. f. (*S. carpinifolia* L. f.) has shorter peduncles (under 1 cm). Rare. Waste places, open areas, thin woods; seSC into Fla and sMiss. May–Oct.

180 Smooth Marsh-mallow
Hibiscus laevis All.
Glabrous perennial to 2 m. Leaves serrate, unlobed, or with divergent basal lobes.
Petals pink with a purple eye, to 8 cm. Seed pod glabrous or nearly so. Seeds covered with dense reddish-brown hairs. Occasional. Marshes and other open places along streams; Fla into e and cnTex, sMinn, cO, sePa, csKy, seAla, CP of SC, and eVa. June–Sept.

   H. grandiflorus Michx. also has lobed leaves but the lobes are usually larger and not as divergent. The blades are soft hairy beneath. Petals pink with a purple to red center, 12–15 cm. Seed pod to 5.5 cm, coarsely hairy outside. Rare. Marshes, ditches, other wet places in open; Fla into sMiss; coastal Ga. July–Aug.

181 Swamp Rose-mallow
Hibiscus moscheutos L.
Perennial to 2 m. In ssp. *moscheutos* the leaves are soft hairy beneath, glabrous above, unlobed or the lower ones shallowly lobed. Petals white or less often pink but always with a purple to reddish center, 10–12 cm. Seed pod glabrous. Occasional. Open swamps, fresh and brackish marshes; Fla into eTex, Tenn, sInd, and Md. May–Sept.

   In ssp. *lasiocarpos* (Cav.) O. J. Blanchard the leaves are grayish stellate, hairy above, and the seed pod is hairy. Occasional. Open swamps and marshes; Fla into e and nwTex, cIll, Pied of Ga, and neNC. June–Oct. *H. incanus* Wendl. f.

Pineland Hibiscus; Comfort-root
Hibiscus aculeatus Walt.
Perennial to 2 m. Stems and leaves rough with short stout hairs. Leaves with 3–5 palmate lobes or clefts. Corolla with a purple eye, otherwise cream-colored at first, turning to deeper yellow, and finally pink as the petals wither. Petals 5–6 cm. Common. Low open pinelands; upland bogs; Fla into seLa, cCP of Ga, and ceNC. June–Sept.

   In *H. coccineus* Walt. the plant is glabrous; the leaves have 5–7 narrow palmate often cleft lobes. The corolla is crimson to deep red. Rare. Open swamps and fresh or brackish marshes; c and nFla into sAla; seGa. June–Sept.

182 Rose-mallow
Hibiscus palustris L.
To 2.5 m. Leaf blades at midstem to 18 × 11.5 cm, broadly ovate to rounded, commonly 3-lobed, upper side glabrous, underside conspicuously hairy; branches of style abundantly hairy. Capsule with a depressed or broadly rounded summit, occasionally abruptly short-tipped. Sometimes considered a var. of *H. moscheutos*. Occasional. Fresh to saline marshes; neIll, nInd, sOnt, wNY, eMass, Va, and eNC. July–early Oct.

183 Seashore Mallow
Kosteletzkya virginica (L.) C. Presl ex Garay
Perennial to 1.5 m, with somewhat rough stellate hairs. Leaves densely hairy, 6–14 cm, cordate-orbicular to lanceolate, with or without hastate lobes at base;

petioles 2–9 cm. Base of calyx tube with 8–10 bracts 1–3 mm across, calyx lobes 4–7 mm; petals pink, lavender, or white; stigmas, styles, and carpels 5 each. Fruit prominently 6-angled, top flattened, covered with hairs 1.5–2 mm. Seeds 5, smooth. Common. Salt, brackish, and fresh marshes, thin swampy woods, sloughs; along the coast from Tex into Fla and NY. July–Oct.

STERCULIACEAE: Cacao Family

184   Chocolate-weed
Melochia corchorifolia L.
Annual with tough stems to 1.5 m and a taproot. Leaves 25–75 × 15–50 mm, doubly serrate. Flowers mostly in a dense terminal head, flower parts 5 each; sepals partially united, persistent, with 3 or more narrow bracts at base; petals short-clawed, pink to purple or white, 4–7 mm; filaments partially united; styles separate. Fruit a smooth capsule 4–4.5 mm; seeds smooth, brown with blackish markings, 2 in each carpel, or 1 by abortion. Occasional. Stable dune areas, roadsides, thin woods and scrub; sSC into Fla, eTex, seArk. Aug–Oct.

In *M. spicata* (L.) Fryxell the flowers are sessile in leaf axils and not in dense terminal heads. Styles partially united. Rare. Thin woods, scrub. Fla into sGa. Aug–Oct. *M. hirsuta* Cav.; *M. villosa* (Mill.) Fawc. & Rendle.

CLUSIACEAE: St. John's–wort Family

185   Common St. Peter's-wort
Hypericum perforatum L.
Members of this genus are herbaceous or woody with entire opposite leaves. The blades ordinarily have translucent internal glands visible with transmitted light; a lens is usually needed to see them. Hypanthium absent; sepals 4–5, separate; petals 4–5, separate, yellow to orange-yellow; ovulary superior. Fruit a many-seeded capsule. Some species cause sensitivity to light if eaten. *Hypericum* has been used as a drug since ancient times in Greece, and probably before. It contains numerous compounds with documented biological activity including anti-inflammatory, antiulcerogenic, antimicrobial, antiviral, cardiotonic, wound-healing, antioxidant, antidepressant, and sedative. Purified components of known strength are routinely prescribed in Eurasia; they are not approved for prescription in the US. *Hypericum* is one of the leading drugs for treatment of mild to moderate depression; 66 million daily doses were prescribed in Germany for this purpose in 1996. Use of unpurified extracts of unknown strength is strongly discouraged.

This species is a perennial to 1 m from a taproot, with many branches. Leaves sessile, spreading; blade linear-oblong, those at midstem 1.5–4 cm, with translucent dots. Petals 5, often black-dotted along margin only, 7–12 mm; stamens many, united at base into 3–5 groups; carpels 3, styles 3–5 mm. Seeds 1–1.3 mm, coarsely roughened by a network of ridges. Native of Europe. Occasional. Weedy in open places; meadows, roadsides, fields, between stable dunes; throughout much of the US and Can (to Nfld). June–Sept.

H. punctatum Lam. is similar but is sparingly branched and leaves are dark-dotted, petals 4–7.5 mm, seeds under 1 mm. Common. Similar places; Tex into eOkla, Minn, sOnt, sQue, Me, Va, and Fla. June–Sept.

186   Pineweed; Orange-grass

Hypericum gentianoides (L.) B.S.P.
Annual to 50 cm. Leaves subulate, usually less than 5 mm, appressed. Flowers sessile or nearly so. Stamens in clusters, fastened together at their bases. Mature capsule to 55 mm, over twice as long as the sepals. Common. Fields, pastures, and roadsides, usually in sandy or poor soils; Fla into e and cTex, sWisc, seOnt, and sMe. July–Oct. *Sarothera g.* L.

In *H. drummondii* (Grev. & Hook.) Torr. & Gray the leaves are linear-subulate, 6–18 mm, and ascending. The sepals are 4–7 mm, about as long as the fruit. Occasional. Dry sandy or clay soils, in open pastures, fields, thin woods; nFla into se and cTex, seKan, sInd, wPa, wNC, and seVa. July–Sept. *Sarothera d.* Grev. & Hook.

VIOLACEAE: Violet Family

187   Green Violet

Hybanthus concolor (T. F. Forst.) Spreng.
Erect perennial to 90 cm, main stem glabrous or its upper portion may be hairy. Leaves 9–17 × 3–6 cm, elliptic to elliptic-oblanceolate. Flowers in leaf axils, those of upper leaves quite small and not opening. Sepals, petals, and stamens 5 each; sepals linear, 3–5 mm, corolla about as long; stamens united at base. Fruit a 3-celled capsule 15–20 mm; seeds ca 4 mm. Occasional. Usually on basic to about neutral soils; wooded slopes, rich woods, ravines, stream terraces; Miss into Ark, Kan, Wisc, Mich, sOnt, Vt, Conn, Va, cNC, ceGa, and cnFla. Apr–June.

188   Field-pansy

Viola bicolor Pursh
All of the violets are said to be emetics, and many are laxative. The English have used a syrup of violets made from the flowers in the treatment of consumption.

This species is an annual, 4–30 cm. Stems slender, erect or decumbent at base, single or branched. Sepals one-half to two-thirds as long as petals. Petals white to blue or reddish-purple, the lower 3 petals with purple veins. Occasional. Fields, pastures, roadsides, thin woods; cwSC into ce and cnTex, cKan, seNeb, ePa, and cwNJ. Mar–May. *V. rafinesquei* Greene.

V. arvensis Murr. is similar but the stems not as slender, the sepals as long as or longer than the petals, and the petals yellowish with purple veins. Rare. Scattered localities; eSC into nMiss, nMich, and Me. Mar–June.

189   Spear-leaved Violet

Viola hastata Michx.
Perennial to 25 cm from an elongate brittle whitish rhizome. Leaves 2–5 near the top of the stem, elongate and with a protruding lobe at each side of the base of the blade. Occasional. Rich deciduous woods; upper Pied of Ga into neAla, e and cwVa, and cSC; nWVa into cnO and cPa. Mar–May.

Two other stemmed violets have yellow corollas. In *V. pubescens* Ait. (*V. pensylvanica* Michx.; *V. eriocarpa* Schwein.) the rhizome is short, dark, and scaly. The leaf blades are cordate and about as wide as long. Common. Rich deciduous woods; nGa into neTex, sMan, NB, and cnSC. In *V. tripartata* Ell. (*V. glaberrima* Ging.) the rhizomes are slender, knotted, and dark. The leaf blades are entire to deeply 3-lobed. Common. Well-drained areas in deciduous woods; swGa into neMiss, neTenn, cnNC, and cnSC; wWVa into sO and swPa. Mar–May.

190   Long-spurred Violet
Viola rostrata Pursh

Perennial to 25 cm from a prominent rhizome. Stems tufted, erect or nearly so. Corolla lilac-purple with a narrow and curved spur 7–20 mm. Occasional. Rich woods; nwSC into nAla, eKy, eWisc, swQue, cNH, and Conn. Mar–May.

Two other species with aboveground stems and purplish flowers but with shorter and broader spurs are: *V. walteri* House, which has prostrate stems that root at the nodes. Spur 3–5 mm. Occasional, but locally common or absent. Rich woods, often well-drained places; nFla into eTex, cwVa, and cNC; csO. Mar–May. *V. conspersa* Reichenb. has erect tufted stems. Spur 4–5 mm. Rare. Moist places, meadows, woods; cnAla into O, seND, sQue, NS, and wNC; nwSC. Mar–May.

191   Lance-leaved Violet
Viola lanceolata L.

Perennial without aboveground stems. Petals white. In ssp. *vittata* (Greene) Russell, which is illustrated, the leaf blades are linear, 6–15 times as long as broad. Common. Wet sandy soils, in thin woods or open; Fla into seTex, CP of Ga, and seVa; seOkla and nwGa. Mar–May.

In ssp. *lanceolata* the leaves are lanceolate, 3.5–5 times as long as broad. Rare. Sandy soils, open areas; nwFla into eTex, eMinn, and NS. Mar–May.

192   Florida Violet
Viola floridana Brainerd

First recognized at localities in Florida, hence its name, but now known from 7 states. Perennial with a thick horizontal rootstock. Some petals hairy. Leaf blades cordate not lobed, glabrous above and below. Occasional. Rich woods, often in swamps; Fla into La, Miss, SC, and ceNC. Mar–Apr.

Other similar species include: *V. affinis* Le Conte, which has very short stiff hairs on the upper surface of the basal lobes of the leaf, and the petioles and peduncles about equal. Considered by some as part of *V. floridana*. Common. SC into nGa, Ark, Ind, neIll, Wisc, wVt, and Mass. *V. cucullata* Ait. has peduncles longer than the petioles, otherwise much like *V. affinis*. Common. NC into Tenn, Mich, eMinn, sOnt, and Nfld. *V. sororia* Willd. has leaf blades hairy on both surfaces. Common. Moist places, meadows and woods; cFla into eTex, eND, swQue, and seMe.

193   Bird-foot Violet
Viola pedata L.

A stemless perennial with a vertical rootstock. Leaves glabrous, with 3 principal divisions, these palmately divided into 5–11 narrow lobes. Pedicels longer than the leaves. Plants often in tight circular clusters having as many as 30 flowers open at once. Flowers large, corolla 3–4.5 cm across, the petals without hairs. Plants with the upper 2 petals dark purple, as seen in the picture, are less common than those having 5 light purple petals. Some books consider the two kinds as separate varieties, the former being var. *pedata* and the latter var. *lineariloba* DC. Common. Roadside banks, thin woods, open places, often rocky, well-drained places; swGa into eTex, eKan, ceMinn, seNH, and NC. Mar–May.

194   Northern Wetlands Violet
Viola nephrophylla Greene
Plants with a thick rhizome and no stolons. Leaf blade at an angle with the petiole, first ones about as wide as to wider than long; margin crenate-serrate, base cordate. Flowers moderately open; sepals blunt to round-tipped; corolla purple (or white), lateral petals directed forward at an angle and bearded on inside, spurred petal hairy; stamens conspicuous. Occasional. A largely wUS and Can species extending E into ND, nIa, Wisc, Mich, sOnt, nNY, NB, and Nfld. May–July.

195   Wild White Violet
Viola macloskeyi Lloyd ssp. *pallens* (Banks ex DC.) M. S. Baker
Plants lacking an aboveground stem, with slender rhizomes under 2 mm across. Leaves orbicular to orbicular-ovate, 1–5 cm, margin shallowly crenate, base cordate. Petals white, under 1 cm. Occasional. Bogs, seepage areas, along cold streams, even in shallow water; Ind into Ia, BC, Lab, Del, and in mts to nGa.

TURNERACEAE: Turnera Family

196   Piriqueta
Piriqueta cistoides (L.) Griseb. ssp. *caroliniana* (Walt.) Arbo
Perennial to 50 cm, spreading by sprouts from the roots. The petals become easy to knock off before the day is over. Flowers heterodistylous, i.e., one kind of flower with the stamens longer than the pistil, as in the picture, and in the other kind the pistil is longer. The species varies considerably in leaf shape and in amount and kind of hairs on the stem and leaves. Some forms have been treated as species or varieties but present information is insufficient to make sound conclusions. Occasional. Sandy soil in open or in thin woods; Fla into swGa and swSC. Apr–Sept. *P. caroliniana* (Walt.) Urb.

PASSIFLORACEAE: Passion-flower Family

197   Passion-flower, May-pop
Passiflora incarnata L.
Climbing or trailing tendril-bearing perennial. Leaves deeply 3-lobed. Sepals and petals 5 each, behind the prominent purple-marked elongate fringes of the crown. Stamens 5, the filaments united into a tube that surrounds the long stalk of the ovary. Stigmas 3 on underside of the tips of the 3 spreading styles. Fruit fleshy, when ripe yellow and edible, ellipsoid, to 7 cm, the many seeds fastened to inside

of outer wall. There are forms with completely white petals and crown. These and the common form are used as ornamentals. Common. Fields, roadsides, thin woods; Fla into eTex, eOkla, sInd, swPa, and seVa. May–Aug.

P. lutea L. has similar-shaped but much smaller and greenish-yellow flowers. Leaves only shallowly lobed, and fruit to 15 mm and black. Common. Woods and thickets; Fla into e and cTex, seKan, cInd, and sePa. June–Sept.

LYTHRACEAE: Loosestrife Family

198   **Toothcup**
Rotala ramosior (L.) Koehne
Annual to 30 cm; stems glabrous. Leaves opposite, simple, entire, base tapered 1.4 cm, linear-lanceolate to oblanceolate. Flowers 1 in each leaf axil; sepals united, lobes acute-tipped; petals 4, separate, white to pink. Style ca 0.5 mm. Fruit a capsule borne free within a subglobose hypanthium. Occasional. Freshwater and brackish marshes, ditches, pond and lake margins, sloughs; eTex into Okla, Ark, Minn, Mich, O, Mass, and wFla; Calif; Mex and tropical Amer. Apr–Nov.

Two species of *Ammannia* are similar but there are 2–several flowers in each leaf axil; petals are purple to pink, and bases of midstem leaves usually auricled to cordate. In *A. latifolia* L. the style is 0.5 mm and tip of calyx lobes obtuse. Common. Similar places; CP—NJ into Fla and Tex, S into tropical Amer. July–Oct. *A. teres* Raf. In *A. coccinea* Rottb. the style is 1.5–3 mm and tip of calyx lobes acute. Rare. Similar places; ND into Pa, NJ, wFla, and Tex; Calif; S into tropical Amer. July–Oct.

199   **Purple Loosestrife**
Lythrum salicaria L.
Stout erect perennial herb to 1.5 m, often in large dense colonies. Leaves opposite or whorled, sessile, lanceolate to nearly linear, 3–10 cm. Flowers in a terminal spikelike cluster 10–40 cm; corolla red-purple, 7–12 mm; stamens mostly 12, alternately longer and shorter; ovulary 2-celled. Fruits capsules with thin walls. Native of Eurasia. Common. Wet places; neArk into Kan, Mo, ND, Que, Nfld, Va, and nwNC; occasionally W to the Pacific. June–Sept.

MELASTOMACEAE: Meadow-beauty Family

200   **Smooth Meadow-beauty**
Rhexia alifanus Walt.
This is our only genus of this large mostly tropical family of about 4000 species, three-fourths of which are American. The 13 species are confined to the US except 1 that also occurs in the W. Indies. Species of this genus are perennials with opposite leaves, an urn-shaped hypanthium that is fused around the ovary and continues above it, asymmetrical petals, and 8 stamens with anthers opening by pores at the apex.

This species is a sparingly branched perennial to 1.2 m and has a completely hairless stem, prominently curved anthers, and a glandular-hairy hypanthium.

The leaves are mostly ovate-lanceolate to elliptic or lanceolate, to 75 mm at midstem, where they are the largest. The petals are spreading and to 25 mm. Common. Sandy and peaty soils of low pinelands, bogs, and savannas; nFla into seTex and CP of Ga and NC. May–Sept.

201   **Pale Meadow-beauty**
Rhexia nashii Small
This species has prominently curved anthers. Plants to 1.5 m, often forming extensive colonies from shallow elongated rhizomes. Stems hairy, 4-angled, 4-sided, with 1 pair of opposing sides much narrower than the other pair. Midstem leaves mostly lanceolate to narrowly ovate to elliptic. Hypanthium 10–18 mm, smooth; petals 20–27 mm, usually light lavender. Common. Sandy or peaty swamps, bogs, roadside ditches, swales, thin flatwoods; seVa into seGa, Fla, and cMiss. May–Oct.

Of other similar species 2 have unequal faces at midstem. In *R. mariana* L. var. *mariana* leaves linear-filiform to ovate or ovate-lanceolate. Petals are white to pale lavender, 12–25 mm; hypanthium 6–10 mm. Common. Wet pinelands, bog margins, ditches; seOkla into Mo, swInd, seKy, seNY, csVa, NC, Fla, and eTex. May–Oct. In *R. cubensis* Griseb. midstem leaves mostly linear, oblong, or spatulate. The hypanthium is glandular-hairy; petals 15–20 mm, bright rose-lavender. Occasional. Similar places; ceNC into Fla and sMiss. May–Oct.

202   **Yellow Meadow-beauty**
Rhexia lutea Walt.
Perennial to 40 cm with a stout woody taproot or rootstock. Stems hairy. Leaves oblong, linear, or spatulate. The hypanthium is 6–7 mm with a narrow neck. Anthers about 2 mm and almost straight. Petals yellow, broadly ascending at pollen-shedding. Common. Low pinelands and savannas, seepage areas, bogs; nFla into seTex, seGa, and ceNC. May–July.

Two other species have stout taproots or rootstocks, ascending petals, almost straight anthers about 2 mm, and ovate to suborbicular leaves. *R. petiolata* Walt. has acuminate-aristate sepal lobes and lavender-rose petals. Occasional. Fla into seTex, cAla, seGa, and seVa. June–Sept. In *R. nuttallii* C. W. James the sepal lobes are blunt to acute and the petals lavender-rose. Occasional. Low pinelands, bogs; Fla into seGa. June–July.

ONAGRACEAE: Evening-primrose Family

203   **Primrose-willow**
Ludwigia decurrens Walt.
Annual to 2 m, usually with several ascending branches. Leaves lanceolate, sessile. Stems with narrow wings. Petals 6–12 mm. Stamens 8. Fruit about twice as long as broad. Common. Open wet places, shallow water; Fla into eTex, seKan, sInd, cwWVa, Tenn, NC, and seVa. July–Oct. *Jussiaea d.* (Walt.) DC.

Two related species have creeping and floating stems and can be troublesome weeds, often forming dense floating mats. The fruits are elongated but the stems are not winged and there are 5 sepals and petals. In *L. peploides* (Kunth.) Raven the stems and leaves are glabrous. Rare. Fla into Tex, seKan, Mo, cwInd, and wKy;

scattered from Ga into NJ. *J. repens* L. In *L. uruguayensis* (Camb.) Hara the stems and leaves are hairy and the petals 12–20 mm. Rare. Fla into Tex, swMo, Miss, Ga, and seNY.

204   False-loosestrife
Ludwigia linearis Walt.
There are over 25 species of *Ludwigia* in the eUS. Few have conspicuous flowers, some have opposite leaves, others have no petals, and some are entirely prostrate. All occur in moist to wet habitats. Usually the scars left by the petals are easily seen on the top of the fruits.

This species is a perennial to 80 cm. Leaves linear, alternate. Sepals narrowly deltoid. Petals 4, longer than the sepals. Stamens 4. Fruit sessile, narrowly conical. Common. Wet pinelands, swamps, bogs, edges of ponds; Fla into seTex, cTenn, nw and swGa, cSC, and sNJ. June–Sept.

L. linifolia Poir. is similar but the sepals are nearly linear and the capsule narrowly cylindrical. Rare. Wet places in pinelands, swamps, and in open. Fla into sMiss, sGa, and seNC. June–Sept.

205   Seedbox; Square-pod Water-primrose
Ludwigia alternifolia L.
Erect perennial to 120 cm, glabrous or nearly so. Leaves alternate, sessile or with a petiole 2 mm; blades to 10.5 × 2 cm, elliptic to lanceolate. Flowers on pedicels 1.4 mm; petals ca 6 mm; stamens 4. Capsules 4-angled or winged, cubical, 4.5–6 × 4.6 mm. Common. Marshes, ditches, savannas, low woods, swamps; Fla into eTex, Kan, sNeb, Ia, sIll, sOnt, NY, Mass, and Va. May–Oct.

206   Showy Evening-primrose
Oenothera speciosa Nutt.
Members of this genus have 4 petals and 8 stamens. The hypanthium is peculiar in that it not only surrounds and adheres to the sides of the ovulary but is prolonged into a tube beyond the ovulary.

This species is an erect to spreading perennial to 70 cm. Leaves oblong-lanceolate to linear, the wider ones irregularly dentate or narrowly lobed, especially near the base. Flower buds nodding. Petals white to dark pink. Anthers 1.2–2 mm. Fruit an ellipsoid to subglobose capsule. This species is hardy and drought-resistant and makes a showy ornamental. Occasional. Dry places, fields, roadsides, waste places, prairies; Fla into Tex, cnKan, Ill, Tenn, and Va. Mar–July.

O. triloba Nutt. is unusual in having no aboveground stems, the leaves and flowers arising from a stout rootstock. Fruit 20–35 mm. Occasional. Dry often calcareous soils, open places; nAla into csTex, cnKan, cTenn, seInd, and seTenn. Apr–May.

207   Sundrops
Oenothera fruticosa L. ssp. *glauca* (Michx.) Straley
Variable perennial to 80 cm. Leaves narrowly ovate or elliptic, to lanceolate or narrowly linear. Flower buds erect, petals 1–2 cm, anthers 4–8 mm. Capsule ellipsoid to nearly oblong, with a short stalk, glabrous or with erect hairs. Com-

mon. Dry to wet often rocky places, thin woods or in open; Fla into La, eOkla, cwMo, sMich, and NS. Apr–Aug.

O. *linifolia* Nutt. is an annual with linear-filiform stem leaves. Petals 3–4 mm. Rare. Dry rocky places, sandy barrens; Fla into eTex, seKan, sIll, and nAla; csNC. Apr–July.

208   Evening-primrose
Oenothera biennis L.
A biennial to 2 m, usually branching only near the top. Petals 1–2.5 cm. Capsules nearly cylindric. Seeds horizontally arranged. Occasional. Fields, roadsides, waste places, prairies; throughout most of the US. June–Oct.

In O. *grandiflora* Ait. the petals are 3–6 cm. Rare. Woods and waste places; CP of Ala; scattered escape from cultivation elsewhere. Two other species have columnar fruits. In O. *laciniata* Hill plants usually branch from near the base, are rarely over 70 cm, flowers sessile in the axils of leaves, bases of leaves deeply cut, and seeds ascending in the capsules. Common. Fields, gardens, waste places; Fla into Tex, ND, and Me. Mar–July. O. *humifusa* Nutt. is similar to the latter species but the upper stems and leaves are closely and densely hairy, and leaves at base of the flowers are entire or obscurely toothed. Occasional. Coastal sands; Fla into La and NJ. Apr–Oct.

209   Small-flowered Evening Primrose
Oenothera parviflora L.
Biennial or short-lived perennial to 1.5 m, highly variable. Hypanthium 1.3–4 cm; petals yellow, 1–2 cm; anthers 4–7 mm. Common. Disturbed open places; Minn, into WVa, cw and nVa, and Nfld. July–Oct.

210   Gaura
Gaura angustifolia Michx.
Slender annual to 1.8 m, usually unbranched near base, principal leaves narrowly elliptic to lanceolate. Flowers and fruits sessile or nearly so. Hypanthium prolonged beyond the ovulary as in Oenothera, this part about the same length as the ovulary; sepals 3–4 and 2.5–8 mm; petals 3–4, white to pink, ca 5 mm. Fruit without a stipe, indehiscent, hairy, acutely 3-ribbed, narrowly ellipsoid to ellipsoid, 1–4 seeded. Common. Dune areas, roadsides; NC into Fla, and Miss. May–Oct.

G. *longiflora* Spach is a similar annual or biennial, with thinly hairy stems, well-branched above, to 4 m. Flowers opening near sunset; sepals 3.7–14 mm; petals 7–13.6 mm. Occasional. Shell deposits along bay shores, thin woods, open sandy areas; Miss into Tex. June–Oct. G. *filiformis* Small.

211   Alpine Enchanter's-nightshade
Circaea alpina L.
To 30 cm; with a tuberous-thickened rhizome; stems weak and soft. Leaves thin, 2–6.5 cm, ovate to deltoid-ovate, apex acute, margin sharply and coarsely wavy-dentate, base cordate to broadly truncate. Flowers to about 15 clustered at top of a terminal raceme, racemes 1.5–7 cm; hypanthium short, tubular, falling soon after pollen shedding; sepals 1–2 mm, white to pinkish, glabrous; petals 1–2.5 mm,

bilobed to about the middle or less. Fruit 2–3 mm, 1-celled, 1-seeded. Common. Moist or wet woods; circumboreal; Alas into Lab; S into Wash, Utah, Colo, SD, Ia, sNY, mts of Tenn, Ga, and NC. June–Aug.

ARALIACEAE: Ginseng Family

212   **Dwarf Ginseng**
Panax trifolius L.
Perennial to 20 cm from a globose root. Leaves 3, once palmately compound, in a whorl on summit of stem; leaflets 3–5, sessile, to 6 × 2 cm, finely serrate. Flowers in an umbel on a solitary peduncle 2–8 cm; pedicels 3–7 mm; sepals very small or absent; petals 1–1.5 mm; styles usually 3. Fruit a yellow berrylike drupe with 2–3 pits. Occasional. Rich woods and bottomlands; Minn into Que and NS; S into Ia, Minn, Ind, Pa, Va, cnNC, and nGa. Apr–June.

APIACEAE: Carrot Family

Members of this family are usually quite distinct from species of other families. All are herbaceous; leaves are alternate or basal and usually compound. Flowers and fruits are in umbels although in some species this is not obvious as flowers are sessile and thus in heads. Ovulary inferior. Fruits are dry and have 2 1-seeded sections that split apart at maturity, a schizocarp.

213   **Seaside Pennywort**
Hydrocotyle bonariensis Comm. ex Lam.
Seaside Pennywort is a glabrous succulent perennial, rooting from nodes on slender creeping or floating stems. Leaves simple; blades peltate, 3–10 cm across, orbicular or nearly so. Peduncles equal or exceed the petioles. Inflorescence a compound umbel of many flowers, usually continuing for some time to develop new sections with numerous additional flowers; fruits and flowers thus usually present at the same time and branches may extend to 10 cm. Calyx lobes lacking. Petals 5, white to light yellowish-green. Fruit flattened, ca 3 mm across, composed of 2 seedlike sections. Common. Primary and stable dunes and swales, sandy marshes, swamps, sand flats; sloughs; NC into Fla and eTex. May–Nov.

214   **Asiatic Pennywort**
Centella asiatica (L.) Urb.
Perennial from thin rhizomes, resembling some *Hydrocotyle* species but petioles are attached to base of the cordate to truncate blades, these ovate to oblong and 1.5–4(10) cm. Flowers 2–9 in simple umbels. Petals white, soon falling. Peduncles 1–5 per node and 1–10 mm, usually shorter than petioles. Fruits 3–4 mm across. Common. Sloughs, swales, meadows, edges of freshwater marshes, yards; CP—sNJ into Fla and Tex; Mex; tropical Amer. May–Oct.

215   **Rattlesnake-master**
Eryngium yuccifolium Michx.
Perennial to 1.8 m, the flowers many in each of 3–30 heads. The narrow parallel-veined leaves suggest it is a monocot instead of a dicot, and the flowers in heads those of a family other than the APIACEAE. Plants were once thought to be effective in the treatment of snakebite and were used as an emetic. Occasional. Thin woods, meadows, and prairies; Fla into eTex, seNeb, Minn, O, Ky, Va, and NJ. June–Aug.

216   **Meadow-parsnip**
Zizia aptera (Gray) Fern.
Perennial to 70 cm, the lower leaves simple, ovate to suborbicular, rarely 3-parted. Terminal leaflet of upper leaves entire near the base. Fruit ribbed and smooth. Common. Dry to moist situations in open or woods; cnFla into nArk, Ia, BC, and RI. Mar–July.

In *Z. aurea* (L.) W. D. J. Koch the lower leaves are compound, leaflet blades with 5–10 teeth per cm. Occasional. Moist habitats in open or in woods; nFla into eTex, Sask, and Me. *Z. trifoliata* (Michx.) Fern. is similar to the latter species, but with coriaceous leaves instead of membranaceous and the blades with 2–4 teeth per cm. Occasional. Rich woods; nFla into eTenn, WVa, Va, and cSC.

217   **Caraway**
Carum carvi L.
Plants 20–75 cm, smooth and glabrous. Leaves mostly on main stem, compound with filiform or narrowly linear divisions. Flowers symmetrical, bisexual. Fruits without prickles. Native of Europe. Occasional. Old fields; Alta into Nfld; S into Ill, Pa, NS.

218   **Lovage**
Ligusticum canadense (L.) Britt.
Stout branched perennial to 2 m from a taproot; root with odor of celery. Leaf blades deeply divided, the leafletlike segments lanceolate to ovate, 5–11 cm × 2.5–6 mm, primary veins of blade directed toward the teeth, margin dentate; upper petioles not expanded. Flowers in peduncled compound umbels, peduncles 5–18 cm; petals white; ovulary and fruit glabrous. Fruit ovoid-oblong, 5–7 mm, one-half to two-thirds as wide as long; with elevated and narrowly winged ribs. Common. Rich soil on wooded slopes, margin of woods, banks and bottomlands of streams; nArk into sMo, Ky, sPa, WVa, csVa, cNC, cwGa; chiefly in mts. June–July.

L. scothicum L. (Scotch Lovage) is the only other species of the genus in eUS. Stems single or branched, 30–60 cm. Leaves thick and fleshy, the largest leaves twice-compound with 9 leaflets. Rays of umbel 10–20, 2–5 cm. Fruit oblong, 6–10 mm, one-third as wide, ribs prominent and narrowly winged. Occasional. Sandy or rocky seashores, saline marshes; Greenl into Lab and Long Isl., NY: n Europe. June–Sept.

219   **Angelica**
Angelica venenosa (Greenway) Fern.
Perennial to 150 cm from a deep stout taproot. Leaves 1–3-divided. Leaflets serrate, several to many and broad, the petioles prominently winged toward their bases. At least the upper part of the stem, the peduncles, and the pedicles are very finely hairy. The snow-white flowers are in compound umbels without bracts. Fruits about as broad as long, finely hairy, flattened, with 3 ridges on each flattened side and 2 wings on each of the edges. Reported many years ago to be poisonous but was probably confused with the poisonous *Cicuta*, Water Hemlock. Angelica is not listed as poisonous in current poisonous plant books. In *Cicuta* the veins of the leaflets end at the notches; in *Angelica* they do not. Common. Thin upland woods, and dry places in open; Fla into Miss, eOkla, sMich, swMass, and eConn. June–Aug.

220   **Wild Carrot; Queen-Anne's-Lace**
Daucus carota L.
Erect, usually branched, hairy biennial to 2 m from a strong taproot. Leaves pinnately dissected, often so deeply as to appear compound, the segments narrow. Inflorescence a compound umbel, rounded when in flower but concave earlier and when in fruit, with a whorl of narrowly once-pinnate leaves at the base. Wings of fruit each with 12 or more prickles. The cultivated carrot is a race of this species. Wild carrot is edible when young but the root, especially the center, soon gets tough. It is a troublesome weed of some crops, in gardens, lawns, and pastures. Common. Throughout the US and sCan. May–Sept.

   D. pusillus Michx. is similar but an annual, smaller. Leaves more finely dissected, those in the whorl at the base of the umbels twice-compound. Spines on the wings of the fruit 10 or less each. Occasional. Open places; Fla into Cal, BC, Mo, Ala, SC, and seVa. Apr–June.

PYROLACEAE: Shinleaf Family

221   **One-flowered Shinleaf**
Moneses uniflora (L.) Gray
Plants 3–10 cm. Leaves basal, orbicular, 1–2 cm, nearly entire to finely toothed; petiole 5–10 mm. Flowers fragrant, 12–20 mm across, solitary and terminal. Rare. Damp woods, bogs; circumboreal; S in Amer into Mich, NY, and Conn; NM. July–Aug.

MONOTROPACEAE: Indian-pipe Family
Herbs without chlorophyll and living on organic matter, probably with the aid of fungi. Flowers nodding, regular, bisexual. Fruits erect.

222   **Pine-sap**
Monotropa hypopithys L.
Plants of this genus turn black on drying and have separate overlapping petals and erect fruits.

This species grows to 30 cm, often clustered; plants at time of flowering may be yellow, tawny, pink, or reddish; they remain colorful, usually in shades of pink, until after fruit has matured. Flowers few to several in a raceme; style shorter than ovulary. Occasional. Moist or dry woods, usually in acid soil; disjunctly circumboreal; in Amer S into neCalif, nFla, and Mex. Apr–Oct.

M. uniflora L., Indian-pipe, is waxy-white, rarely pink to red, to 20 cm, solitary to several clustered. Leaves scalelike. Flowers solitary, to 15 mm. Occasional. Rich woods; Calif into seAlas, Nfld, Fla, eTex, and Mex; Columbia; eAsia. Apr–Sept. *Monotropsis odorata* Ell., which has a spicy odor, is somewhat similar to *Monotropa hypopithys* but to only ca 10 cm and flowers ca 10 mm with petals united. Fruits nodding instead of erect. Rare. Dry woods, often sandy soils; Ala into cWVa, Md, Me, Del, nSC, and nGa. Mar–June.

DIAPENSIACEAE: Pixie Family

223 Diapensia
Diapensia lapponica L.
Plants mat-forming. Leaves opposite, evergreen; blade spatulate, 6–15 mm. Flowers solitary on peduncles 1–4 cm; bracts below flowers ovate; sepals deeply 5-parted; corolla 7–10 mm, bell-shaped; stamens 5, borne on inside of corolla tube alternate with the lobes; fruit subglobose, 5 mm. Rare. Circumboreal, S on alpine summits into Me, NH, Vt, and NY. June.

224 Galax
Galax urceolata (Poir.) Brummitt
The shiny leathery cordate leaves of this evergreen perennial may be 15 cm wide; size is dependent on the chromosome number of the plant and on environmental conditions. Individual plants may have twice as many chromosomes as others. Leaves of the former may reach a width of 15 cm, those of the latter only 10 cm. The small flowers are on a leafless stem to 50 cm. Petals are white or sometimes light pink or light blue. The leaves turn reddish to bronze in winter and are often used in Christmas decorations. Common. Moist to dry woods chiefly in the mts; nGa and nAla into Ky, Md, seVa, and nSC. Apr–July. *G. aphylla* auct. *non* L.

PRIMULACEAE: Primrose Family

225 Water Pimpernel
Samolus parviflorus Raf.
A slender glabrous perennial to 50 cm; small plants sometimes unbranched as in the photograph, the larger ones freely branched, usually with prominent basal leaves. Flowers in terminal racemes, pedicels bearing a small bract about midway; petals white, united, the tube very short; a hypanthium fused to base of ovulary. Fruit a globose capsule 2–3 mm. Not easily confused with any other kind of plant. Common. Moist to wet places, fresh or brackish; swales, sloughs, pond and marsh margins, roadside ditches. Tropical Amer N into BC, sMich, sOnt, sQue, NB, and PEI. Apr–Oct.

226   **Whorled Loosestrife**
Lysimachia quadrifolia L.
Members of this genus are perennials, usually with leaves entire and opposite or whorled. Petals united at base; stamens 5, borne on inside of corolla tube. Fruits capsules, placentas central, seeds few to many.

In this species, plants are erect, to 90 cm, glabrous. Stems 4-angled, rarely branched. Leaves in whorls of 4, uncommonly less or more; blades narrowly or broadly lanceolate, 5–10 cm, lateral veins obscure; petioles ciliate only near the base. Flowers 10–15 mm across, on long slender spreading pedicels from leaf axils; corolla yellow, inner base streaked with black and reddish-purple, lobes 10–15 mm. Capsule subglobose, 2.5–3.5 mm, seeds few. Occasional. Moist to dry places; thin woods or scrub, bluffs, shores of freshwater ponds and marshes; cnKy into cnArk, Man, eMinn, Wisc, sOnt, seNY, cwVa, cnNC, and cGa. May–Aug.

L. ciliata L. (Fringed Loosestrife) also has flowers on slender axillary pedicels, but it has opposite petioled leaves and the petiole is ciliate throughout its length. Plants with long slender creeping rhizomes. Flowers 15–28 mm across; corolla yellow, unspotted; lobes 6–8 mm. Common. In NM and Ore; BC into Alas, Que, NS, Mass, cNC, ceGa, and La. May–Sept. *L. tonsa* (Wood) Wood ex Pax & R. Knuth also has leaves of midstem opposite but rhizomes are very short. Leaves mostly ovate, to 8 × 3.5 cm; petioles without cilia or rarely sparsely ciliate near stem. Plants to ca 1 m, in small to large clumps with as many as 70 stems from 1 crown. Thin upland woods, bluffs; nAla into cnArk, cn and eKy, WVa, cs and swVa, cn and csNC, and cwGa. May–Aug.

227   **Lance-leaved Loosestrife**
Lysimachia lanceolata Walt.
Plants to 70 cm from rhizomes. Leaves sessile, lacking punctate dots, main leaves to 15 cm, linear to narrowly oblong, tapering to base, lateral veins evident, underside pale. Flowers to 20 mm across; petal margins sharply toothed. Common. Dry to moist habitats; woods, thickets, or in open; prairies; nFla into eTex, eOkla, Ia, Wisc, sMich, Pa, Va, cn and swNC, and cwSC. June–early Aug.

228   **Swamp Loosestrife; Swamp-candles**
Lysimachia terrestris (L.) B.S.P.
Perennial to 1 m, rhizomatous. Leaves opposite, those on about the lower third of stem scalelike, the others 3–10 × 0.7–2 cm, lanceolate to narrowly elliptic. Flowers in terminal racemes with conspicuous narrow bracts at base of each whorl of flowers. The species reproduces vegetatively by means of axillary buds 1–2 cm that are produced late in the season and drop to the ground in the autumn. Occasional. Moist to wet places—swamps, thin woods, freshwater marshes, pond margins; Tenn into Ia, Minn, Nfld, Va but absent from w portion, neNC, and cSC. May–Aug.

229   **Sea-milkwort**
Glaux maritima L.
Perennial, usually much branched, spreading to erect, succulent, often forming dense masses to as much as 1 m across. Leaves opposite, crowded, larger ones 3–12 × 1.5–6 mm, entire. Flowers solitary in leaf axils, sessile or nearly so; corolla

absent; sepals petallike, white to pink or red, united by bases. Occasional. Beaches, brackish or salt marshes; circumboreal; in Amer S into NM and Va. June–July.

230   Shooting-star
Dodecatheon meadia L.
Perennial to 60 cm with thick fleshy roots. Leaves all basal, entire, elliptic-lanceolate to oblanceolate, 10–30 cm long, up to 8 cm wide. Flowers few to many in a terminal umbel. Young pedicels erect, flowers nodding when fully opened, fruits on erect pedicels. Petals united at the base, the lobes curved backward. Ovulary 1-celled with central placentation. Occasional. Rich woods, meadows, prairies; csSC into nAla, eTex, seKan, sWisc, and DC. Mar–May.

LOGANIACEAE: Logania Family

231   Indian-pink
Spigelia marilandica L.
Erect perennial to 70 cm with 4–7 pairs of sessile leaves. Fruit 4–6 × 6–10 mm, with 2 distinct lobes. Seeds few. Extracts of the roots, which contain an alkaloid, have been used in medicine to eliminate intestinal parasites. Misuse has caused poisoning. Common. Rich woods; Fla into eTex, seOkla, swInd, nwGa, and eSC. Apr–June.

BUDDLEJACEAE: Butterfly-bush Family

232   Rustweed; Polypremum
Polypremum procumbens L.
Glabrous perennial with several to many radiating prostrate to ascending branches, often forming a circular mass, to 70 cm across in vigorous plants. Leaves entire, 10–25 mm, awl-shaped to linear, opposite, the pair connected at base by a fine line. Flowers axillary and sessile in leaf axils and at ends of branches; corolla regular, basal ⅔ united, 4-lobed, bearded inside throat; stamens 4, very short; ovulary superior with 2 carpels. Fruit a capsule, notched at apex; seeds yellow, many, 1.5–2.5 mm, more or less square. Common. Meadows, gardens, roadsides, waste places; eTex into seMo, sIll, ePa, Del, seNY, Fla; Mex. May–Sept.

GENTIANACEAE: Gentian Family

233   Marsh-pink
Sabatia dodecandra (L.) B.S.P.
Perennial to 100 cm, with slender to robust rhizomes. Branches alternate. Upper leaves wider than diameter of stem. Flowers pedicelled. Corolla lobes 8–13, pink or rarely white. Occasional. Var. *dodecandra* usually occurs in brackish habitats; cw and nwFla; SC into Conn. June–Sept. Var. *foliosa* (Fern.) Wilbur occurs on riverbanks and beside ponds and streams; Fla into seTex and lower CP of SC. June–Sept.

S. bartramii Wilbur is similar but the upper leaves are narrower than the stem. Occasional. Moist open places, savannas, pinelands, ditches, often in water; Fla into seMiss and sGa. June–Sept. *S. calycina* (Lam.) Heller also has many similar characteristics but is shorter, to 50 cm, and has only 5–7 corolla lobes. Occasional. Edge of marshes, lowland hardwoods, ditches; Fla into seTex, swGa, and seVa. June–July.

234   Rose-pink Sabatia
Sabatia angularis (L.) Pursh
Erect biennial to 90 cm without rhizomes. Branches opposite. Lower part of stem strongly 4-angled. Pedicels usually 15–35 mm. Corolla pink, rarely white. Common. Usually moist open places, pinelands, roadsides, wooded ravines, granitic outcrops; nwFla into eTex, eKan, swMich, Conn, and SC. July–Aug.

    S. brachiata Ell. is similar but smaller (to 60 cm) and the lower part of the stem is terete. Occasional. Dry places, fields, savannas, open pine and oak woods; La into seMo, NC, and seVa. June–July. *S. quadrangula* Wilbur also has many similar characteristics but is smaller, to 60 cm, the petals are white, and the pedicels are shorter than 5 mm. Occasional. Wet to dry places, fields, savannas, thin woods; nc and nwFla into cPied of Ga, cs and ceVa. June–July.

235   Upland Sabatia
Sabatia capitata (Raf.) Blake
Perennial to 60 cm, usually with few branches. Flowers sessile and with many petals. The basal and stem leaves are similar in shape, the upper ones usually over 10 mm and always over 5 mm wide. Calyx lobes linear, erect. Rare. Thin hardwoods, hillsides, ridges, occasionally in moist lowland woods; c and neAla into nwGa and adjTenn. June–Aug.

    S. gentianoides Ell. is similar but usually unbranched and with the basal leaves broad, the upper leaves narrowly linear, and under 5 mm wide. Calyx lobes subulate, the tips usually reflexed. Occasional. Bogs, wet meadows, savannas, ditches; se and ceTex into Fla, sGa, and eNC. Aug–Oct.

236   White Sabatia
Sabatia difformis (L.) Druce
Erect perennial to 1 m, with a gnarled stout branched rhizome. Upper stem angular, branches opposite. Flowers with long pedicels. Corolla lobes over 7 mm. Common. Usually moist savannas and pinelands; CP—Fla into seAla and NC. May–Aug.

    S. macrophylla Hook. is similar but the stems are terete and corolla lobes are to 7 mm. Occasional. June–July. In var. *macrophylla* calyx lobes are erect or only slightly curved at tip, equaling or less than the calyx-tube in length. Savannas, pine barrens, flatwoods, swampy places; cnFla into seLa and cwGa. In var. *recurvans* (Small) Wilbur the calyx lobes are strongly curved backward and exceed the calyx-tube. Savannas, pine barrens, seepage areas; neFla into CP of Ga.

237 **Pennywort**
Obolaria virginica L.
A purplish-green fleshy perennial to 20 cm. Roots brittle. Stem simple or with a few erect branches. Leaves opposite. Flowers dull white or tinted with light purple, 7–15 mm. Stamens separate and of equal size. Seeds very small, shorter than 0.25 mm and about 1600 per capsule. Rare. Rich woods, usually moist places; cAla into nwMiss, sInd, NJ, and NC; nwFla; seLa. May–June.

238 **Pale Gentian**
Gentiana villosa L.
Glabrous perennial to 60 cm. Leaves in 5–12 pairs. Flowers in a terminal cluster with 2–6 leaves at their bases, sometimes other flowers in the upper 1–6 pairs of leaf axils. Calyx lobes linear, mostly longer than the tube. Corolla greenish-white, often tinged with purple, the lobes with an obliquely triangular appendage on the lower edge of 1 side of each lobe. Anthers sometimes united. Occasional. Rich woods; nwFla into seLa, Ky, sO, and sePa. Sept–Dec.
   G. alba Muhl. ex Nutt. is similar but has deltoid-ovate to ovate-lanceolate calyx lobes. Occasional. Moist woods, prairies, and meadows; nArk into cMinn, Pa, and neNC. Sept–Oct. *G. flavida* Gray. In *G. decora* Pollard the calyx lobes are very narrow, pointed, and shorter than the tube. Rare. Rich woods and openings; nwSC into neGA, neTenn, and cWVa. Sept–Nov.

239 **Soapwort Gentian**
Gentiana saponaria L.
Glabrous or finely hairy perennial to 65 cm. Leaves in 7–15 pairs, resembling those of *Saponaria,* thus the name. Flowers in terminal clusters, with leaves at their bases, sometimes other flowers in the upper 1–10 pairs of leaf axils. Calyx lobes shorter than to about as long as the tube. Corolla purple or blue, or if lighter colored then with purple or blue lines. Appendages centrally located between the corolla lobes and with 2 erect lobes. Anthers united. Common. Moist, often open places; nwFla into eTex, seOkla, cTenn, Ill, seNY, and Va. Sept–Nov.
   G. catesbaei Walt. has similar flowers but the calyx lobes are longer than the tube. Common. Moist places, thin woods, pinelands; cn and neFla into NJ. Sept–Nov.

240 **Stiff Gentian**
Gentienella quinquefolia (L.) Small
Annual to 80 cm. Stem and branches wing-angled. Flowers in terminal clusters, rarely single, the pedicels short to as long as the flowers. Corolla lobes with no appendages between them. Anthers not united. Our plants are var. *quinquefolia.* Common. Rich woods, road banks, moist open places; cnGa into eTenn, WVa, sOnt, wNY, mts of Va, and nwSC. Aug–Nov. *Gentiana q.* L.

241 **Columbo**
Frasera caroliniensis Walt.
Stout biennial or triennial to 3 m. Leaves in whorls of 3–9, lanceolate to oblance-olate, to 45 cm. Flowers in a pyramidal panicle sometimes with smaller clusters from the leaf axils below. Flowers occasionally occupy the top third of the plant.

Sepals, petals, and stamens 4 each. Corolla 20–35 mm broad, light greenish-yellow often marked with small purple-brown dots, each petal bearing a large fringed greenish gland. Fruit a flattened, 1-celled capsule; the seeds large, flat, and borne on the outer wall in 2 rows. Rare. Rich woods, dry open places often in calcareous habitats; nwSC into seLa, seMo, Ind, sWisc, sOnt, wNY, O, and cTenn. May–June. *Swertia c.* (Walt.) Kuntze.

MENYANTHACEAE: Buck-bean Family

242 **Buck-bean**
Menyanthes trifoliata L.
Leaves all basal, emergent, 3-foliate; petiole 5–30 cm, base conspicuously expanded. Flowers in a raceme on a leafless peduncle 10–30 cm; calyx 3–5 mm, deeply cleft; corolla 1.5–2 cm across, whitish, usually purple-tinged, petals united, tube about twice as long as the calyx, lobes 5–7 mm and ovate-lanceolate. Only species of genus. Rare. In quiet shallow water; circumboreal; S into Calif, Wyo, Neb, Mo, Ind, cO, NJ, and nVA. Apr–July.

243 **Floating-heart**
Nymphoides aquatica (J. F. Gmel.)
Perennial from a thick rhizome. Leaf blades floating, nearly circular, 5–20 cm across, veins prominent, undersurface pebbly and purple. What seem to be long petioles are mostly stems, on the upper end of which develop an umbel of flowers and 1 leaf with a short petiole. After flowering, tubers to 20 × 4 mm usually develop and hang downward under the umbel. Occasional. Ponds, lakes, and slow streams; Fla into eTex, CP of Ga, and s half of CP of NC; scattered localities N to Del. Apr–Sept.

    N. cordata (Ell.) Fern. has ovate, usually smaller leaf blades that are not purple-pebbly beneath. The tubers are much elongated and very slender. Occasional. Fla into La, CP of Ga, Conn, NY, and Nfld; locally into Minn, Wisc, and sOnt. Apr–Aug. *N. lacunosa* (Vent.) Kuntze of some books.

APOCYNACEAE: Dogbane Family

244 **Blue-star**
Amsonia tabernaemontana Walt.
Perennial to 1.1 m, little-branched or unbranched, the stems glabrous, leaves alternate, juice milky. Corolla finely hairy on outside. Pods erect, 2 per pedicel, about 130 × 3 mm, each splitting on 1 side. Seeds cylindrical and packed into 1 row. Common. Rich deciduous woods; Fla into neTex, eKan, swInd, and sVa. Mar–May.

    A. rigida Shuttlew. ex Small is similar but has no hairs on the outside of the corolla, and the leaves are usually shorter and more abundant. Rare. Moist situations in thin woods or in open; nFla into s and neMiss, sAla, and sGa. The corolla is also glabrous outside in *A. ciliata* Walt. but the leaves are only 1–5 mm wide although 3 – 8 cm long. Common. Sandy areas in thin woods or in open; cNC into CP of Ga, Fla, and cs and cn Tex; then N into sMo.

245   Dogbane
Apocynum androsaemifolium L.
Members of this genus are reported to be poisonous when eaten.

This species is a perennial to 9 cm, juice milky. Leaves spreading or drooping. Corolla 5 – 10 mm, bell-shaped, the lobes spreading or curved backward. The fruit of any one flower consists of a pair of slender pods 12 – 22 cm that split along 1 side. Common. Open areas and thin woods; mts of Ga into eTenn, WVa, O, cInd, neOkla, Colo, cTex, sCalif, Alas, and NS. June – Sept.

A. *cannabinum* L., which is similar, has spreading or ascending leaves, a cylindrical corolla 2 – 5 mm and with erect or only slightly divergent lobes. Fla into sCalif, Wash, cAlba, Que, and Vt. May – Aug.

ASCLEPIADACEAE: Milkweed Family

246   Red Milkweed
Asclepias lanceolata Walt.
Members of this genus are perennials and the juice is milky except in 2 species. The flowers are in umbels and are complicated, although some characteristics are easily understood and are quite helpful in identification to species. Flowers consist of an outermost calyx with 5 small reflexed lobes, usually hidden by the 5 reflexed lobes of the corolla, which is red in the photograph. The orange structures above the red corolla are hoods that together form the crown. The crown may be elevated on a column as seen in the photograph. Each hood contains a cavity from which a needlelike horn may protrude. (Horns are visible in the picture of *A. perennis.*) In most species the fruits are erect and on deflexed pedicels. Seeds, except in 1 species, bear a tuft of silky hairs. Several species are known to be poisonous when eaten raw; it is likely that most, if not all, species are toxic. Young fruits and shoots of some species are reported to be excellent for food when boiled and the first water discarded.

This species may grow to 120 cm. The stem is erect and rarely branched. The leaves are narrowly lanceolate and in 3 – 6 pairs. There are 1 – 6 umbels; corolla is red, crown usually orange, and horns hidden. The fruits are erect on deflexed pedicels. Occasional. Wet savannas and pine barrens, fresh or brackish marshes; Fla into seTex, cCP of Ga, and coastal NC into NJ. May – Sept.

A. *rubra* L. has similar flowers, except the corolla is sometimes purplish-red. The leaves are lanceolate. Occasional. Bogs, marshes, low pine barrens; nwFla into eTex, c upper CP of Ga, and seNY; ePied of Ga. June – Aug.

247   White Milkweed
Asclepias variegata L.
Stem unbranched, to 1 m. Leaves broad, in 2 – 5 pairs. Peduncles 1 – 7 cm, pedicels 1 – 2 cm. Corolla lobes 7 – 8 mm. Crown 4 – 7 mm wide. The horns are exposed but are turned in toward the top center of the flower. Mature fruits erect on deflexed

pedicels. Common. Upland woods; cnFla into eTex, ceOkla, neO, Md, sConn, and SC. Apr–June.

248   Curly Milkweed
Asclepias amplexicaulis Sm.
Plants with a single stem, to 1 m. Leaves sessile, clasping the stem, in 4–6 pairs, wavy-margined, thus the common name. The single umbel, rarely 2 or 3, is on a stout peduncle 10–60 cm above the uppermost leaves. Crown 5–8 mm broad and prominently darker than the corolla. Fruits erect on deflexed pedicels. Occasional. Thin woods, open and often sandy places, usually dry habitats; nFla into eTex, Okla, Ia, cWisc, O, Md, ePa, cVt, and Mass. May–June.

249   Aquatic Milkweed
Asclepias perennis Walt.
Sometimes growing to 50 cm, the stems slender, usually branching only from the base. All leaves opposite, lanceolate. Peduncles thin, 2–5 cm. Pedicels 5–15 mm. Umbels 1–few. Horns longer than the hoods. Crown 2–3 mm wide. Mature fruits drooping. Seeds hairless. Common. Wet woods, especially in swamps and along rivers; Fla into seTex, sInd, wTenn, cAla, and CP of SC. May–Sept.

250   Whorled-leaf Milkweed
Asclepias verticillata L.
Stem usually simple, sometimes branched near the top, to 80 cm. Leaves very narrow, whorled. Umbels 2–14 from the upper nodes. Peduncles slender, 15–25 cm. Corolla lobes 3.5–4.5 mm, the crown 3–5 mm wide. Horns prominently exposed. Fruits erect, on erect pedicels. Common. Dry thin woods, sandhills, rocky places; Fla into eTex, ND, seMan, cwWisc, and Mass. May–Sept.

   In *A. cinera* Walt. also the leaves are quite narrow but they are opposite and are longer, 5–9 cm. The umbels are 1–4, terminal or axillary from the upper nodes, on slender peduncles 5–22 mm. Corolla and crown lavender to almost white, the horns protruding from the hoods. Fruits erect on erect peduncles. Occasional. Pinelands, sandy ridges, sometimes in moist open places; nFla into lower half of CP of Ga and swSC.

251   Common Milkweed
Asclepias syriaca L.
Plants to 2 m. Leaves opposite, widely elliptic to ovate-elliptic, finely hairy below. Flowers are heavily fragrant. Corolla 8–10 mm, greenish to purplish; hoods 6–8 mm, pale purple; horns shorter than hood. Fruit erect on drooping pedicels, surface with many slim to conic structures 1-3 mm, a character unique solely to this milkweed species in our area. Common. Open usually dry places; fields, roadsides, waste places, stable dune areas; nwGa into nwKan, cnNeb, eSD, seMinn, Mich, NS, nVa, and neNC. June–Aug.

   A. exaltata L. is similar. Plants with a single upright stem to 1.3 m. Leaves opposite but otherwise much like those of Pokeweed (Pursh named the plant *A. phytolacioides,* apparently because of the similarity). Umbels 2–7 from the upper nodes. Peduncles slender, 3–9 cm. Longest pedicels 3–6 cm. Horns prominently longer than the hoods. Fruits 12–14 cm. Occasional. Deciduous woods, meadows; nwSC into neAla, cWisc, sMe, Mass, N into seVa. June–Aug.

252   Butterfly-weed; Chigger-weed
Asclepias tuberosa L.
Plants roughish-hairy. Stems 1–several from a thick root; erect, ascending, or decumbent; usually branching at the top into 2–5 parts, these sometimes elongated and spreading and each bearing 1–6 umbrels. The sap is not milky. Leaves abundant, alternate, linear to elliptic, obovate, oblanceolate, or hastate. Corolla and crown usually orange but varying to yellow or red. Either the crown or the corolla may be the darker. Common. Dry places, thin woods or open; Fla into nMex, Ariz, sUtah, seNeb, cwMinn, sNH, and Mass. May–Aug.

This highly variable species has been divided into intergrading subspecies. The leaves of ssp. *tuberosa* are typically obovate to oblanceolate; of ssp. *rolfsii* (Britt. ex Vail) Woods. hastate; of ssp. *interior* Woods. lanceolate to ovate.

253   Fragrant Milkweed
Asclepias connivens Baldw.
Erect unbranched perennial to 95 cm. Leaves opposite, sessile or nearly so, blades narrowly lanceolate to oblong-elliptic. Inflorescences terminal, or terminal and lateral in the upper 1–5 nodes. Flowers very unusual: large, the corolla lobes 12–15 mm, hoods 7–9 mm and converging over the stigma. Fruits erect on declined pedicels, long and narrowly elliptical. Rare. Low pinelands and savannas, margins of cypress swamps; Fla into seMiss; s half of CP of Ga. June–Aug. *Anantherix c.* (Baldw.) Feay.

A. viridis Walt. has equally large and showy flowers but the petals are spreading with ascending tips. The hoods are much smaller, deflexed at their bases, the tips rounded and ascending. Rare. Dry pinelands, thin woods on hillsides, cedar glades, prairies; Fla; seGa; Ala into eTex, seNeb, sIll, nwGa; eKy into seO and nwWVa. Apr–Aug. *Asclepiodora viridis* (Walt.) Gray.

254   Swamp Milkweed
Asclepias incarnata L.
Stems to 1.5 m. Leaves numerous, ovate-elliptic to linear-lanceolate, petioled, pubescent beneath, 1–3 cm wide. Hoods 3–4 mm. Corolla and crown pink to rose-purple. Fruits erect on erect pedicels. Reported as edible when treated like *A. syriaca.* Common. Moist open places, edge of water; cPied of Ga into eTenn, Ky, nwArk, Kan, seMan, sOnt, Me, sNS, and ceNC, also in scattered localities to the S and W. July–Sept.

There are 2 subspecies: ssp. *incarnata* is essentially glabrous to weakly fine hairy and usually much-branched, whereas ssp. *pulchra* (Ehrh. ex Willd.) Woods. is generally conspicuously hairy and infrequently branched to simple. *A. purpurascens* L. is similar but the leaves are 4–8 cm wide and the hoods 5–7 mm. Rare. Wet or dry places in thin woods or open; cnTenn into seOkla, cIa, ceWisc, sNH, and seVa; mostly absent from Appal Mts. June–July.

255   Four-leaved Milkweed

Asclepias quadrifolia Jacq.

Stem unbranched, to 50 cm. Leaves thin and opposite. The middle internode obviously shorter than those above or below, most often reduced until the 2 middle pairs of leaves appear whorled, thus the common name. Peduncles 15–35 mm. Pedicels 15–30 mm. Crown 5–8 mm wide. Corolla lobes about 5 mm. Horns shorter than the hoods. Fruit erect. Common. Upland woods; nwSC into neAla, csKy, cInd, O, NY, and Mass; cArk into csOkla, neMo, seMinn, and c and sIll. Apr–June.

256   Green Milkweed

Asclepias obovata Ell.

Stems usually unbranched, to 55 cm tall, densely hairy, as are the leaves. Umbels 1–8 at the upper nodes, sessile or nearly so. Pedicels stout, 8–10 mm. Petals greenish-yellow, thus the common name. Horns shorter than the hoods. Crown 6–8 mm wide. Rare. Sandy pinelands and ridges; cnFla into seTex, CP of Ga, and seSC; cArk. June–Sept.

In *A. tomentosa* Ell. the pedicels are about 2 cm and the horns longer than the hoods. Rare. Sandy soils, open places; sw into nFla; cSC into cNC; eTex. In *A. viridiflora* Raf. the crown is only 2–3 mm broad and the plant usually taller and not as hairy. Hoods and corolla lobes about equal in length. Occasional. Mostly in prairies but also dry fields, roadsides, and rocky places; Pied of Ga, into neMex, eWyo, cnND, sOnt, Md, ePa, and swMass. June–Aug.

257   Sandhill Milkweed

Asclepias humistrata Walt.

Stems stiff and spreading 1–several from a deep narrowly fusiform root. The spreading habit; 5–10 close pairs of broad, sessile, clasping leaves; and abundant milky juice make this species distinctive. Other prominent features include the tan-colored flower buds, a nearly white crown, and erect fruits on deflexed pedicels. Plants often grow in very hot and dry places without wilting. Common. Sandhills, dry oak woods, pine barrens; Fla into seLa, CP of Ga, and c and ceNC. Mar–July, occasionally into Sept.

258   Spiny-pod

Matelea carolinensis (Jacq.) Woods.

As the common name suggests, fruits of this opposite-leaved perennial twining vine are spiny yet not especially sharp. Fruits are pointed and contain many seeds which at their tips bear a conspicuous tuft of long hairs. Petals are 10–15 mm and spreading. Occasional. Woods and thickets on slopes; Ga into Miss, cTenn, WVa, and Del. Apr–Oct.

M. flavidula (Chapm.) Woods. is similar except the petals are yellow to green. Rare. Similar habitats; eSC into nwFla. *M. floridana* (Vail) Woods. has maroon or green petals, 4–8 mm. Rare. Similar habitats; nFla. *M. decipiens* (Alexander) Woods. has maroon, ascending petals up to 16 mm. Occasional. Similar habitats; Ga into eTex, sMo, sInd, and NC. *M. obliqua* (Jacq.) Woods. has rose petals that are 4–6 times longer than wide. Occasional. Similar habitats with limy soil; nGa into seMo, sePa, and wNC. Apr–Oct. *M. baldwyniana* (Sweet) Woods. has white or cream ascending petals. Similar habitats with limy soil; wAla into eOkla and sMo.

259 Climbing-milkweed

Matelea gonocarpa (Walt.) Shinners
Leaves 8–12(20) × 4–10 cm. Calyx lobes glabrous on outerside; corolla 3–4 times as long as calyx, innerside glabrous; petals often are spreading, color can be yellow, green, brown, or black or even 2 colors with the darker one toward center of the flower, lobes linear-lanceolate, 7–10 mm. Pods smooth with 5 winged angles, 8–13 × 2–3 cm. Occasional. Woods and thickets on slopes and upper terraces of lowland woods; eTex into eOkla, sMo, sIll, sInd, Tenn, Ga, c and seVa, and Fla. May–Aug.

CONVOLVULACEAE: Morning-glory Family

260 Dodder; Love-vine

Cuscuta gronovii Willd.
Dodders all lack chlorophyll, are annuals, and are yellowish to orange. Mature plants are without roots and instead attach to other plants by rootlike suckers (haustoria). In this species the calyx is 5-lobed but not 5-angled and is shorter than corolla tube; petals united below, lobes wide-spreading, ovate, obtuse. Fruit a capsule breaking open along irregular lines; seeds about 1.1 mm. Common. Parasitic on a variety of species usually growing in moist to wet places; fresh and brackish marshes, sloughs, swamps, pond margins; Ariz into Tex, Mo, Man, Que, NS, Fla, and La. July–Oct.

261 Pony-foot

Dichondra carolinensis Michx.
Finely hairy prostrate perennial 1–12 cm, the spreading stems rooting at nodes, often forming dense masses and sometimes used as a ground cover. Leaves 1–3 cm across. Flowers axillary, solitary; sepals 2–3 mm at pollen shedding; corolla white, somewhat bell-shaped, shorter than calyx. Fruit a 2-lobed 2-seeded capsule. Common. Usually moist to wet places, tolerating short-time flooding; broadleaf woods, pinelands, roadsides, pond margins, swales, lawns; seVa into neNC, ceGa, Fla, eTex, and sArk. Mar–May.

262 Stylisma

Stylisma patens (Desr.) Myint
A prostrate or spreading vine with no tendency to twine. Flowers usually solitary. Subspecies *angustifolia,* shown in the picture, has glabrous sepals and narrow leaves; in ssp. *patens* the sepals are hairy. Common. Dry sandy soils, rarely sandy loams; CP—NC into cFla and sMiss. May–Sept.

S. *humistrata* (Walt.) Chapm. is similar but the flowers are mostly 2–4 together, the sepals glabrous, and the stem tips have a tendency to twine. Common. Sandy soils in open or in woods; seVa into nFla, eTex, and sArk. S. *villosa* (Nash) House is like the previous species and can be identified by its hairy sepals.

Occasional. Dry sandy soils; pen Fla and CP of Tex. *S. aquatica* (Walt.) Raf. (CP—seNC into neFla, also La into seTex and seArk) has pink, maroon, lavender, or red corollas.

263   Jacquemontia
Jacquemontia tamnifolia (L.) Griseb.
Erect or reclining and usually twining annual. Plants, especially the inflorescence, tawny hairy. Leaf blades entire, cordate-ovate to elliptic-ovate, 5–12 cm, the upper ones increasingly smaller. Flowers in heads on long axillary peduncles equal to or longer than the adjacent leaves; style 1; stigmas 2, flattened, elliptic or oblong; ovulary 2-celled; fruit 4-seeded. Occasional. Fields, gardens, roadsides, waste places; Fla into eTex, Pied of Ga, eNC, and seVa. June–frost. *Thyella t.* (L.) Raf.

264   Low Bindweed
Calystegia spithamaea (L.) R. Br.
Members of this genus are perennial twining vines, the calyx is concealed by 2 large bracts, and the fruits are 1-celled.
   This species has 1–4 flowers which are from the axils of the lower and/or medial leaves only. Occasional. Thin woods, open areas, often in sandy or rocky soils; cnFla into cwMo, sOnt, sQue, swMe, and neSC. May–July. *Convolvulus spithamaeus* L.
   Calystegia sepium (L.) R. Br. is similar but flowers are more abundant and come from axils of leaves along a greater part of the stem. The stem and leaves are glabrous to short hairy. Occasional. Fields, roadsides, waste places, thin woods; Fla into sTex, Ore, BC, and Nfld. Apr–July. *Convolvulus sepium* L.; *C. americanus* (Sims) Greene; *C. repens* L. The leaves are felty hairy in *Calystegia catesbiana*. Rare. Thin woods or in open, usually on slopes; nGa, nwSC, and swNC. June–July. *Convolvulus sericatus* House.

265   Common Morning-glory
Ipomoea purpurea (L.) Roth
Our species of this genus are twining vines, the calyx is not hidden by bracts, and the 1 stigma is 2-lobed in some species. The mature fruits are dry, 2–4-celled, and 2–6-seeded.
   This species is an annual with 1 main stem from a taproot. Leaves unlobed. Pedicels with reflexed hairs. Sepals acute or acuminate, the points shorter to slightly longer than the base, glabrous or hairy. Corolla funnel-shaped and purple, red, bluish, white, or variegated. Stigma 3-lobed. Mature fruit usually 3-celled and 6-seeded. Common. A weed of fields, fencerows, waste places; Fla into Ariz, Wisc, and NS. June–frost.

266   Fiddle-leaf Morning-glory
Ipomoea imperati (Vahl) Griseb.
Glabrous, trailing, fleshy perennial. Stems rooting at the nodes. Most leaf blades lobed near base, sometimes deeply so. Flower stalks about as long as the leaves. Occasional. Coastal dune areas; seNC into Fla and Mex. June–Oct. *I. stolonifera* (Cyrillo) J. F. Gmel.

267 **Railroad-vine**
Ipomoea pes-caprae (L.) Mey.
Trailing fleshy glabrous perennial rooting at the nodes. (We have seen stems as long as 31 m trailing across dunes.) Leaf blades unlobed but summit notched, 4–11 × 4–10 cm. Peduncles 1–several flowered; sepals elliptic-orbicular, corolla funnel-shaped. Common. Dunes, drift area of beaches, overwash flats; sSC into Fla and Tex; warm regions around the world. June–Nov. *I. brasiliensis* (L.) Sweet.

268 **Red Morning-glory**
Ipomoea coccinea L.
Glabrous annual. Leaves ovate, the tip acuminate, basal lobes rounded or with 1–3 angular projections. Calyx 6–8 mm. Corolla with a long tube. Stamens and pistil longer than the tube. Seeds 4. Occasional. Fields, thickets, roadsides, waste places; Fla into Ariz, cwIll, sMich, sPa, and seMass. May–Oct. *Quamoclit c.* (L.) Moench.

    I. hederifolia L. is similar but some or all leaves are 3-lobed on most plants and the calyx 4–5.5 mm. Rare. Fla into seTex, CP of Ga, and seSC. *Quamoclit h.* (L.) G. Don. *I. quamoclit* L. has similar flowers but the leaves are pinnately divided into many narrowly linear segments about 1 mm wide or less. Rare. Fla into eTex, Mo, Miss, CP of SC, and seVa. June–Nov. *Quamoclit vulgaris* Choisy.

269 **Ivyleaf Morning-glory**
Ipomoea hederacea Jacq.
Hairy annual from a taproot. Leaves usually lobed, sometimes resembling those of English Ivy, hence its common name. Pedicels with reflexed hairs. Sepals prominently hairy at the base, with long tapering tips that are spreading or curved backward. Corolla funnel-shaped, light blue when fresh, then turning to light purple, or rarely white. Stigma 3-lobed. Mature fruit 3-celled and 6-seeded. Common. Weed of fields, fencerows, waste places; Fla into Ariz, cKan, eND, and NY. July–frost.

270 **Wild Potato-vine**
Ipomoea pandurata (L.) Mey.
Leaves and calyx glabrous or nearly so, anthers 5–7 mm, and the corolla always with a reddish-purple center. Stems arise from a deep vertical perennial tuberous root that sometimes weighs as much as 30 lbs. Indians are reported to have roasted the roots for food, but caution should be taken as the fresh root is said to be purgative. Starch extracted from the root is probably safe. Common. A troublesome weed, fields, roadsides, waste places, in open; Fla into cTex, eKan, sOnt, seNY, and eMass. June–Sept.

    In *I. macrorhiza* Michx. leaves are felty hairy beneath, the corolla is smaller, 5–8 cm, and lacks the reddish-purple center, and the sepals are prominently but finely hairy. Rare. Sandy open places; Fla to sMiss and coastal SC. June–Aug.

271 **Coastal Morning-glory**
Ipomoea cordatotriloba Dennst. var. *cordatotrilobata*
Perennial from a branched root. Leaves usually lobed. Pedicels glabrous. Sepals ciliate on margin and hairy at base. Corolla funnel-shaped, 28–55 mm, pink to

purple or rarely white. Anthers 1.5–3.2 mm. Stigma 2-lobed. Mature fruit 2-celled and 4-seeded. Common. Open places; roadsides, fencerows, thickets, abandoned fields; Fla into eTex, CP of Ga, and seNC. May–Oct. *I. trichocarpa* Ell.

I. lacunosa L. is similar but the flowers are only 15–23 mm and the corolla is white. Common. Similar places; Fla into cTex, eKan, cwIll, sO, and sNJ. Aug–frost. *I. sagittata* Poir. has narrowly sagittate leaves, anthers are 5–7 mm, the corolla 55–75 mm and rose-lavender. Common. Bogs, fresh and brackish marshes, interdune areas, occasionally in dry habitats; along and near the coast of Fla into Tex and NC. May–Sept.

POLYMONIACEAE: Phlox Family

272 **Blue Phlox**
Phlox divaricata L.
A perennial with spreading basal shoots. Flowering stems to 50 cm, with a few well-separated pairs of oblong to lanceolate or narrowly ovate leaves. Flowers loosely arranged, on glandular hairy branches. Corolla blue to purple, often with a reddish-purple eye, stamens shorter than the tube. Occasional. Rich deciduous woods; cnFla into eTex, eSD, swQue, nwVt, ne and swNC, and ne CP of Ga. Mar–Apr.

Other species with loosely arranged flowers and a few well-separated pairs of leaves on the flowering stem are: *P. latifolia* Michx. with oblanceolate to elliptic or obovate leaf blades and stamens projecting beyond the corolla tube. Occasional. Thin deciduous woods or in open; cnSC into nAla, eInd, sePa, and Pied of NC. May–June. *P. stolonifera* Sims with abundant creeping stems, obovate to spatulate leaf blades, and stamens longer than the finely glandular hairy corolla tube. Rare. Deciduous woods, especially stream terraces; cnSC into neGa, e edge of Tenn, sO, cPa, and swVa. May–June.

273 **Thick-leaf Phlox**
Phlox carolina L.
This species is one of several tall Phloxes; most make excellent garden plants. Plants are often difficult to name, partly because they often hybridize. This species is a perennial to 1 m with 6–25 nodes. Leaves 4–12 cm, the lateral and marginal veins indistinct, the margins entire. Inflorescence a somewhat cylindrical corymb with the lower branches peduncled. Calyx cylindrical or nearly so. Corolla tube glabrous. Stamens project beyond the corolla tube. Common. Thin deciduous or mixed woods or open places, wet or dry; swGa into Miss, neTex, seMo, sInd, swVa, eMd, and SC. May–Oct.

P. maculata L. is similar but to 1.5 m with 18–35 nodes; leaves 70–150 × 10–50 mm; inflorescence cylindrical, to 40 cm, the lower branches few if any, longer than the upper. Stamens just inside rim of the corolla tube. Occasional. Moist places, thin woods or in open; nwSC into seMo, seMinn, swQue, swConn, and ceNC. June–Sept.

274   Smooth Phlox; Summer Phlox
Phlox paniculata L.
Perennial to 1.5 m, nodes 15–40. Leaf blade margin finely serrate and ciliate, lateral veins prominent. Inflorescence compact and glabrous. Calyx cylindrical or nearly so; corolla tube hairy; some stamens equal or exceed corolla tube. Occasional. Rich moist places in thin woods or open, stream banks; cArk into neKan, Mo, nIa, sWisc, O, cNY, Va, ce and swNC, nGa, and nMiss. July–Sept.

Two other species that are tall are: *P. glaberrima* L., leaves 4–20 cm, blade margin entire, lateral and marginal veins indistinct. Inflorescence corymblike with 7–15 nodes below. Stamens project beyond the corolla tube. Occasional. Moist or wet places in woods; swales, meadows; Ark into seMo, wKy, seWisc, swO, cw and sVa, ceNC, nwSC, se and cGa, nwFla, and eTex. Apr–Aug. *P. amplifolia* Britt. has a broad open inflorescence with glandular hairy bracts and glabrous corolla tube. Rare. Hardwoods; nAla into cMo, sInd, swVa, and cwNC. July–Sept.

275   Annual Phlox
Phlox drummondii Hook.
This species has a great variety of cultivated forms differing especially in color and shape of the corolla. Many of these have escaped and become established in numerous places. Different forms are often found in close proximity, as seen in the picture, which was taken from directly above the plants. All plants are annuals with the upper leaves mostly alternate. Stems, leaves, calyx, and the corolla tube bear glandular hairs. Common locally. Open areas, usually sandy and well drained; native to Tex but escaped E into Fla and seVa, and northward at least into seMo. Mar–June.

276   Jacob's-ladder
Polemonium reptans L.
Glabrous to hairy perennial with 1–several loosely clustered, ascending to erect, branching stems to 50 cm. Leaves alternate, pinnately compound. Leaflets 3–8 pairs and a terminal one. Petals united. Carpels 3. Common. Rich moist woods; cnNC into Ky, nwGa, eOkla, seMinn, and sNH. Apr–May.

HYDROPHYLLACEAE: Waterleaf Family

277   Waterleaf
Hydrophyllum canadense L.
Rhizomatous essentially glabrous perennial to 70 cm with fibrous roots. Upper leaves with 5–7 palmate lobes, the lower ones often partly or wholly pinnately compound. Flower buds in short coiled clusters that straighten after the flowers open. Petals united. Filaments prominently long hairy. Young parts of this and other species are probably edible raw or cooked. Rare. Rich woods; nGa into neArk, eMo, sOnt, and swVt; cNC. May–June.

H. appendiculatum Michx. is similar but is a biennial and finely hairy. Rare. Tenn into neArk, ceKan, seMinn, sOnt, and swPa. Two species have pinnately compound leaves. In *H. virginianum* L. there are 3–7 leaf segments and the plant is glabrous to slightly hairy. Occasional. Rich, usually moist woods; cNC into nArk, sMan, and sNH. In *H. macrophyllum* Nutt. larger leaves have 9–13 segments. Stems are conspicuously hairy. Rare. Rich moist woods; nAla into seIll, nWVa, and cwNC.

278   Phacelia

Phacelia bipinnatifida Michx.

Members of this genus are hairy, annual or biennial herbs, the leaves are pinnately lobed to divided, and the flower buds are in coiled clusters that straighten after all flowers open.

This species is an upright biennial to 60 cm, corolla lobes are not fringed, the lower two-thirds of the filaments are hairy, stem and branches of the inflorescence have a few to many small glandular hairs, and fruiting pedicels are recurved. Occasional. Rich woods, often in rocky places; cnGa into nAla, neArk, cIll, cInd, sWVa, and eNC. Mar–May.

P. ranunculacea (Nutt.) Const., in contrast, has weak stems to 25 cm, the corolla 2–4 mm broad, the filaments glabrous. Rare. Rich often alluvial woods; eTenn into neArk, seMo, sIll, and swInd; cNC; DC and vicinity. Apr–May.

279   Fringed Phacelia

Phacelia purshii Buckl.

Weak-stemmed branched annual to 50 cm. Stems and pedicels with appressed hairs. Leaves pinnately lobed or divided. Corolla lobes fringed across their ends. Stamens 5. Carpels 2. Fruit 1-celled. Seeds minute and numerous. Occasional. Moist places in woods or open; cnGa into nAla, seMo, ceIll, neO, neWVa, and cwNC. Apr–May.

P. fimbriata Michx. is similar but the hairs on the stem and pedicels are spreading and the petals are usually cream or white and have longer fringes. Rare, sometimes abundant locally. Rich mt woods; swNC into ceTenn and WVa.

280   Phacelia

Phacelia dubia (L.) Trel.

This species, like most others of the genus, often grows in showy masses. An annual with weak, usually much-branched, stems to 40 cm. Stem and branches of the inflorescence are hairy but with few glands. Pedicels spreading to ascending. Marginal hairs of the sepals not spreading. Corolla blue to white, the lobes entire. Occasional, though abundant locally. Usually thin woods, fields, sterile soils, on and around granitic outcrops; csSC into nwCP of Ga, cTenn, cO, swPa, and NC. Mar–May.

P. maculata Wood is similar but the marginal hairs on the sepals are spreading. Occasional. On and around granitic rocks, fields, sandy margins of creeks; Pied of Ga and adjAla into cnSC and adjNC. Apr–May.

281   Tall Hydrolea

Hydrolea corymbosa J. F. Macbr. ex Ell.

Erect slender perennial to 70 cm, with few spines, if any. Leaves elliptic to elliptic-lanceolate, alternate, entire. Flowers at top of the stem. Petals light violet to purplish-pink, united at the bases. Stamens borne on the corolla tube. Styles 2, separate and much longer than the ovulary, which is glandular-hairy and 2-celled. Fruit a capsule with many small seeds. Common locally. Aquatic habitats; sw and csGa, adjFla and southward. July–Sept. *Nama corymbosa* (J. F. Macbr.) Kuntze.

H. ovata Nutt. ex Choisy is also erect but stouter, to 1 m and armed with spines. Leaves are ovate. Flowers to 28 mm wide and at the top of the stem. Locally abundant in aquatic habitats; swGa into eTex, csMo, and cwAla. June–Sept. *Nama ovata* (Nutt.) Britt.

282   Hairy Hydrolea
Hydrolea quadrivalvis Walt.
Perennial with succulent, hairy, spiny stems ascending from a creeping or decumbent base. Flowers 1–8 in short axillary clusters. Petals united at their bases. Stamens 5, attached to corolla tube. Styles 2, separate. Ovulary and fruits 2-celled. Fruits dry, splitting into 2 halves. Seeds small, many, longitudinally ribbed. Occasional. Swampy woods, edges of ponds and lakes, marshes, stream banks; Fla into seLa, cMiss, CP of Ga, and seVa. June–Sept.

H. uniflora Raf. is similar but is not hairy. Occasional. Similar habitats; ceMiss into La, eTex, seMo, and sIll. June–Sept. *H. affinis* Gray; *Nama affinis* (Gray) Kuntze.

BORAGINACEAE: Borage Family

283   Bluebells; Virginia-cowslip
Mertensia virginica (L.) Pers.
Ascending to erect plants to 70 cm, stems and leaves glabrous; stems tender, usually several from the perennial root. Leaves simple, alternate; blade 5–15 cm, mostly rounded at apex. Flowers hanging like bells, the buds pink; corolla 18–25 mm, tube densely hairy inside base; stamens 5, anthers yellow. Fruit 4 ovoid nutlets. Occasional, locally abundant, especially in mid part of range. Rich woods, often in moist places, sometimes in open; neArk, into seMinn, seMe, cVa, Tenn, and nwGa; cNC. Mar–Apr.

M. maritima (L.) S. F. Gray, Seaside Bluebell, is also glabrous but main leaves only 2–6 cm and corolla only 6–9 cm. Plants to 1 m. Rare. Along the seacoasts; Greenl into Mass; Alas into BC; nEurope. June–Aug. In *M. paniculata* (Ait.) G. Don., Northern Bluebell, plants to 1 m, the leaves and calyx are hairy, main leaves 5–14 cm and corolla 10–15 mm. Occasional. Damp woods; Ore into Alas, Hudson Bay, nMinn, nWisc, and neIa. June–July.

284   Puccoon
Lithospermum carolinense (Walt.) MacM.
Perennial to 80 cm. Stems erect, simple or with 2–several branches, from a prominent taproot. Roots with a reddish-purple content, once used as a dye for cloth. Leaves with rough hairs. Inflorescence coiled when flowers are in bud. Fruiting calyx lobes 9–15 mm. Corolla 15–25 mm wide at end. Corolla rich yellow to orange-yellow. Occasional. Dry soils, especially sandy ones, thin woods or in open; sw and cwSC into nwFla, c and nTex, swOnt, wNY, cwIll, eTenn, and sAla; Mex; some western states. Mar–June. *Batschia caroliensis* (Walt.) J. F. Gmel.

L. canescens (Michx.) Lehm. is similar but the hairs on the leaves are soft, the fruiting calyx lobes 6–8 mm, the corolla 10–15 mm wide. Occasional in the seUS; common in the nwUS. Dry places, often sandy; prairies, open woods; nwGa into eOkla, eKan, sSask, swOnt, cVa, and Tenn. Apr–May.

VERBENACEAE: Vervain Family

285   Moss Verbena
Glandularia pulchella (Sweet) Tronc.
Abundant along roadsides, producing conspicuous expanses of flowers. The stems branch abundantly and are prostrate or decumbent, root at the nodes, and are largely unaffected by mowing. The leaves are opposite, deltoid in outline, and deeply and narrowly segmented. The corolla is lavender to purple, or white. It can probably serve as an ornamental much more than is generally appreciated, seemingly free of pests and diseases. Common. Roadsides, fields, waste places; Fla into s and seTex, Pied of Ga, and seNC. Feb–frost. *G. tenuisecta* (Briq.) Small; *Verbena t.* Briq.

286   Stiff Verbena
Verbena rigida Spreng.
The flower of many Verbenas resembles that of Phloxes except that it is slightly irregular. Also, the fruit is an aggregate of 4 nutlets held tightly by the calyx.

This species is an erect perennial from coarse elongate rhizomes, often forming large patches that are conspicuous when in flower. The stems and especially the leaves are rough-hairy. Leaves lanceolate to oblanceolate, coarsely serrate, clasping the stem. The flowers overlap in dense short stiffly erect spikes, rarely to 6 cm, the axis with short-stalked glands. This makes a hardy, drought-resistant ornamental. Occasional. Waste places, roadsides, fields, pastures; Fla into eTex, Pied of Ga, and eNC. Mar–Oct.

287   Lippia
Phyla nodiflora (L.) Greene
Perennial, usually with prostrate or decumbent rooting stems. Leaves opposite, blunt- or round-tipped, and with 1–7 teeth on each edge above the middle. Flowers many in a tight head that is longer than broad as the seeds become mature. Common. Sandy, usually moist and open habitats; seVa and mostly along the coast into Fla, Tex, and Mex; N into seMo. Apr–frost. *Lippia n.* (L.) Michx.

The similar *P. lanceolata* (Michx.) Greene has taller ascending to erect stems, and acute leaves with 5–11 teeth on each edge to below the middle. Wet places; occasional along the coast from sNJ into Fla and Tex; more scattered into sCalif and from Tex N into Neb, Minn, Ont, and cPa. June–frost. *Lippia l.* Michx.

LAMIACEAE: Mint Family
Members of this family have opposite simple leaves, usually square stems, petals united and irregular, ovulary superior, style arising from the central depression of the 4 lobes of the ovulary, fruit usually 4 nutlets nestled in the persistent calyx tube.

288   **Blue-curls**
Trichostema dichotomum L.
This species is an annual to 80 cm. Leaves to 7 cm, less than 5 times as long as wide; the lower ones often fall during dry periods. Lower lip of the corolla narrow but prominent and drooping. The 4 stamens are strongly arched and bluish, thus the common name. Common. Dry places in open, or thin woods; Fla into eTex, Ill, sMich, and Me. Aug–frost.

T. *setaceum* Houtt. is quite similar but with narrower leaves, more than 5 times as long as wide. Occasional. Fla into eTex, seMo, sO, and Conn. *T. lineare* Nutt. Aug–frost.

289   **Wood-sage; Germander**
Teucrium canadense L.
Erect rhizomatous perennial to 150 cm. Leaves lanceolate to ovate-lanceolate. Flowers in 1 or more terminal spikes. Corolla purplish, pink, or cream-colored. Upper 4 lobes of the corolla nearly equal, oblong, turned forward, so that there seems to be no upper lip. Lower lip very prominent. Anther-bearing stamens 4. Ovulary 4-lobed, not deeply 4-parted as in most mints. Fruit of 4 nutlets joined at their sides. Often confused with *Stachys*, which has a distinctly 2-lipped corolla. Occasional. Moist to wet places in thin woods or open; prairies; Fla into Tex, BC, and NS. June–Aug. *T. nashii* Kearney; *T. virginicum* L.

290   **Hairy Skullcap**
Scutellaria elliptica Muhl. ex Spreng.
In members of this genus the calyx is enlarged when in fruit and has a helmetlike protuberance on the upper side.

This species is a perennial to 75 cm. Stems hairy, sometimes finely so, with 2–7 nodes below the inflorescence. Upper leaves crenate-serrate, the first pair below the inflorescence less than 3 times as long as wide. Corolla 12–21 mm, blue to violet to rarely white. Common. Usually in deciduous woods on slopes; Ga into seTex, sMo, swMich, and seNY. May–July.

S. *incana* Biehler is similar but is taller and often has more racemes of flowers below the terminal one. Occasional. Pied of Ga into eOkla, Ind, swNY, NJ, and NC. June–Aug.

291   **Narrow-leaved Skullcap**
Scutellaria integrifolia L.
Perennial to 75 cm, upper leaves linear-lanceolate to narrowly elliptic, more than 3 times as long as broad. The bases of lower leaves obtuse, often cordate. Racemes 1–7, or rarely many. Corolla 13–28 mm. In var. *integrifolia* lower internodes have small hairs that curve upward and usually inward. Common. Open, usually wet situations, rarely in thin woods; Fla into eTex, Ark, sO, and Mass. Apr–July. In var. *hispida* Benth. the lower internodes have prominent divergent hairs and few or no small incurving hairs. Occasional. Fla into eTex, Ark, cwTenn, CP of Ga, and seVa. Apr–July.

In S. *multiglandulosa* (Kearney) Small ex Harper the lower leaf bases are less than 90°, usually broadly cuneate, and there are spreading glandular hairs on the lower internodes. Rare. Dry soil, usually sandy; Ga, upper CP into ePied. May.

292   Large-flowered Skullcap

Scutellaria montana Chapm.
Perennial to 50 cm. Upper leaves crenate to crenate-serrate, the first pair below the inflorescence less than 3 times as long as wide. Corolla 25 mm or longer. Perhaps not distinct from *S. elliptica*. Rare. In deciduous woods or in open; nGa into eTenn, nAla, and nMiss. May–July.

S. parvula Michx. is small, growing to 35 cm, has flowers in the axils of leaves only, and bears underground tubers. The stem is simple or branching from the base. The upper leaves are narrowly ovate-lanceolate to ovate, to 17 mm, and entire to shallowly toothed. Occasional. Various habitats, but usually in the open; nFla into e and ncTex, eND, cME, and nSC. Apr–June.

293   Heal-all

Prunella vulgaris L.
Perennial to 80 cm, with short branches anywhere below the central flower cluster. Leaves variable. Flowers in dense almost globose spikes, or later cylindrical, the spikes over 1 cm wide excluding corollas. Corolla white to pink and purple. Pollen-bearing stamens 4. Common. Natzd. in a variety of habitats; Fla to Calif and N into sCan and Alas. Apr–frost.

P. laciniata L. is smaller, the margins of the upper leaves usually with projections, the flower clusters under 1 cm wide excluding the corollas. Rare. Natzd. in moist habitats; nGa into NY. June–Sept.

294   False Dragon-head

Physostegia virginiana (L.) Benth.
Perennial to 1.5 m, the stems single but sometimes with short branches at the top. Leaves in 13–22 pairs, the lower ones often falling early, the blade margins of the larger leaves with sharp teeth. Flowers few to many in 1–7 spikes. Corolla from almost white to rose-purple, 1.8–3 cm. Common. Moist to mesic situations in open or woods; nGa into cTex, ND, sOre, sMe, and SC. May–Oct. *P. angustifolia* Fern., *Dracocephalum virginianum* L.

In *P. purpurea* (Walt.) Blake (*Dracocephalum denticulatum* Ait.), the leaves have rounded teeth or are undulate. Fla into sMiss, sGa, and seVa. Apr–Aug. *P. intermedia* (Nutt.) Engelm. & Gray has corollas shorter than 1.8 cm. Fla into sGa and sTex; N into seMo, sIll, and eKy. Apr–July.

295   Hemp-nettle

Galeopsis tetrahit L.
Taprooted annual to 0.7 m; stems bear long spreading or slightly reflexed hairs; swollen under nodes. Calyx shaped like a dilated funnel, with ca 5 veins, and 5 somewhat equal spiny teeth. Fertile stamens 4, extending beyond the corolla, upper pair shorter than the lower, both pairs parallel and ascending under upper lip of corolla. Ovulary deeply 4-parted; summit of nutlets rounded. Native of Eurasia. Common. Waste places, roadsides, fields; Alta to Nfld, NS, S into Ia, Wisc, Mich, and NY. June–Sept.

296 Henbit

Lamium amplexicaule L.

Annual or winter-annual to 35 cm, branching from the base, most branches arched and ascending. Leaves beneath the flower clusters sessile, the blades horizontal or ascending. Corolla 12–18 mm. Pollen-bearing stamens 4. A troublesome weed, especially in yards and gardens. Has been used as a cooked green when young, but possibly poisonous if used in large quantities. Common. Natzd. in a variety of open habitats; Fla into Calif and N into Can. Throughout freeze-free winter–May.

In *L. purpureum* L. the leaves beneath the flower clusters are stalked, horizontal to drooping, and deeper green or purplish. Corolla 10–16 mm. Occasional. Natzd. in open places; nSC into Tex and cCalif, and N into sCan. Feb–May.

297 Stachys; Hedge-nettle

Stachys tenuifolia Willd.

Members of this genus have a distinctly veined calyx tube and nearly equal ascending calyx teeth; stamens longer than corolla tube, 2 pairs, upper pair shorter than the lower pair, opening in anthers longitudinal; ovary deeply 4-parted; summit of nutlets rounded.

Plants of Hedge-nettle to 1 m. Upper part of the stem glabrous although the corners may have hairs. Leaves acuminate. Calyx teeth two-thirds to as long as the tube. Variable and perhaps not separable from the next species below. Common. Rich woods and open places; Pied of Ga into Tex, Minn, sQue, and NH. May–Aug.

S. hyssopifolia Michx. has narrowly linear to narrowly oblong leaves with acute to obtuse tips. The underground tubers are crisp and nutty and good to eat from fall into early spring. Occasional. Moist places; CP of SC into CP of Va, mts of Va, and seMass; also swMich and nwInd. *S. latidens* Small ex Britt. is separated by some on the basis of its having calyx teeth about half as long as the tube. Occasional. Mts of Ga into eTenn, WVa, and DC. June–Aug.

298 Blue Sage

Salvia azurea Michx. ex Lam.

In our members of this genus the flowers are in terminal spikelike clusters, the calyx without a cap or protuberance on the tube, and the corolla 2-lipped.

Perennial to 1.5 m. Stems 1–several from the base, simple or with a few branches above. Leaves above the base linear to elliptic-lanceolate, at the base cuneate. Calyx 2-lipped. Corolla blue to almost white, rarely purplish. Pollen-bearing stamens 2. Occasional. Dry places in thin woods or open; seNC into Fla, Pied of Ga, Tex, eColo, Neb, Minn, and eTenn. Aug–Oct.

299 Lyre-leaved Sage

Salvia lyrata L.

The leaves of this perennial are all basal or nearly so. They are often lyrate in shape, prompting the common and scientific name. There may be several stems from the base although there is usually 1. The main stem may have 2 leafless branches from the upper part. Pollen-bearing stamens 2. Common. Various habitats in thin woods or open, sometimes weedy; Fla into Tex, sIll, sPa, and Conn. Feb–May, sometimes in fall.

300   Bee-balm
Monarda didyma L.
Our members of this genus have flowers in dense headlike clusters or whorls, the calyx lobes are nearly equal, and there are 2 pollen-bearing stamens.

This species is a perennial to 1.8 m. Leaves petioled, the blades ovate-lanceolate, 6–15 cm. Flowers in 1 tight head, or rarely a second above the first, those on the inside opening last, as seen in the picture. Corolla scarlet. The leaves have been used as a mint flavoring in cooking and in making tea. Occasional. Moist woods or in open; mts of Ga into mts of Va and WVa, Mich, Me, and NJ. July–Sept.

Individual flowers of *Salvia coccinea* P. J. Buchoz ex Etlinger are of similar color and shape but are arranged loosely in 3–9 whorls. Occasional. Sandy soils in open, or thin woods; Fla into e and sTex, and coastal Ga and SC. May–frost.

301   Horse Mint
Monarda punctata L.
Perennial to 1 m. Flowers in 2 or more tight clusters on the end of each flowering stem, each cluster with several wholly or partially pink to lavender leaflike bracts beneath. Calyx teeth acute to acuminate, to 2 mm. Corolla yellow, spotted with purple. Common. Dry places in open or in thin woods; Fla into Tex, Minn, eVt, and Miss. July–Sept.

M. citriodora Cerv. ex Lag. is similar, but is an annual with aristate calyx teeth 3 mm or longer and a white to pink corolla. Common. Open areas; La into Tex, Kan, and Mo; adventive E to Mich, Tenn, eSC, and nFla. May–July.

302   Blephilia
Blephilia ciliata (L.) Benth.
Perennial resembling some species of *Monarda* but the calyx is 2-lipped and has many hairs. In this species the petioles of the leaves beneath the lowest cluster of flowers are shorter than 10 mm. The corolla may be almost white. Stamens 2. Occasional. Dry places in woods or open; cGa into neTex, eIa, Wisc, sVt, and cSC. Mostly absent from the Appal Mts. Apr–July.

In *B. hirsuta* (Pursh) Benth. the petioles of the leaves beneath the lowest flower cluster are 10 mm or longer. The corolla is pale with purple dots. Occasional. Moist woods; mts of wNC and Ga into Tenn, neTex, Minn, wQue, Vt, and cnNC. June–Aug.

303   Rose Dicerandra
Dicerandra odoratissima Harper
Annual to 45 cm. Lower calyx teeth subulate, over 2.5 mm. Corolla pink to rose-purple, longer than the stamens. Stamens 4, the anther sacs tipped with blunt to acute horns. Rare. Sandy soils in open, or thin scrub or live oak woods; csGa into s tip of SC. Sept–Oct.

In the similar *D. densiflora* Benth. the stamens are longer than the corolla and the lower calyx teeth are 2–2.6 mm. Rare; n pen of Fla into Long Co., Ga. Sept–Oct.

304   White Dicerandra

Dicerandra linearifolia (Ell.) Benth.

Annual to 50 cm. Lower calyx teeth triangular, 1–1.7 mm. Corolla white to light lavender, the upper lip lined and spotted with purplish-red on the inside. Stamens longer than the corolla. Anther horns acuminate. Plant quite aromatic. Abundant locally, often forming spectacular colonies. Occasional. Sandy soil in open or thin pine, scrub oak, or live oak woods; Fla into swAla and CP of Ga. Sept–Nov.

305   White Horse-mint

Pycnanthemum incanum (L.) Michx.

A perennial to 2 m. The calyx teeth are distinctly unequal in length, the longest ones are less than half as long as the calyx tube and bear long hairs at their tips. Corolla white to pink-tinged and spotted with purple. Occasional. Dry habitats in open or in thin woods; Pied of Ga into nAla, sIll, sNH, and Pied of NC. June–Sept.

In *P. pycnanthemoides* (Leavenw.) Fern. the corollas are deeper colored and the longest calyx lobes more than half as long as the calyx tube. Pied of Ga into sIll, sInd, WVa, and SC. June–Sept. *P. albescens* T. & G. lacks long hairs on the tips of the calyx teeth. Dry habitats; cFla into sAla, eTex, sMo, and eTenn. June–Sept. *P. montanum* Michx. differs by having glabrous but ciliate bracts below the flowers. Rare. Woods or in open in mts; nwSC into nGa, eTenn, and cWVa. June–Sept.

306   Mountain Horse-mint

Pycnanthemum montanum Michx.

To 0.8 m. Blade of main leaves broadly lanceolate to lance-elliptic or narrowly ovate, mostly 6.5–12 × 2–4.5 mm. Inflorescence dense, headlike, no branches or only the lower ones evident; bracts long-ciliate, otherwise glabrous or nearly so. Calyx with scattered long hairs, at least toward the summit, the teeth bristle-tipped; corolla yellowish to off-white spotted with purple. Occasional. Mt woods, balds, roadsides; eTenn into sWVa into seVa, wNC, neSC, and nGa. Sept–Oct.

307   Dittany; Stone Mint

Cunila origanoides (L.) Britt.

Perennial to 45 cm, usually with several branches from the upper half of the main stem. Leaves ovate or nearly so. Calyx lobes nearly equal. Corolla rose-purple to nearly white. Stamens 2. Indians and early settlers used this plant to treat colds and fevers. Occasional. Dry habitats in open or in woods; Pied of Ga into cnTex, cMo, cPa, and seNY. Aug–Sept. *Mappia o.* (L.) House.

308   American Bugleweed

Lycopus americanus Muhl.

Members of this genus have flowers in tight axillary clusters; corolla regular or nearly so, pollen-bearing stamens 2, stamens and style longer than corolla tube, stamens not resting on lower lip. Ovulary deeply 4-parted.

Plants of this species to 0. 9 m; rhizomes not ending in a tuber; stems with low blunt angles. Most leaf blades taper to a short petiole, lower and medial ones with bases cut or pinnately lobed, blade underside with hairs 0.1–0.5 mm. Calyx 2–3.3 mm, its 5 lobes narrow, firm, tip subulate, and longer than the nutlets; corolla 4-lobed; anthers 0.25–0.5 mm. The set of 4 nutlets concave at top. Common. Marshes, wet woods, pond margins, ditches; cArk into BC, Que, Nfld, Fla, Tex, NM, and Calif.

In *L. asper* Greene (Rough Bugleweed) rhizomes are tuberous. Leaves scabrous, lower and medial leaves sessile, larger blades with 6–12 teeth on each edge. Occasional. Marshes, wet shores; Mo into Calif, Alas, Mich, and Ill; w end of Lake Erie. July–Aug.

309   Field Mint
Mentha arvensis L.
Erect, to 0.8 m, retrorse hairs on stem angles. Leaf blade with teeth scattered along margin; upper leaves slightly to no smaller than midstem ones, 2–8 × 0.6–4 cm. Flower clusters arranged along the stem in leaf axils. Calyx tube and lobes hairy, lobes more or less equal and similar; corolla 4-lobed, nearly regular; anther-bearing stamens 4. Fresh and dried leaves have been used for tea. Highly variable, several varieties recognized. Common. Circumboreal; moist low ground along streams, around ponds, alluvial woods; Alas, Lab, Va, Ind, Kan, Ariz, Calif, and Wash; swNC. July–Sept.

310   Horse-balm
Collinsonia canadensis L.
Perennial to 1.5 m. Leaves at flowering time 6 or more in separated pairs, the largest with 15–40 teeth on each edge. Lower teeth of the calyx with subulate tips. Corolla yellow, the lower lip fringed. Stamens 2. Common; rare in CP. Moist rich woods; nwFla into seMo, Wisc, and sNH. Aug–frost. *C. punctata* Ell.

C. tuberosa Michx. has 5–15 teeth on each leaf edge. Rare. Pied of Ga into seLa and swTenn; also into SC and cNC. Aug–Oct. In *C. serotina* Walt. the lower calyx teeth are subacute to acuminate. Stamens 4. Occasional. Rich woods, nFla into seLa, Pied of Ga, and cNC. July–Oct. *C. anisata* Sims. In *C. verticillata* Baldw. there are usually 4 closely crowded leaves. Flowers are in close-set groups of 3–6. Stamens 4. Occasional. Rich woods; cGa into neAla, eTenn, swNC, and nwSC. Apr–May.

311   Hyptis
Hyptis alata (Raf.) Shinners
Perennial to 3 m. Stem single or sparsely branched in the upper portion. Flowers in dense heads, 1 head per peduncle. The peduncles are up to 8 cm and arise from the upper leaf axils. Calyx nearly regular. Corolla almost white, spotted with purple, the lower lip with a sac-shaped lobe. Stamens 4 protruding from the corolla tube, 2 longer than the others. Common. Moist situations in open or in thin pine or cypress woods; Fla into seTex, CP of Ga, and ceNC. June–Oct. *H. radiata* Willd.

In *H. mutabilis* (A. Rich) Briq. there are up to 18 tight clusters of flowers on each of usually several spikes, the spikes terminal and on axillary peduncles. Occasional. Various habitats; Fla into swAla and sGa. July–Oct.

SOLANACEAE: Nightshade Family

312   Ground-cherry; Husk-tomato
Physalis heterophylla Nees
Members of this genus are annuals or perennials and the fruit is enclosed in a considerably larger papery sac with only a small opening at the tip. The sepals, united and enlarged, form the sac. Fruits are berries. Ripe fruits of all our native species are probably edible; those of some of the exotic species, several of which have become naturalized, may be poisonous, especially when green.

This species is a perennial with simple spreading sticky hairs. Leaves cordate to broadly rounded at the base. Anthers 3.5–4.5 mm on wedge-shaped filaments. Common. Thin woods, fields, waste places, dry situations; Fla into eTex, seND, sQue, and swME. Apr–Oct.

P. viscosa L. is also a perennial but has very small branched hairs. Common. Sandy soil of coastal dune areas, thin woods; Fla into eNM, csKan, and sAla; coast of Ga into seVa. Apr–Sept.

313   Horse-nettle; Bull-nettle
Solanum carolinense L.
Erect, simple to branching perennial to 80 cm from deep vigorous horizontal rhizomes. Stems and undersides of leaves with straw-colored prickles. Leaf blades generally ovate in outline with 2–5 large teeth or shallow lobes on each edge. Stem and leaves also loosely covered with small 4–8-rayed hairs. Corolla light purple to white. The fruits, green at first and later yellow, are much like little tomatoes but are poisonous when eaten. Common. Troublesome weed in gardens, pastures, fields, and other places. Fla into eTex, eNeb, Minn, sOnt, and Vt. Apr–Sept.

S. rostratum Dunal is similar vegetatively but is an annual, larger, the spines stouter, and the fruits (berries) partially or wholly covered by the enlarged prickly calyx; petals yellow. Rare. Weedy; about same habitats and distribution.

314   Sticky Nightshade
Solanum sisymbriifolium Lam.
Much-branched coarse very prickly annual; prickles short to long, flat, yellow to orange, on all vegetative parts and flower stalks and calyx. Corolla white to violet. Mature fruits are red berries 12–16 mm across. Occasional. Roadsides, waste places; eTex into Fla and Ga (scattered in CP), O, NY, and Mass. May–Sept.

315   Bittersweet
Solanum dulcamara L.
Rhizomatous unarmed perennial climbing or scrambling to 3 m. Leaves 2.5–8 × 1.5–5 cm, many blades with lobed basal portion or even a pair of leaflets, underside with simple hairs only. Corolla light blue to violet. Fruit bright red, POISONOUS when eaten. Native of Eurasia. Occasional. Thin woods, clearings, thickets, fencerows; Mo into Ill, Nfld, nVa, S into mts of NC and Tenn; Ida into Wash and Calif. May–Sept, to Nov in sUS.

SCROPHULARIACEAE: Figwort Family

Members of this family have weakly to strongly irregular corollas and superior
2-carpelled ovularies with a terminal style.

316   Woolly Mullein; Flannel-plant

Verbascum thapsus L.

This species is a densely woolly biennial to 2 m. Stem upright, stout, unbranched.
Leaves have the feel of thick flannel, in a basal rosette only during the first year;
those of the second year are as long as 40 cm with bases of upper leaves extending
down sides of the stem. Flowers and fruits in a very dense elongated cylindrical
spike. We leave a few of the volunteers as ornamentals each year. Corollas yellow,
or rarely white. Filaments with yellow hairs. Seeds many and long-lived. Common. Fields, pastures, roadsides, and other open places; Fla into Calif, BC, and
NS. June–Nov.

   V. phlomoides L. is similar but the upper leaves merely clasp the stem, and some
of the flower and fruit clusters are separated by distinct spaces. Rare. Open places;
seSC into WVa, Minn, and Me; Ia. May–Aug.

317   Moth Mullein

Verbascum blattaria L.

Erect annual or biennial to 1.2 m, with small scattered glandular hairs. Stems
sometimes branched in upper part. Leaves coarsely toothed, those of the first
season in a basal rosette. Flowers and fruits in elongated racemes, on pedicels
10–17 mm. Petals yellow or rarely white. Filaments bear prominent purple hairs.
Common. Fields, roadsides, other open places; Fla into eTex, neMo, sOnt, and
Me. May–July.

   V. virgatum Stokes is similar but the pedicels are only 3–5 mm. Rare. Roadsides, weedy places; sSC; cNC; eTex; scattered localities in the neUS. Apr–May.

318   Toadflax

Nuttallanthus canadensis (L.) D. A. Sutton

Slender glabrous biennial or winter annual to 75 cm, with a rosette of prostrate
stems to 10 cm. Upper leaves alternate, linear; those on the prostrate stems opposite or nearly so, and wider. Flowers in 1–several racemes. Corolla blue to purple
or rarely white, with a 5–9 mm spur at base. Plants with corollas over 1 cm excluding the spur are separated by some as *Linaria texana* Scheele. Others treat
these as a variety. Common. In fields especially, roadsides, waste places; Fla into
eTex, eSD, seMinn, swQue, and NS; Calif to BC and scattered localities. Mar–
May. *Linaria c.* (L.) Dum.-Cours.

   N. floridanus (Chapm.) D. A. Sutton has smaller flowers, the spur is under
1 mm, and plants are only to 40 cm. Occasional. Dry sandy places in open or thin
woods; Fla into sMiss; seGa. Mar–Apr. *Linaria floridana* Chapm.

319   Butter-and-eggs

Linaria vulgaris Mill.

Strong-scented perennial to 1 m, spreading by rhizomes. Leaves abundant, blade
pale green, linear, 20–50 × 3–4 mm, narrowed to a petiolelike base. Flowers
numerous in a compact spike; corolla yellow with an orange center and bearing a

stout spur. Native of Europe. Occasional. Fields, pastures, roadsides, along railroads, borders, waste places; nTex into Calif, BC, Nfld, Va, n and swNC, nGa, and
nArk. Mar–Sept.

320 **American Figwort**
Scrophularia lanceolata Pursh
Erect perennial to 2 m. Leaves opposite, petioled, 8–20 cm. Bracts below flowers
small; sepals separate almost to base; corolla dull reddish-brown, lacking a spur,
upper lip not forming a hood; stamens 4 plus a sterile yellowish-green filament
that is often wider than long. Thin woods, roadsides, fencerows; BC into Que, NS,
ce and cwVa, Ind, wMo, Okla, NM, Utah, and Calif. May–July.

321 **Turtlehead; Snakehead**
Chelone glabra L.
Members of this genus are perennials with opposite leaves, flowers in spikelike
racemes, calyx lobes longer than the tube, a tubular corolla, and densely hairy
anthers.
    This species is a perennial to 1.6 m. Leaves mostly widest at or near the middle,
acuminate, serrate, and with an acute base. Petiole up to 10 mm. Flowers in a
dense cluster. Petals white or pink to rose-purple toward the tip. Pollen-bearing
stamens 4, the sterile filament green. Common. Stream banks, wet woods, moist
pastures, other moist habitats; swGa into nwMo, Minn, sOnt, Nfld, and c coast
of SC. Aug–Oct. *C. montana* (Raf.) Pennell & Wherry; *C. chlorantha* Pennell &
Wherry.

322 **Turtlehead; Snakehead**
Chelone obliqua L.
Perennial to 1 m. Leaves widest mostly at or near the middle, serrate, and with an
acute base. Petioles slender, 5–15 mm. Flowers in a dense cluster. Corolla purple.
Sterile filament white. Occasional. Stream banks, wet woods, swampy meadows,
margins of springs; swGa into nMiss, eMo, sMinn, cInd, and eTenn; eSC and
swNC; seNC into c and eMd. Sept–Oct.
    C. cuthbertii Small is similar but the leaves are sessile and widest toward the
base. Sterile filament purple. Occasional. Swampy woods, bogs, wet meadows; mts
of NC; seVa. July–Sept. In *C. lyonii* Pursh the leaves are ovate, rounded at their
bases, the petioles 1.5–6 cm. Rare. Rich woods in mts; nwSC into wNC, eTenn,
and swVa; Conn into Mass. July–Sept.

323 **Southeastern Beard-tongue**
Penstemon australis Small
Members of this genus are opposite-leaved with 4 fertile stamens and a prominent
bearded one (staminode). Basal leaves in a rosette and shaped differently from the
cauline ones. A genus of about 300 species, perhaps 20 occurring in the eUS. Identification to species is often difficult.
    In this species the axis of the inflorescence, midstem blades, and stems are
finely hairy. Staminode golden and projecting beyond the corolla tube. Corolla
20–24 × to 6 mm. Common. Dry pinelands, thin upland woods, sandhills, dry
fallow fields; Fla into s and ceAla, SC, and seVa. Apr–July.

In *P. smallii* Heller the axis of the inflorescence is also hairy but the leaves just below the inflorescence are broader and bigger, at least two-thirds as large as the midstem leaves. Occasional. Woods, cliffs, roadbanks, usually shaded; nwSC into ce and neTenn; nwGa. Apr–June.

324   Appalachian Beard-tongue
Penstemon canescens (Britt.) Britt.
Plants erect, to 80 cm. The axis of the inflorescence, midstem blades, and stems of this species are also finely hairy, but the staminode is yellow and does not project outside the corolla tube. Corolla violet purple to pinkish, 25–32 mm and over 6 mm wide at the widest section. Common. Thin woods, rocky places, dry sandy soils, fallow fields, roadsides; nwSC into neAla, neTenn, seInd, sWVa, cPa, and cnVa. Apr–July. *P. brittonorum* Pennell.

   P. dissectus Ell. is unique in having finely dissected stem leaves. The basal ones are entire or merely few-toothed. Rare. Gravelly soil and thin soil on rocks; c and neCP of Ga. Apr–May. *P. multiflorus* Chapm. ex Benth. is unique among se species in having each anther sac opening by a short slit in the end joining the filament. Fla into cCP of Ga. May–July.

325   Beard-tongue
Penstemon laevigatus Ait.
Plants erect, to 1 m. Leaves entire. Axis of inflorescence glabrous or with glandular hairs separated by more than their lengths. Sepals 3–6 mm, ovate-lanceolate. Corolla 15–22 mm, at first nearly white, later purplish. Anthers brown, staminode yellow. Common. Thin woods or in open, usually moist places, meadows, along streams, cedar barrens; neFla into ceMiss, neTenn, sPa, sNJ, and SC. May–June. *P. pentstemon* (L.) MacM.

   P. calycosus Small is similar but the leaves are serrated. Sepals are 5–12 mm and linear-attenuate. Occasional. Thin woods, meadows, thin soil on limestone, stream banks; nwSC into nAla, eTenn, cw and neIll, swO, cKy, and w edge of NC; scattered localities from sePa into se and wNY, neO, sMich, and Me. *P. tubiflorus* Nutt. is different in having microscopic glandular hairs over the inner surface of the corolla tube. Occasional. Open woods, fallow fields, prairies; ceMiss into eTex, ceNeb, and sOnt; ePa into Me and Mass.

326   Monkey-flower
Mimulus ringens L.
Erect glabrous perennial to 1.3 m. Stems 4-angled. Leaves opposite, serrate, and sessile or clasping, the blades lanceolate to ovate-lanceolate. Flowers solitary from leaf axils, calyx lobes 5, 3–5 mm, shorter than the tube. Corolla strongly irregular. Stamens 4. Common. Marshes, edges of ponds and slow-moving streams and other wet places in partial shade or open; cAla into cOkla, swMan, James Bay in Ont, NB, NS, and ceNC. June–Sept.

   M. alatus Ait. is similar but the leaves petioled, calyx lobes under 2 mm, and fruiting pedicels under 2 cm. Common. Similar habitats; cnFla into eTex, ceKan, cnNY, Conn, and SC. July–Oct.

327   Creeping Gratiola
Gratiola ramosa Walt.
Members of this genus have opposite leaves, 5 nearly equal sepals with usually 2 small bracts below, and 2 pollen-bearing stamens, the 2 sterile stamens absent or minute.

This species is a slender perennial to 35 cm from branching rootstocks, often forming conspicuous colonies. Leaves glabrous or minutely glandular, linear-lanceolate, 1–2 mm wide. Flowers solitary in leaf axils. Sepals 5, 3–8 mm, sometimes with 1–2 bracts just below. Pollen-bearing stamens small, on inside of the corolla tube. Common. Moist to wet places, pinelands, swamps, edges of quiet water, ditches; Fla into eTex, seOkla, cCP of Ga, seSC, and seNC. Apr–Aug.

All plants with sepals 5–7 mm and 1–2 sepallike bracts just below are considered by some to be a separate species, *G. brevifolia* Raf. Mostly w part of range E into nFla and sGa; nwGa into neAla and adjTenn. Apr–Aug.

328   Florida Gratiola
Gratiola floridana Nutt.
A perennial to 40 cm from slender whitish rhizomes. All flowers axillary and solitary, on pedicels 20–45 mm. Easily recognized by its relatively large flowers. Corolla 15–20 mm, later ones sometimes smaller, lobes white or the lower one yellowish. Corollas of other species are 6–15 mm. Rare. Wet places in open or in woods, edges of streams, depressions, swamps; nwFla into nwAla, seTenn, and cGa. Mar–July.

G. aurea Pursh is also easily recognized. The corollas are golden yellow and 10–15 mm. Leaves are intermediate in width between those of *G. ramosa* and *G. neglecta*. Rare; nFla into seNC; W to sAla; also Nfld S to neVa and scattered W into neIll and eWisc. May–Sept.

329   Gratiola
Gratiola neglecta Torr.
Annual to 40 cm. Stems minutely glandular hairy. Leaves linear to elliptic, 2–12 mm wide. Flowers solitary in leaf axils, the pedicels slender, 10–30 mm. Corolla 8–10 mm, lobes white, the tube yellowish with fine dark lines. Occasional. Moist to wet places in open or woods, depressions, edges of ponds and lakes; nGa into eTex, eKan, ND, BC, seQue, cMe, and NC. Apr–May.

G. virginiana L. is similar but the pedicels are stout and usually 1–5 mm, single pedicels rarely to 10 mm; stem usually glabrous. Common. Fla into eTex, cKan, nwInd, cWVa, csVa into NJ. Mar–May. In *G. viscidula* Pennell the mature fruits are 2–2.5 mm across and the plants are perennial. Occasional. NC into Pied of Ga and W of Appal to sO, and E of them to Del. June–frost.

330   Blue Water-hyssop
Bacopa caroliniana (Walt.) Robins.
Creeping or floating perennial to 30 cm. Lemon-scented when crushed. Young stems hairy. Leaves opposite, ovate to broadly elliptic, with 3–7 palmate veins. Flowers solitary from leaf axils, on short pedicels bearing 2 small bracts below the sepals. Corolla bell-shaped, 9–11 mm, the lobes slightly different in shape and

size. Stamens 4. Common. Shallow water and moist edges of ponds, streams, and marshes; ditches; Fla into seTex, CP of Ga, and seVa. Apr–Oct. *Hydrotrida c.* (Walt.) Small.

B. rotundifolia (Michx.) Wettst. is similar vegetatively but not lemon-scented, the pedicels bractless and 2–3 times longer than the calyx, the corolla white and only 6–8 mm. Occasional. Shallow water of ponds and pools; swMiss into se and cTex, cMont, seND, and sInd. May–Nov. *Macuillamia r.* (Michx.) Raf.

331   Smooth Water-hyssop
Bacopa monnieri (L.) Pennell
Glabrous and not lemon-scented. Stems prostrate, decumbent, or ascending to 30 cm; often forming mats. Leaves opposite, spatulate to cuneate-obovate, with 1 vein. Pedicels, at least in fruit, exceed leaves. Sepals unequal. Corolla bell-shaped, white to light purple, the lobes slightly unequal. Stamens 4, small. Occasional. Margins of streams, ponds, ditches, and fresh or brackish marshes; often on sandy soils; Fla into sMiss, e and sTex; also sGa and along the coast into seVa. Apr–Nov. *Bramia m.* (L.) Pennell.

B. innominata (G. Maza) Alain has a similar growth form but stems are finely hairy, pedicels are shorter than leaves, petals are white, and there are only 2 stamens. Rare. Drainage ditches, muddy banks, marshes; cnFla and S, NE along the coast into seNC. June–Nov. *Herpestis rotundifolia* C. F. Gaertn. *B. cyclophylla* Fern.

332   Amphianthus
Amphianthus pusillus Torr.
This diminutive annual is restricted to shallow flat-bottomed pools in granitic rocks. In Nov or Dec rooted seedlings can be seen under the water, forming small rosettes of leaves. In late winter when the water warms slightly a slender scape arises to the water surface, supported by 2 floating leaves. The first flowers, 6–8 mm, are borne above the water between these 2 leaves. Soon flowers that do not open but bear seeds form at the rosette. The seeds from both types of flowers rest in the dry thin soil, withstanding summer temperatures often unbearable to humans. Sometimes a wet autumn combined with a delayed freeze may allow plants to complete a life cycle before winter. Rare. Pied of Ga; cnSC; ceAla. Jan–May.

333   False-pimpernel
Lindernia monticola Nutt.
Members of this genus have opposite leaves, single flowers in upper leaf axils, no bracts below the 5-lobed calyx, 4 stamens of which 2 are sterile, and 2 separate stigmas.

This species is a perennial to 30 cm, with a basal rosette of leaves, cauline leaves usually small, the upper ones bract-size. Stem strongly angled and glabrous. Fruits 3–5 mm. Seeds as wide as long. Occasional, common locally. Shallow soil at margins of granitic and sandstone outcroppings, moist pinelands, swamp margins; neFla into ce and neAla, Pied of Ga, and seSC; nwSC into nwNC; cNC. Mar–June, sporadically until freezing. *Ilysanthes m.* (Nutt.) Benth.

L. saxicola M. A. Curtis is rarely over 10 cm, upper leaves slightly smaller than those of the rosette, and fruits only 1–2 mm. Rare. On rocks in Hiawassee and Tallulah Rivers; neGa and swNC. July–Sept.

334   Bird's-eye Speedwell
Veronica persica Poir.
In members of this genus the flowers are solitary in the axils of alternate leaves or bracts, the other leaves being opposite. Sepals 4. Corollas 2–12 mm broad, the lobes nearly alike. Stamens 2.

This species is a decumbent annual; the stems bear scattered glandless hairs. Leaves wider than long. Flowers in the axils of leaves that are the same size as those without flowers. Corolla 7–12 mm broad. Fruits hairy, strongly flattened, and on pedicels longer than the leaves. Occasional. Fencerows, fields, lawns, waste places; Fla into Calif, sAlas, Man, and Nfld. Feb–Aug.

V. hederifolia L. is similar vegetatively but the leaves are longer than wide, the corolla only 2–2.5 mm broad, and the fruit glabrous. Rare. Similar habitats; cSC into eTenn, eWVa, eO, sNY, and neNC. Mar–May.

335   Thyme-leaved Speedwell
Veronica serpyllifolia L.
Stems closely hairy, often creeping at base. Leaf blade entire, elliptic to broadly ovate, 10–25 × 5–15 mm. Bracts much smaller than the leaves below. Main stem terminates in a racemelike inflorescence. Corolla off-white with dark blue to deep purple lines. Fruits longer than wide. Our var. is *V. s. humifusa* (Dickson) M. Vail; var. *serpyllifolia* is a native of Europe. Occasional, but common locally. Irregularly dist; moist woods, alpine meadows, lawns, roadsides, waste places; neArk into sMinn, nMich, NS, nNY. May–Aug.

336   Gerardia; False-foxglove
Agalinis fasciculata (Ell.) Raf.
About 30 species of this genus occur in the eUS. They have opposite filiform to linear leaves, often with leaf clusters in the axils; 5 sepals, the lobes alike and shorter than the tube; a pink to purple, or rarely white, corolla with a bell-shaped tube and nearly equal lobes, the lower lobe being on the outside when in bud; and 4 stamens in 2 unequal pairs. Identification of most species is difficult. About half are entirely or essentially confined to the CP.

This species is an annual to 120 cm with prominent clusters of leaves in axils of the main leaves, stems finely rough, and pedicels shorter than the calyx. Corollas are 20–35 cm. Plants usually blacken in drying. Common. Fields, thin pinelands, savannas, usually in sandy soils; Fla into cTex, swMo, cAla, CP of SC, and ceNC. Aug–Oct. *Gerardia f.* Ell.

337   Gerardia; False-foxglove
Agalinis tenuifolia (Vahl) Raf.
Annual to 80 cm. Stems slender and smooth. Upper leaves narrow, to 1 mm wide. Pedicels longer than the calyx, usually longer than entire flower. Corolla tube at inner base of 2 upper lobes glabrous; upper lobe arches forward over the stamens

and sometimes nearly closes the throat. Capsule globose, 3–5 mm. Seeds dark. Common. Thin woods or in open, slopes, sandhills, fallow fields, prairies; nwFla into eTex, cWyo, seMan, cMich, sQue, and sMe. Aug–Oct. *Gerardia t.* Vahl.

A. setacea (Walt.) Raf. is similar but on average smaller; upper leaves quite narrow, about ⅓ mm wide; and all corolla lobes spreading, with long hairs at the inner base of the 2 upper lobes. Occasional. Dry sandy pinelands, poor upland soils, thin woods, or in open; ceAla into neFla, SC, and seNY. Sept–Oct. *Gerardia s.* (Walt.) J. F. Gmel.

338   Hairy False-foxglove
Aureolaria pectinata (Nutt.) Pennell
Members of this genus have opposite leaves; 5 calyx lobes; yellow corollas 3 cm or longer, the lobes spreading and nearly equal; and 4 fertile stamens, the anthers hairy and awned at the base.

This species is a much-branched densely glandular hairy annual to 1 m, leaves deeply dissected, and the calyx lobes toothed. Common. Dry places, thin woods, sandhills, open places; Fla into seTex, La, sMo, sKy, Ga, csVa, and ceNC. May–Oct. *Gerardia p.* (Nutt.) Benth.

A. pedicularia (L.) Raf. is similar but the upper parts of the plant are almost glabrous, the lower part of the stem glandular hairy. Occasional. Similar habitats; neGa into seKy, NY, neIll, ceMinn, Mass, and sMe. July–Oct. *Gerardia p.* L.

339   Downy False-foxglove
Aureolaria virginica (L.) Pennell
Perennial to 1.5 m. Stems finely hairy, little-branched. Lower leaves with 1–2 pairs of large lobes near base of blade, upper leaves smaller and less lobed, the uppermost often entire. Fruits finely but densely hairy. Common. Thin woods, usually in dry places; neFla into seAla, Ind, cMich, sNH, and Mass. May–Sept. *Gerardia v.* (L.) B.S.P.

A. laevigata (Raf.) Raf. is similar but the fruits glabrous, stems usually glabrous, and leaves lanceolate and rarely lobed. Occasional. Thin to rich woods; nGa into sO, cPa, and cMd. July–Sept. *Gerardia l.* Raf.

340   Smooth False-foxglove
Aureolaria flava (L.) Farw.
Perennial to 2 m; main stem, axis of inflorescence, peduncle, and calyx glabrous. Stems glaucous, lower leaves rarely lobed, pedicels stout, upcurved at pollen-shedding; corolla 3.5–5 cm. Common. Upland woods; se and nArk into csMo, sMinn, Wisc, Mich, Ont, swMe, ceNC, Fla, and La. June–Sept. *Gerardia f.* L.

341   Blue-hearts
Buchnera americana L.
Perennial to 80 cm, rarely branched, from a small rootstock, probably root-parasitic, and turning very dark in drying. Leaves 3-veined and opposite, or the uppermost alternate. Corolla purple or white, lobes shorter than 5 mm. Flowers in gradually elongating spikes up to 15 cm. Fruits many-seeded. Common. Dry to wet places in open, or thin woods; Fla into eTex, cMo, sMich, sOnt, wNY, and NJ. May–frost.

Some verbenas are similar but their fruits consist of 4 readily separable nutlets. *B. floridana* Gand.

342 Indian Paint-brush

Castilleja coccinea (L.) Spreng.
Hairy, usually biennial, rarely an annual, to 70 cm, with 1 unbranched stem from a basal rosette of leaves. Stem leaves varied, from entire to more commonly 3–5-cleft. Flowers in the axils of scarlet, or rarely yellow, leaflike bracts; each lateral half of the calyx with a rounded or barely notched tip. Corollas yellow to greenish-yellow. Occasional. In thin woods, around rock ledges and cliffs, meadows; nwFla into nLa, eOkla, Mo, seMan, and sNH. Apr–July.

There are perhaps 200 species of *Castilleja* but only 2 more occur in the eUS. *C. sessiliflora* Pursh (Downy Paintbrush) also has some or all of its leaves lobed or cleft but each lateral half of the calyx is distinctly 2-lobed at tip. Occasional. Dry prairies or plains; Wisc into nIll, Sask, Mo, Tex, and Ariz. May–July. *C. septentrionalis* Lindl. is similar but all foliage and bracteal leaves are entire. Occasional. Damp rocky soil; Lab into Nfld, and Vt; nMich. July–Aug.

343 Cow-wheat

Melampyrum lineare Desr.
An erect annual to 50 cm, usually branched in the upper half, the stems finely hairy. Leaves opposite. Flowers 1 per leaf axil. Calyx with 4 lobes. Corolla 2-lipped. Stamens 4. Seeds 1–4. Fruit an asymmetrical capsule. Occasional. Woods, bogs, open places, peaty soils; nwSC into cnGa, e edge of Tenn, wNC, Pa, neO, neIll, Man, BC, Lab, and NS. May–July.

344 Lousewort; Fernleaf

Pedicularis canadensis L.
Members of this genus are perennials with pinnately-lobed leaves. Sepals are strongly unequal and united. The corolla is extremely irregular, the lower lobes outermost in the bud. Stamens 4, anthers glabrous, the 2 pollen sacs equal and parallel. Fruit an asymmetrical flattened capsule. Seeds numerous.

This species has hairy stems with alternate leaves. Corollas of various combinations of yellow, red, rusty red, and purplish. Common. Rich woods and soils, in open, usually well drained; Fla into eTex, csMan, Wisc, sOnt, cQue, and cMe. Mar–May.

P. lanceolata Michx. has a similar inflorescence and leaves, but the stem is glabrous and the stem leaves opposite. Occasional. Moist open places, often in calcareous areas; swNC into ceWVa, ceVa, Pa, csMo, Ill, seMan, Wisc, sOnt, and Mass; e half of Neb. Aug–Oct.

345 Yellow-rattle

Rhinanthus minor L. ssp. *minor*
To 0.8 m, stem and branches green, simple or few-branched. Leaves firm, lance-triangular to oblong, 2–6 cm × 4–15 mm, serrate to crenate-serrate. Calyx somewhat inflated at pollen-shedding, very conspicuously inflated with fruit. Corolla 9–14 mm. Common. Fields, thickets, in open, and other various habitats. Circumboreal; S into Ore, Colo, NY, and sMe. May–Sept. *R. crista-galli* L.

OROBANCHACEAE: Broom-rape Family

346   **Squaw-root**
Conophilis americana (L.) Wallr. f.
Parasitic on the roots of several kinds of trees, mostly oaks and beeches. Usually several stems in a tight cluster. As the plants grow older they darken and often blend with surrounding leaves. Occasional. In various types of woods; Fla into nAla, Wisc, and NS. Mar–June.

A related species, *Epifagus virginiana* (L.) Bart., Beech-drops, is parasitic on beech trees. The stems are slim and usually branched. The flowers are mostly separated from each other, often by 1 cm or more. Occasional. Fla into eTex, Wisc, and NS. Sept–Nov.

347   **Cancer-root**
Orobanche uniflora L.
These nongreen plants are parasitic on roots of various kinds of plants. Stems are underground or nearly so with a few overlapping scales and 1–few erect finely hairy bractless pedicels to 16 cm. There is, therefore, 1 flower per stalk. Corollas are 12–20 mm, white to yellowish, sometimes tinged with light purple. They wither with age but persist, capping the capsule. Rare. Rich woods; cFla into Calif; BC, Yukon, and Nfld. Apr–May.

In *O. minor* Sm. the flowers are sessile and several on a prominent spike. Corollas shorter, 10–15 mm. Very rare. Parasitic on clover, tobacco, and other crops; cNC; Va into NS.

LENTIBULARIACEAE: Bladderwort Family

348   **Yellow Butterwort**
Pinguicula lutea Walt.
All species of *Pinguicula* bear short glandular hairs. Their leaves are clammy and sticky, trapping small insects that provide nourishment.

This species is a perennial to 50 cm. Leaves broad, 1–6 cm, succulent, in a basal rosette. Flowering stems 1–8. Corolla, including spur, 20–35 mm. Common. Moist places, thin pinelands, savannas; seLa into Fla, CP of Ga, and seNC. Feb–May.

349   **Dwarf Butterwort**
Pinguicula pumila Michx.
Perennial to 18 cm. Leaves broad, 1–2 cm, succulent, in a basal rosette. Flowering stems 1–10. Corolla, including spur, less than 20 mm, almost white to lavender, pink, or light yellow. Occasional. Moist places, sandy soil, pinelands and savannas; seNC into csGa, Fla, and seTex; cLa. Feb–May.

P. planifolia Chapm. differs from other species by its narrow and deeply cleft corolla lobes. Corolla violet, 10–20 cm including the spur. Rare. Pond margins, ditches, other moist places; cnFla into coastal Miss. Mar–Apr.

350   Horned Bladderwort

Utricularia juncea Vahl

Flowers of this genus are on erect leafless stems. Leaves dissected or forked into filiform segments and on stems underwater, or at the surface of wet soil or muck, in which case they may be small and even simple. They bear bladders, or traps (see picture), that catch small aquatic life for food. Petals united, irregular, and usually yellow. There are 2 pairs of stamens.

Plants to 45 cm. Leaves quite small, little-dissected or simple, with only a few small bladders, and at the surface of the soil or muck in which the plant grows. Bracts at the base of pedicels attached at their bases. Lower flower does not reach basal end of the pedicel of the one above. Corolla, including the spur, 6–15 mm. Occasional. Wet sand, peat, or mud; Fla into seTex, seArk, CP of Ga, and seNY. July–Sept. *Stomoisia j.* (Vahl) Barnh.

U. cornuta Michx. is quite similar but the lower flower reaches above the base of the one above and the corolla including spur is over 15 mm. Occasional. Similar habitats; Fla into sTex, cnSC, eWVa, neIll, eMinn, and eNfld. Apr–Sept. *Stomoisia c.* (Michx.) Raf.

351   Floating Bladderwort
352

Utricularia inflata Walt.

Besides finely dissected leaves, this species has some partially swollen floating, radiating leaves over 5 cm. Corolla about 2 cm broad. Occasional. Ponds, lakes, pools, sluggish waters; Fla into eTex, CP of Ga into NC, and sNJ. Feb–Nov.

U. radiata Small is quite similar but the radiating leaves are under 5 cm and the corolla about 15 mm broad. Occasional. Similar habitats; Fla into eTex, CP of Ga into NC, and NS; nwInd. Feb–Nov. *U. inflata* var. *minor* (L.) Chapm.

353   Purple Bladderwort

Utricularia purpurea Walt.

Flowering stems to 15 cm from long immersed stems with whorled finely dissected leaves, some with bladders at the tips. Corolla purple to deep pink, 10–13 mm broad. Rare. Ponds, lakes, pools; Fla into seTex, sGa, eNC, Mass, nInd, sMich, eMinn, and NS. Apr–Sept. *Vesiculina p.* (Walt.) Raf.

The Dwarf Bladderwort, *U. olivacea* C. Wright ex Griseb., is delicate, the flowering stem under 1 cm and bearing 1 flower about 2 mm. Petals yellowish but almost white. Leaves alternate, usually of only 1 segment, and 1 bladder. Fruit reported to have only 1 seed. Plants are easily overlooked and may be more common than records indicate. Very rare. In mats floating on water, frequently among the leaves of other Bladderwort species; Fla into sGa and seNC. Sept–Oct; Mar. *Biovularia o.* (C. Wright ex Griseb.) Kam.

354   Wiry or Slender Bladderwort

Utricularia subulata L.

Plants to 18 cm. Stems filiform, wiry, and bearing 1–several minute widely separated peltate bracts 1–2 mm, similar ones at base of each flower stalk. The leaflike branches are all filiform, unbranched, under the sandy or sandy-peaty substrate, and seldom collected. Common. Wet places; thin pinelands, bogs, depressions, roadside ditches; Fla into eTex, seArk, Tenn, seMass, swNC, seVa, and wNS. Mar–Oct.

U. gibba L. is similar but the bracts are joined at their base; plants 16–35 cm and usually in floating tangled branches or mats. Leaves all bladder-bearing. Ponds, swamps, drainage ditches, canals; Fla into NM, seNeb, and Okla; CP of Ga into NJ and seMass. Mar–Nov. *U. biflora* Lam.

ACANTHACEAE: Acanthus Family

355   Dyschoriste

Dyschoriste oblongifolia (Michx.) Kuntze
A dull green perennial to 50 cm. Leaves opposite. Calyx lobes slender, alike, less than half as long as the corolla. Petals united, slightly irregular, narrow part of the tube shorter than the wider part. Stamens 4, anther sacs sharp-pointed at the base. Ovulary superior, carpels 2, seeds 1 per carpel and borne on a hooked projection. Occasional. Sandhills, dry pinelands, occasionally in pine flatwoods; Fla into CP of Ga and adjSC. Apr–Aug.

D. humistrata (Michx.) Kuntze is the only other species of this genus in our area. The corolla is shorter, under 12 mm and about as long as the calyx. Rare. Woods in low places; c and cnFla into neCP of Ga, and vic of Charleston, SC. Apr–June.

356   Ruellia

Ruellia caroliniensis (J. F. Gmel) Steud.
A perennial to 80 cm with petioled opposite entire leaves. Flowers in sessile axillary clusters, usually only 1–2 open on any given day. Calyx lobes linear, alike, less than 2 mm wide. Corolla light purple, 25–50 cm, the lobes as a unit not quite at right angles to the slender corolla tube. Seeds borne on hooked projections. Quite variable and several varieties have been named. Common. Open woods, usually in dry upland soils; Fla into eTex, sIll, cWVa, and NS. Apr–Sept.

R. humilis Nutt. is similar but leaves are sessile or nearly so, and stems are often branched. Occasional. Dry woods; cNC into WVa, nwGa, eTex, eNeb, sMich, and csPa. June–Sept.

357   Water-willow

Justicia americana (L.) Vahl
Perennial to 1 m from a coarse rhizome. Leaves opposite, shaped like those of the black willow tree. Flowers in dense axillary long-peduncled spikes. Corolla united, 2-lipped, upper lip 2-notched, lower one 3-lobed and about as long as the tube. Stamens 2, anther sacs of each separated. Fruit a club-shaped capsule, smooth, mostly 4-seeded, 15–20 mm. Occasional. In and at edges of streams and occasionally lakes, often in rocky or sandy places; Pied of Ga into eTex, ceKan, Wisc, swQue, nwVt, and neNC. June–Oct. *Dianthera a.* L.

J. ovata (Walt.) Lindau is slender, to 50 cm, from slender rhizomes. Leaves in 5–8 pairs, ovate to elliptic or oblanceolate. Flowers in loose axillary spikes. Corolla similar to that of the above species but thinner. Capsule about 1 cm. Occasional.

Wooded bottomlands and swamps, pond and stream margins, occasionally in marshes; Fla into La, seMo, sIll, wTenn, sMiss, CP of Ga, and seVa. May–July.

PLANTAGINACEAE: Plantain Family

358 **Common Plantain**
Plantago major L.
Members of this genus are annuals or perennials with simple mostly parallel-veined leaves. Each flower is sessile or nearly so in axil of a bract aggregated into spikes or heads; sepals 4, in 2 slightly different pairs, one pair adjacent the axis, the other adjacent the bract; petals regular, united, long-persistent, corolla tube covering summit of the fruit. Fruit a 2-celled capsule splitting open in a circular line releasing the 2–many seeds.

This species is a perennial; leaves mostly lying on the ground, blade ovate, 2.5–10 cm across. Bract at base of flower broadly ovate; bracts and sepals glabrous or inconspicuously ciliate; corolla lobes under 1 mm. Fruit broadest at the middle and splitting open there. Common. Around buildings, roadsides, fields, lawns, beside paths; Greenl, Lab, irregularly scattered S throughout Can and US. June–frost.

P. rugelii Dcne. is similar but leaves more erect, petioles reddish, bracts narrowly lance-triangular. Fruit broadest below the middle and splits open there. Occasional. Similar places; Tex into ND, Mont, swQue, NS, Va, and Fla. June–frost.

359 **Hoary Plantain**
Plantago virginica L.
Winter annual to 15 cm with a taproot. Leaf blades hairy, oblanceolate to obovate, 5–40 mm across, longest ones 2–16 mm. Flowers in dense spikes; bracts hairy on outer side. Common. Lawns, borders, roadsides, fields, swales, waste places; eTex into SD, Ia, Wisc, sMe, Mass, and Fla. Mar–June.

P. aristata Michx. (Buckhorn Plantain) is another winter annual with a taproot, to 25 cm. Leaf blade linear to narrowly oblanceolate, to 20 × 0.8 cm. Flowers in spikes to 15 cm, a conspicuous bract at base of each flower. Common. Yards, borders, roadsides, waste places, adapted to dry areas. Native from Ill, La, and eTex. Now natzd. throughout most of US and sCan. Apr–Nov.

360 **English Plantain**
Plantago lanceolata L.
Perennial to 60 cm, roots fibrous (no taproot). Leaves all basal; blade narrowly elliptic to lanceolate, 7–50 mm across, to 30 cm long with 3–several prominent parallel veins. Spike 1–8 cm, less than one-fourth as long as supporting stem. Difficult to eradicate from lawns; if any of the many radiating roots are left in the soil they usually will form a sprout at the cut end. Native of Europe. Common. Lawns, gardens, fields, roadsides, waste places; throughout moister portions of US and Can; Greenl, Lab. May–Oct.

P. maritima L. (Seaside Plantain) is also a perennial with basal leaves; leaf blade thick, fleshy, entire or nearly so, 5–20 cm. Corolla tube hairy on outside. Fruit splits open near the middle. Common. Salt marshes, beaches, coastal rocks; circumboreal; S in eUS to NJ.

RUBIACEAE: Madder Family

Members of the Madder Family have opposite or whorled entire leaves with stipules in between on the stem. The stipules are sometimes as large as the leaves, and the combination often appears whorled. Flowers regular; petals united, the lobes 4–5, or rarely 3; ovulary wholly or partly inferior.

361   Bluets; Quaker-ladies

Houstonia caerulea L.

This species is a matted perennial with erect stems to 20 cm, with very slender fragile rhizomes; basal leaves 5–15 mm; stipules membranous and entire. Peduncles filiform, arched in bud, erect when flower opens. Corolla pale blue, or rarely white, with a yellow eye; the tube 5–10 mm and glabrous inside. Fruit a capsule 2–3.5 mm across. Common. Thin woods or open areas; fields, meadows, swales, grassy areas, roadsides; usually moist soils; se and cArk into Wisc, sOnt, sQue, NS, Mass, NC, cGa, and Ala. Apr–June. *Hedyotis c.* (L.) Hook.

H. pusilla Schoepf (Small Bluet; Star-violet) has similar flowers but is an erect annual to 10 cm with leaves mostly near the base. Corolla violet to deep purple with a reddish eye, the tube 2–5 mm. Capsule 4–5 mm across. Common. Thin woods, fields, pastures, usually in well-drained areas; Fla into eTex, seNeb, csMo, sIll, Pied of Ga, and seVa. Feb–Apr. *Hedyotis crassifolia* Raf.

In *Houstonia serpyllifolia* Michx. stems are creeping, the tips and some branches erect and each bearing a flower; lower leaves petioled, blades suborbicular, 2–7 mm across. Corolla tube hairy within. Occasional. Margin of streams, wet slopes, other moist places; nwSC into neOkla, swPa, and mts of Va. Mar–June. *Hedyotis michauxii* Fosberg.

362   Trailing Bluet

Houstonia procumbens (Walt. ex G. F. Gmel.) Standl.

Creeping perennial, some stems decumbent. Leaf blade ovate to suborbicular. Flowers solitary on erect pedicels; corolla white, the tube 5–7 mm and glabrous inside. Fruits 2-carpelled capsules on curved pedicels. Common. Dunes, swales, thin scrub, thin woods; sSC into Fla and Miss. Feb–Apr. *Hedyotis p.* (Walt. ex J. F. Gmel.) Fosberg.

363   Summer Bluet

Houstonia purpurea L.

Erect perennial to 40 cm. Stems 1–many from a common base. Stem leaf blades ovate to ovate-lanceolate. Flowers several to many in terminal clusters or from upper leaf axils. Petals white to light or reddish purple. Common. Deciduous woods, rocky places, roadsides; swGa into seTex, eOkla, sIll, swPa, sNJ, and csSC. *Hedyotis p.* (L.) Torr. & Gray.

Other species with similar flowers but narrower leaves include *H. canadensis* Willd. ex R. & S., the only species with basal leaves ciliate and present at flowering time. Rare. Rocky woods on slopes; nwGa into nAla, Ark, cKy, Minn, eO, sOnt, NY, Me, and cw and swVa. Apr–May. *Hedyotis c.* (Willd. ex R. & S.) Fosberg. In *Houstonia longifolia* Gaertn. the stipules are rounded to deltoid. Common. Dry, usually open places; nFla into eOkla, nIll, seSask, Me, Va, and cSC. May–July. *Hedyotis l.* (Gaertn.) Hook.

364 Mexican-clover
Richardia scabra L.
Weedy annual with branching and spreading often decumbent stems. Leaves opposite, nearly glabrous but outer parts and lower midrib rough, with bristle-bearing stipules between the petiole bases. Flowers and fruits sessile in dense terminal clusters. Corolla tubular, with 6 lobes, white, 5–6 mm. Ovulary inferior. Fruit covered with small tubercles, separating into 4 indehiscent 1-seeded units. Common. Open places, fields, roadsides, waste places; Fla into e half of Tex, sArk, cGa, and seVa; rarely in Pied of Ga into NC. June–frost.

  R. brasiliensis Gomes is similar but the leaves are appressed-hairy on both surfaces and the fruits hairy. The roots are used in S. Amer as an emetic. Occasional. Similar places. Fla into coastal Tex, CP of Ga, and seVa. Apr–frost.

365 Buttonweed
Diodia virginica L.
Prostrate, ascending, or erect branching perennial. Stems almost glabrous to quite hairy. Base of leaf blades narrowed to nearly cordate; stipules membranous at base, tip consisting of 3–5 linear projections. Flowers 1, or rarely 2, per leaf axil, sessile; sepals 2, rarely 3, persistent; corolla tube 7–9 mm; ovulary inferior; stigmas 2, filiform. Fruit 5–9 mm, leathery, splitting into 2 indehiscent 1-seeded segments. Sometimes a weed in lawns; difficult to eradicate. Common. Usually in open moist places, shallow pools, pond margins, marshes, swamps, ditches, swales; Fla into eTex, sMo, WVa, and sNJ. May–frost. *D. tetragona* Walt.; *D. hirsuta* Pursh.

366 Rough Buttonweed
Diodia teres Walt.
A usually much-branched annual with erect to spreading hairs. Leaf blade linear-lanceolate to elliptic-lanceolate; stipules with about 5 threadlike projections. Sepals 4, under 4 mm, persistent; corolla 2–6 mm; stigma shallowly 2-lobed. Fruit 2.5–4 mm, splitting into 2 indehiscent 1-seeded segments. Common. Dry open areas; fields, thin woods, lawn borders, roadsides, waste places, especially in sandy soils; Fla into csTex, cKan, sMich, sePa, and Conn. June–frost.

367 Field Madder
Sherardia arvensis L.
Much-branched annual, often forming dense masses. Leaves 5–15 mm, in whorls of 4–6. Flowers in terminal heads surrounded by a whorl of leaves fused near their bases. Corolla with a tube about 3 mm. Occasional. Fields, lawns, waste places; Ga into eTex, swMo, Tenn, neO, swQue, and NS. Mar–Aug.

  Some species of *Galium* are similar to Field Madder, but there are no fused leaves surrounding the flower clusters. Similar species include *G. virgatum* Nutt., which has solitary flowers and bristly fruits. Occasional. Dry places; La into Tex, sMo, swIll, and cTenn.

368   Bedstraw; Catchweed

Galium aparine L.

In members of this genus leaves and stipules are alike and combined into whorls. For simplification we shall refer to all as leaves. Flowers are small; petals 3–5, the lower portions united into a tube. Fruit when mature separates into 2 seedlike indehiscent 1-seeded parts.

This species is an annual with a small root system and weak slender stems to over 1 m, often forming tangled masses; stems little-branched, angles armed with stiff retrorsely hooked bristles. Leaves usually in whorls of 8, blades 1–8 cm, bristle-tipped. Flowers solitary or in groups of 2–3. Fruits with many hairs, these hooked at the tip. Common. Moist or rich soils; scrub areas, woods, meadows, around buildings, waste places; Alas into Nfld and Greenl; S to near tropics. May–Aug.

369   Dye Bedstraw

Galium tinctorium (L.) Scop.

Perennial with weak reclining to ascending stems, often forming tangled masses; stems sharply 4-angled. Leaves glabrous, those of main stem in whorls of 4, 5, or 6, blades broadly oblanceolate to oblong-spatulate, margin retrorsely scabrous, rounded at tip. Flowers solitary; corolla greenish-white, 1.5 mm or less across, with 3(4) obtusely tipped lobes. Mature fruits dry, black, smooth, the pair 2–3 mm across. Common. Moist to wet places; pond margins, ditches, sloughs, swamps, floating mats; Tex into Neb, SD, Ont, Nfld, NC, and cnFla. Apr–Sept.

G. obtusum Bigelow is erect, the stem is smooth and angles rounded. Leaves in whorls of 4; flowers 2–2.5 mm across, and corolla with 4 acutely tipped lobes. Common. Similar habitats; Fla into Ariz, Neb, Minn, sOnt, swNS, Va, cNC, and cGa. Mar–July.

370   Purple Galium

Galium hispidulum Michx.

Finely scabrous, usually much-branched perennial with stems to 60 cm. Leaves in whorls of 4; blades elliptic, firm, persistently green into winter. Flowers usually in pairs; corolla white. Fruit purple, smooth, juicy when fresh. Common. Dunes, sandy scrub, pinelands, maritime woods, live oak woods; sNJ into ceGa, Fla, and La. June–Aug.

CAPRIFOLIACEAE: Honeysuckle Family

371   Perfoliate Horse-gentian

Triosteum perfoliatum (L.) Nieuwl.

Erect perennial; main stem with hairs mostly under 0.5 mm. Leaves opposite with bases united. Sepals 5, linear; corolla with a tubular base and expanded upper portion, unequally 5-lobed; stamens 5, filaments very short, borne on inside of corolla tube; ovulary inferior. Fruit a dry drupe with 3 oblong stones. Occasional. Woods, thickets, thin or rocky soils; cArk into eOkla, eNeb, Minn, sOnt, Mass, nSC, nGa, and nAla. May–July.

T. aurantiacum E. Bickn. is similar except leaf bases not joined and hairs on main stem mostly over 0.5 mm. Leaves widest near the base. Occasional. Rich woods, thickets; nArk into eOkla, Minn, wOnt, sQue, wNB, cVa, and Ky. In *T. angustifolium* L. (Yellow-flowered Horse-gentian) the largest leaf blade usually under 5 cm across and widest near the middle. Occasional. Deciduous and mixed woods; usually on basic to neutral soils; nwGa into La, seArk, Mo, Ill, Ind, Pa, cConn, c and swVa, and Pied of NC. Apr–June.

VALERINACEAE: Valerian Family

372 Corn-salad
Valerianella radiata (L.) Dufr.
A succulent annual to 70 cm; stems dichotomously branched; main leaves opposite, sessile. Flowers bisexual, in tight cymes; calyx minute to missing; petals 1.5–2 mm, united, lobes white, 0.4–0.8 mm; stamens 3, extending beyond corolla; ovulary inferior, carpels 3, only 1 developing and this into a leathery nutlet. Common. Fields, roadsides, gardens, borders, woods margins, waste places; Tex into Kan, sIll, O, Pa, Va, NC, and Fla. Apr–May.

CUCURBITACEAE: Gourd Family

373 Creeping Cucumber
Melothria pendula L.
Slender perennial vine, trailing or climbing by tendrils. Leaf blade palmately veined, varying from scarcely to strongly 3–5-lobed, base cordate. Flowers few, unisexual, rarely bisexual, the female solitary and 8 mm across; corolla small, bell-shaped; stamens separate. Fruit green to black, ovoid, ca 1 cm, pulpy, with about 20 white seeds; POISONOUS when eaten. Common. Dunes, swales, scrub areas, edges of marshes, swamps; sTex into eOkla, sMo, sInd, sVa, and Fla. June–frost.

374 Balsam-apple; Wild-cucumber
Echinocystis lobata (Michx.) Torr. & Gray
High climbing annual vine, tendrils 3-forked. Leaf blade orbicular in outline, with (3)5(7) sharp triangular lobes. Male flowers axillary, in long erect racemes; corolla 8–10 mm across with lanceolate lobes. Female flowers few and solitary, short-peduncled from same leaf axil. Fruit green, ovoid, 3–5 cm, weakly prickly, a 4-seeded berry opening by 2 pores at apex. Occasional. Moist soils, thickets, rich soil along streams; Tex into Sask, NB, neVa, and neNC. June–frost.

CAMPANULACEAE: Bellflower Family

375   Bellflower

Campanula divaricata Michx.

Glabrous perennial with 1–many spreading to drooping or erect stems from a rootstock. Leaves elliptic to lanceolate, serrate, tips usually acuminate, bases acute. Flowers in a terminal, mostly drooping, open panicle. Corolla nearly white to blue or light purple, 6–9 mm. Stamens 5, separate. Stigmas 3. Ovulary inferior. Many-seeded fruit opens by basal pores. Common. Rocky woods, cliffs, unstable soil on steep slopes; ceGa into cAla, s and eWVa, wMd, and cNC, but not common in Pied. Aug–frost. *C. flexuosa* Michx.

    Campanulastrum americanum (L.) Small is a biennial and otherwise quite different. Corolla 2–3 cm wide, the lobes wide-spreading. Style curved abruptly upward near the tip. The ovulary is inferior, but the fruit opens by pores near the top. Common. Rich deciduous woods; cnFla into eOkla, seSD, sOnt, wNY, CP of Va, and cNC. July–frost.

376   Harebell

Campanula rotundifolia L.

Glabrous perennial 10–80 cm. Leaf blades on main stem linear or nearly so, 1.5–8 cm and usually under 1 cm across, entire or very minutely toothed. Flowers usually several in a lax racemelike to long paniclelike inflorescence; corolla blue, 15–30 mm, bell-shaped, the lobes much shorter than the tube; style shorter than the corolla. Common. Dry woods, meadows, cliffs, beaches; circumboreal, S in N. Amer into Ia, Ind, and NJ; Mex. June–Sept.

377   Venus' Looking-glass

Triodanis perfoliata (L.) Nieuw.

Erect annual to 1 m but most often half that tall or less, simple or with a few branches, with fibrous roots. Stem leaves strongly clasping. Flowers sessile in leaf axils, those on the lower half of the stem not opening. Petals united, deep purple to pale lavender. Stamens 5, separate. Stigmas 3. Ovulary inferior. Many-seeded fruit opens by 3 small elongate pores at or just below the middle. Common. Fields, waste places, gardens; Fla into sTex, Ariz, Mont, BC, sMinn, sQue, sMe, and NC. Apr–July. *Specularia p.* (L.) DC.; *T. biflora* (Ruiz & Pavón) Greene.

378   Purple Lobelia

Lobelia elongata Small

Some *Lobelia* species have been used medicinally but probably are best considered poisonous. Overdoses of plants or extracts produce adverse symptoms and many cause death.

    This species is a glabrous perennial to 1.6 m, from horizontal basal shoots. Leaves narrowly lanceolate, tapering to both ends. Flowers 20–25 mm. Calyx lobes entire. Corolla tube 8–14 mm, with openings in the side. Inner base of lower corolla lip glabrous. Occasional. Marshes, bogs, swamp woods; seGa into c and eNC and Del. July–frost.

    L. glandulosa Walt. is similar but the inner base of the lower corolla lip is hairy. Common. Damp pinelands, savannas, swamp forests, marshes; c and cnFla into

CP of Ga, nwNC, and seVa. Aug–Oct. *L. puberula* Michx., the most abundant of our large-flowered species, is also similar but has finely hairy stems, at least near the base. Fla into eTex, eOkla, seMo, sIll, wVa, and sNJ. July–Oct.

379   Cardinal-flower
Lobelia cardinalis L.

The brilliant flowers allow this species to be recognized at considerable distances. As is true of all our species of *Lobelia,* the corolla is 2-lipped, the upper lip gener-ally erect and the lower spreading and 3-cleft. The 5 anthers are united. Common. Moist to wet places; Fla into eTex, seKan, Minn, and NB. July–Oct.

Our other species have purple or lavender to blue or white corollas. Most are difficult to name. One species, *L. inflata* L., is easy to recognize when the fruits are fully grown. They are ovoid to subglobose, and on thin pedicels. Common. Open woods, fields, waste places; nGa into eOkla, Minn, seSask, Lab, and cnSC. July–frost.

ASTERACEAE: Composite Family

This family is variable and exceptionally large and widespread, with many struc-tures unique to the Composite, a fortunate circumstance in view of the many genera involved. An understanding of most of the characteristics of the family is essential for identification of all but a few genera and some of the species. We feel confident the following details will serve as a useful tool.

Flowers of all species are sessile in *heads* surrounded by bracts (see illustration on page xxiv). These bracts are called *phyllaries* and as a group the *involucre.* They may be in 1–2 whorls or vary with a few to many phyllaries overlap-ping. Flowers are borne on a common *receptacle,* a disc- to cone-shaped to columnar structure, and are closely compacted with rarely fewer than 5 but varying to a vast number (e.g., in *Erigeron* species). Each individual flower is in the axil of a *receptacular* bract, or the bract is sometimes absent. Flowers in a head may be all bisexual, all unisexual with both sexes in same head, or rarely sexes in heads on different plants. *Sepals* are borne on the apical rim of the hypanthium (defined below) and represented by a *pappus* (possibly absent), which may consist of limber to stiff capillary bristles, scales, a thin and limber to firm or hard ring or crown, or combinations of these; bristles may be simple or plumose (somewhat resembling a bird feather). *Petals* 5, bases united into a tube terminated by either 5 equal lobes or a single strap-shaped structure called both *ligule* and *ray.* Heads may have all flowers with lobed corollas (known as *disc flowers*), all strap-shaped (*ray flowers*), or both, with ray flowers usually occupying the perimeter of the head. *Stamens* 5, borne toward basal inside of corolla tube alternate with corolla lobes; filaments very slender and separate; anthers linear with sides fused into a tube around the style and perhaps its 2 stigmas. *Pistil* with a 2-carpelled, 1-celled ovulary surrounded by and fused with a cylindric tissue (the *hypanthium*); style rises from summit center of the ovulary and extends into anther tube and usually beyond it; style commonly 2-cleft, each half bearing a stigmatic surface. The *fruit* is an achene; the single seed with a tissue-thin covering develops free inside the hypanthium-ovulary wall covering.

380   Ironweed

Vernonia gigantea (Walt.) Trel.

Members of this genus are erect perennials with prominent alternate stem leaves. Heads with 9–34(120) disc flowers but no rays or receptacular bracts, flowers all perfect, corollas a deep reddish-purple; receptacle flat to slightly rounded; pappus consisting of an inner circle of long capillary bristles and an outer circle of short ones.

Plants of this species 1–2(3.5) m; upper portion of main stem hairy. Stem leaves numerous; blades of middle leaves 6–30 × 1–7.5 cm, lanceolate to linear-lanceolate, oblanceolate, or elliptic. Heads with 9–30 flowers; involucre cylindric to cylindric-campanulate, to 7 × 5.5 mm, apical portion of middle phyllaries straight and pressed against those above, inner ones 3.5–5.3 × 1.2 mm, tip acute to obtuse, outerside thinly hairy, margin ciliate. Corollas 9–11 mm. Achenes ribbed; pappus tan to brown, inner bristles ca 6 mm, outer scales 0.2–0.8 mm. Common. Usually in low places, pastures, thin woods, roadsides; nFla to Tex, Okla, Mo, Ind, Pa, Md, and NC. July–Oct.

Plants of *V. flaccidifolia* Small are also tall (1–2.5 m) and stem leaves abundant but stems glabrous. Leaf blades at midstem 10–25(35) × 2–7(8) cm, narrowly to broadly lanceolate, base tapered. Inner phyllaries tightly appressed, 3.2–5 × 1.5–2 mm, apex obtuse to acute. Flowers 16–26 per head, corollas 11–12 mm; pappus straw-colored, inner bristles ca 6 mm, outer scales irregular, 0.2–0.8 mm. Rare. Upland deciduous woods, pastures, roadsides; ce and neAla into nwGa and adjTenn. Sept. *V. altissima* Nutt.

Other tall species with ca 30 or fewer flowers per head include *V. angustifolia* Michx., which has linear to broadly linear leaves and tawny-purplish pappus. Common. Sandy well-drained pinelands; sandy ridges and scrub; Fla into seMiss, cAla, ceGa, and seNC. July–Aug. *V. pulchella* Small grows only to 70 cm and has 20–35 flowers per head. Rare. Sandy scrub and pinelands; seGa into swSC. July–Aug.

381   Ironweed

Vernonia glauca (L.) Willd.

This species grows to 1 m. Stem leaves only, 32–48 flowers per head, and a straw-colored pappus. Occasional. Well-drained soil of deciduous woods; neGa, csVa, eWVa, sePa, and swNJ; cwAla. July–Aug.

Other species with 30–50 flowers per head are: *V. acaulis* (Walt.) Gl., which has basal as well as cauline leaves. Occasional. Well-drained soils of thin woods or open, roadsides; swGa into ePied of Ga, c and seNC, and seGa. July. *V. noveboracensis* (L.) Michx., which is 1–2 m and has a brownish-purple pappus. Common. Moist places in open, fields, pastures, along streams; cnFla into eAla, seKy, Pa, and eMass. July–Sept.

382   Elephant's-foot

Elephantopus tomentosus L.

Perennial to 60 cm. Leaves basal, soft-hairy beneath, over 7 cm wide. Flowers in heads surrounded by 3 conspicuous bracts 10 mm or longer. Petals pink to purple, rarely white. Common. Dry places, evergreen or deciduous woods; Fla into eTex, seOkla, seKy, Ga, and cs and seVa. July–Oct.

Two other species are similar but leaves are narrower and bracts only to 8 mm. In *E. nudatus* Gray the glandular hairs on the bracts are easily seen because other hairs are scanty. Occasional. Thin woods or in open; Fla into seTex, seArk, CP of Ga, and Del. In *E. elatus* Bert. the glands are mostly hidden by long and dense hairs. Occasional. Dry pinelands; Fla into sMiss, sGa, and seSC.

383   Leafy-stemmed Elephant's-foot
Elephantopus carolinianus Raeusch.
Plants 0.4–1 m; stems leafy, with 4 or more nodes before branching. Leaf blades 9–25 × 5.5–9.5 cm, ovate to broadly lanceolate. Heads 4-flowered; involucre cylindric; bracts 3, cordate, 11–30 × 7–18 mm; phyllaries lanceolate, 6–10 × 1.5–2 mm, resin-dotted, hairy, margin membranous, apex abruptly acute. Corolla white to violet, 7–8 mm. Pappus of 5 bristles in 1 series, 4–5 mm. Common. Moist habitats on hillsides and along margins of streams and swamps, deciduous woods; cnFla into eAla, seKy, Pa, and eMass. Aug–Oct.

384   Dog-fennel
Eupatorium capillifolium (Lam.) Small
Our species of this genus are mostly perennials; leaves opposite or whorled or rarely alternate. Flowers bisexual and all disc, receptacular bracts absent; achenes 5-angled, the pappus a single series of capillary bristles bearing small antrorse barbs that are turned toward the tip. Plants often difficult to name to species.

This species is an annual to 2 m with clustered stems; leaves opposite or the upper ones alternate, blades glandular-dotted and finely dissected, the segments under 0.5(1) mm across. Flower heads evenly placed around branches of the inflorescence. Common. Thin woods, fields, pastures, roadsides; Fla into eTex, sTenn, SC, and sNJ Sept–frost, or all year if frost-free.

E. leptophyllum DC. is a rhizomatous perennial to 1.2 m and has similar leaves and flowers but flower heads hang from only 1 side of the branches. Rare. Similar habitats; Fla into sMiss, sGa, and seNC. Sept–frost. *E. compositifolium* Walt. has similar flowers and leaves except leaf segments 1–2.5(4) mm across. To 2 m. Common. Fields, pastures, thin woods; usually sandy soils; Fla into eTex, sw and ceGa, and ceNC. Sept–frost.

385   False-hoarhound
Eupatorium rotundifolium L.
Perennial to 1.2 m, finely and densely hairy. Leaves sessile or with petioles up to 3 mm, blades not more than twice as long as broad and with 2 prominent lateral veins. Branches of inflorescence opposite. Heads 5-flowered, phyllaries obtuse to acute. Corollas white. Common. Rare in the mts, and occasional in the Pied. Thin woods, pine barrens, savannas; Fla into e and cnTex, sO, and seNY. July–Sept.

E. album L. has leaves 3 times or more as long as wide, the blades 15–35 mm wide. The phyllaries are long acuminate, the tips and margins white. Corollas white. Common, rare in higher mts. Dry thin woods, especially in sandy pinelands; Fla into seLa, seArk, sO, and seNY. July–Sept.

386   **Joe-pye-weed**
Eupatorium dubium Willd. ex Poir.
Perennial to 1.5 m; stems speckled or covered with purple. Leaves mostly in whorls of 3–4; blades thick and firm, midstem ones ovate to lanceolate-ovate and with 3 prominent veins. Inflorescence slightly to strongly rounded; involucre 6.5–9 mm, often purplish; flowers (4)6–9(12) per head, corolla purple. Occasional. Freshwater marshes, low meadows, pond shores, sloughs; NS to SC. July–Oct.

    E. maculatum L. (Queen-of-the-meadow; Joe-pye-weed) is similar but to 2.5 m; leaves in whorls of 4–5, those at midstem with 1 main vein. Inflorescence or its segments flat-topped, flowers 9–22 per head, corolla purple to pale lavender. Occasional. Moist places: marshes, meadows; thin woods; calcareous soils; nGa into WVa, Ill, BC, Nfld, and wMd. July–Sept.

387   **Boneset; Thoroughwort**
Eupatorium perfoliatum L.
Perennial to 2 m; main stem little to widely branched, conspicuously long-hairy, especially basal portion. Easily recognized by its opposite sessile broad-based perfoliate or strongly clasping leaves; blades 7–20 × 1.5–4.5 cm. Flowers 9–23 per head; corolla dull white. Common. Moist to wet places in woods or open; pond margins, sloughs, swales, meadows, ditches; nFla into La, Okla, Neb, Minn, Que, and NS. Sept–Oct.

388   **Pink Eupatorium**
Eupatorium coelestinum L.
Petioled leaves, blue to reddish-purple corollas, and 35–70 flowers per head with a conic receptacle serve to identify this species. Often used as a perennial in flower beds although it sometimes becomes a pest because of long white creeping underground stems and abundant distribution of seed by wind. Occasional. Moist woods and meadows, stream borders, and almost any habitat near ornamental plantings; Fla into se and cnTex, cwInd, Va, and sePa. July–Oct. *Conoclinium c.* (L.) DC.

    E. incarnatum Walt. is similar but corollas are pink to purplish, flowers 13–24 per head with a flat receptacle. Plants have an odor resembling vanilla. Rare. Rich woods, swampy places; Fla into cs and cnTex, csInd, sO, wWVa, seVa, and nwGa. Aug–Oct.

389   **Joe-pye-weed**
Eupatorium fistulosum Barr.
Perennial to 2 m. Stems purplish throughout, strongly glaucous and hollow. Midstem leaves mostly in whorls of 4–7, the blades not resin-dotted beneath or only with a few scattered ones. Inflorescence strongly rounded. Flowers 5–8 per head, phyllaries and corollas generally bright pink-purple. Common. Moist places in thin woods and open; Fla into eTex, seIa, swQue, and sMe. July–Oct.

    E. purpureum L. is similar but the stem is usually greenish and not hollow. The bruised fresh plant is strongly vanilla-scented. Corollas generally very pale pinkish or purplish. Common. Thin woods, usually in drier places than the above species; cwSC into eOkla, eKan, sWisc, and sNH. July–Oct.

390   White Snakeroot

Ageratina altissima (L.) King & H. E. Robins.
Members of this genus are perennials, leaves opposite. Flowers bisexual; phyllaries weakly overlapping, the main ones about equal and in a somewhat double series, a few shorter outer ones possibly present also; corollas white. Achenes 5-angled, pappus a single circle of capillary bristles. There are 3 species in the eUS; 2 are common.

This species is a perennial to 1.5 m. Leaves opposite, with petioles over 2 cm. Blades narrowly to broadly ovate, rather thin, acuminate, 6–18 cm. Flowers 15–30 per head. Phyllaries glabrous or lightly hairy, acute to acuminate at the tips, their margins not overlapping. Corollas white. The plant contains a poison that can kill an animal eating it. The poison is also readily transmitted from a grazing animal to suckling young or to people using milk or butter from that animal. Many deaths from snakeroot have been recorded for humans and grazing animals. Common. Rich woods and openings, pastures; nFla into cTex, seND, Minn, sSask, and NS. July–Oct. *Eupatorium urticaefolium* Reichard. *E. rugosum* Houtt.

391   Small-leaved White Snakeroot

Ageratina aromatica (L.) Spach
This species has leaf blades more than 4 times as long as petioles, relatively thick and firm, mostly crenate to crenate-serrate, acute to obtuse, and largest ones 3–7(10) × 2–5 cm. Plants to 80 cm, rarely more. Flowers mostly 10–19 per head. Common. Dry woods, especially in sandy soil; nFla into eTex, Ky, sO. Sept–frost. *Eupatorium aromaticum* L.

392   Climbing Hempweed

Mikania scandens (L.) Willd.
The genus contains about 250 species. They are common in the tropics, especially in S. Amer.

This species is a twining perennial, often forming extensive masses over other low vegetation. Leaves opposite. Flowers per head 4, none with ligules, phyllaries narrowly acute, the corollas nearly white to lilac or light pink. Common. Moist to wet places in open; Fla into eTex, seMo, wTenn, Ala, seTenn, nw and Pied of Ga, and swMe. June–frost.

M. cordifolia (L.) Willd. is quite similar but phyllaries are obtuse or broadly acute. It is also more vigorous. Rare. Low places and hammocks; pen Fla and sLa; abundant in subtropical and tropical localities. Flowers during frost-free periods.

393   Blazing-star

Liatris pilosa (Ait.) Willd. var. *pilosa*
Members of this genus are erect perennials from a tuberous underground base, with alternate entire leaves, the inflorescence a spike or raceme, flowers all bisexual, without rays or receptacular bracts, phyllaries much overlapped, achenes ca 10-ribbed, and a pappus of 1–2 rings of stout barbed or plumrose bristles. Most species are difficult to name.

In this species the flower heads are longer than broad, sessile or on peduncles to 1 cm, and phyllaries obtuse. The basal rosette of leaves is absent, the lower stem leaves linear and to 20 cm, corolla lobes under 3 mm and the tube hairy inside. Common. Dry places, uncultivated fields, thin woods, especially among pines; Fla into seMiss, nw and Pied of Ga, SC, and NJ. Aug–Oct. *L. gramnifolia* Willd.

L. gracilis Pursh is similar but the peduncles are over 1 cm. Rare. Similar places; Fla into csAla and swSC. Aug–Oct.

394   Blazing-star

Liatris aspera Michx.
Plant stiffly erect, to 2 m. Leaves 25–90, the lower ones narrowly elliptic, the upper ones nearly linear. Flowers 16–35 per head. Phyllaries with obtuse to rounded apex; thin, pink to white margins and tip. Occasional. Thin woods or open places, usually dry and sandy or rocky; cwSC into eTex, eND, and sOnt. July–Sept.

L. elegans (Walt.) Michx. also has pink to white phyllaries, these acute to acuminate and broadest above the middle. There are usually 5 flowers per head. Occasional. Similar places; Fla into e and cnTex, seOkla, CP of Ga, and csSC. Aug–Oct.

Plains Blazing-star

Liatris squarrosa (L.) Michx.
To 0.8(1) m from large cormlike rhizome. Leaves near base of stem 6–25 × 0.4–1.3 cm, base clasping stem. Heads few or even 1; involucre 12–25(3) mm; phyllaries firm, green to purple, tips pointed away from the head; flowers 20–45 per head, innerside of corolla lobes coarsely hairy. Common. Dry open places, thin woods; Fla into e and cnTex, seSD, sIND, swWVa, Ky, NC, and Del; absent from Appal Mts. June–Aug.

In *L. microcephala* (Small) K. Schum. (Narrow-leaf Blazing-star) plants are slender, most 30–80 cm, essentially glabrous throughout. Leaves numerous, linear, lower ones 3–16 cm × 1.5 mm, gradually reduced upward. Heads usually numerous, involucre nearly cylindric, 6–8 mm; phyllaries green, narrowly thin-margined, apex blunt. Flowers 4–5(6) per head, most corollas 6–8 mm including the 1.5–2-m lobes; pappus antrorsely barbed, seldom reaching base of corolla lobes. Occasional. Exposed places; glades, thin woods, rock outcrops, sandy shores; Pied and adjCP of Ga and Ala, eKy, wNC. Aug–early Oct.

395   Smooth Blazing-star

Liatris helleri Porter
Glabrous plants 10–50 cm. Leaves numerous and crowded, basal ones 2–20 × 3–7(10) cm. Heads 3–20(30); involucre 7–10 mm; flowers mostly 7–10 per head; corolla 6–11 mm, tube innerside hairy near base; pappus about half as long as corolla tube. Rare. BR of NC; neAla. July–Aug.

396   Deer-tongue; Vanilla-plant

Carphephorus odoratissimus (J. F. Gmel.) Hebert
Glabrous perennial to 2 m, with a distinct odor of vanilla that can sometimes be detected from a distance. Leaves alternate, the basal ones 35 × 7 cm. Flowers all bisexual, rays absent, phyllaries in several overlapping series, achenes 10-ribbed, the pappus of tawny to purplish finely barbed bristles 3–4 mm, entire heads well

under 1 cm. Tons of leaves are collected from the wild and used to flavor smoking tobacco. Common. Pinelands, savannas, thin mixed woods, usually in poorly drained places; Fla into seLa, cAla, CP of Ga, and seNC. July–Oct. *Trilisa odoratissima* (J. F. Gmel.) Cass.

C. paniculatus (J. F. Gmel.) Hebert is closely related. Stems have a dense coat of fine hairs and flower heads are usually in a cylindrical rather than spreading inflorescence. Common. Similar habitats; Fla into s half of NC CP. Aug–Oct. *Trilisa paniculata* (Walt.) Cass.

Carphephorus
Carphephorus corymbosus (Nutt.) Torr. & Gray
Perennial to about 1 m with a single finely and densely hairy stem and spatulate to elliptic-spatulate basal leaves. Inflorescence flat-topped or rounded. Larger heads over 1 cm. Flowers are bisexual, with no rays. Phyllaries obtuse, in several overlapping series. Achenes 10-ribbed, the pappus of finely barbed bristles. Occasional. Thin, dry pinewoods, scrub oak sandhills; cs and seGa and southward into Fla. June–Oct.

In *C. tomentosus* (Michx.) Torr. & Gray the phyllaries are acute and the corollas have small resin droplets on the outside surface. Occasional. Similar habitats; seGa into seVa. Aug–Oct. *C. bellidifolius* (Michx.) Torr. & Gray is also similar but the stem is usually glabrous below the inflorescence and the inflorescence is usually more open and more slenderly branched. Occasional. Scrub oak sandhills, thin pinewoods of uplands, sandy fields; ceGa into CP of SC and seVa. July–Oct.

397   Golden-aster
Chrysopsis mariana (L.) Ell.
In members of this genus the cauline leaves, and often the basal ones, are alternate. Outer flowers of the head have yellow ligules, disc flowers with yellow corollas. Receptacular bracts absent. Phyllaries much overlapping. Pappus of capillary bristles in 2 rings, the outer much shorter than the inner.

This species is an erect perennial to 80 cm, loosely woolly-hairy when young. Lower leaves pinnately veined, elliptic to oblanceolate, to 12 cm, serrate. Upper stem leaves much smaller. Involucral bracts with stalked glands and not woolly-hairy. Common. Thin woods, pine barrens, uncultivated fields; Fla into eTex, sO, and seNY. July–Oct. *Heterotheca m.* (L.) Shinners.

In *C. gossypina* (Michx.) Ell. phyllaries are conspicuously woolly-hairy. Common. Sandy pinelands, scrub oak; Fla into sAla, CP of Ga, and seVa. Aug–Oct. *Heterotheca g.* (Michx.) Shinners.

398   Grass-leaved Golden-aster
Pityopsis pinifolia (Ell.) Nutt.
Glabrous perennial to 80 cm, with grasslike leaves. Rare. Sandy soils, scrub oak, thin pinelands, uncultivated fields; nCP of Ga, and along the Fall Line to halfway across NC. Aug–Sept. *Chrysopsis p.* (Ell.) Nutt.; *Heterotheca p.* (Ell.) Ahles.

Other species with grasslike leaves but with silvery-silky hairs on stems and leaves include *Pityopsis graminifolia* (Michx.) Nutt. with no glands on peduncles and phyllaries, and no stolons. Common. Usually dry places in thin woods or open; Fla into eTex, sO, and Del. June–Nov. *Heterotheca g.* (Michx.) Shinners;

Chrysopsis nervosa (Willd.) Fern. In *Pityopsis graminifolia* var. *tenuifolia* (Fern.) Semple & Bowers there are small stalked glands on the phyllaries and none on the pedicels. Common. Similar places; Fla into sArk, CP of Ga and swSC. July–Nov. *Heterotheca microcephala* (Small) Shinners. In *Pityopsis adenolepis* (Fern.) Semple glands are on peduncles as well as the phyllaries. Common. Thin woods, usually sandy soils; Fla into sMiss, Pied of Ga, and seVa. July–Oct. *Heterotheca a.* (Fern.) Ahles.

399   Camphorweed

Heterotheca subaxillaris (Lam.) Britt. & Rusby

Plants annual or biennial, glandular and sticky, with a camphorlike odor when crushed. Stems to 2 m, erect, ascending, or decumbent, the former type usually inland and W of the Miss R, the latter along the coasts. The tall western form of this species was introduced around Athens, Ga, and Spartanburg, SC, before 1950. It is aggressive, occupying abandoned farmlands, thin woods, and other open areas, and is spreading rapidly. Often abundant in fields the first year after crops. Whereas it occupied only about 1500 square miles around Athens in 1954, it now occurs in much of the Pied of Ga. Heads of flowers numerous. Ray flowers essentially without a pappus. Common. Open areas; Del into Fla, Ga, and Tex, on W into Ariz and N into Kan and Ill. Flowering in frost-free periods but most abundantly July–Oct. Includes *H. latifolia* Buckl.

400   Field Goldenrod

Solidago canadensis L.

Placing some plants in *Solidago* with certainty can be difficult because distinguishing features are often troublesome. Identification to species can be even more difficult because many of them are quite variable, hybrids can be involved, and differences may depend on features confusing to interpret. Usually Goldenrods may be recognized as follows: Perennials; leaves alternate, simple, and entire to variously toothed; phyllaries overlapping and of several lengths; heads with both disc and ray flowers; pappus a single circle of bristles; style with flattened branches that are glabrous inside and finely hairy outside.

Plants of this species to 2.5 m, quite variable, with perhaps 5 var. Stems with sparse to abundant appressed fine hairs. Blade of lower leaves narrowly lanceolate, acuminate at tip, with 2 prominent lateral veins and short petioles. Common. Moist to dry open places and thin woods; Fla into Tex, Ariz, eND, nMinn, swQue, Lab, and Nfld. Aug–frost. *S. altissima* L.

In *S. stricta* Ait. plants are 0.3–2 m, with long stolonlike rhizomes, glabrous. Blade of basal leaves thick and firm; lowest ones oblanceolate to elliptic-oblanceolate, sometimes very narrow, 6–30 × 0.3–2(3) cm; stem leaves abruptly reduced and sessile; upper ones numerous, small, and pressed against the stem. Flowers in narrow panicles; rays 3–7, disc flowers 8–12. Common. Sandy usually moist places; open pinelands, savannas; Fla into eTex, CP of Ga, and pine barrens of NJ. Aug–frost.

401   Wreath Goldenrod

Solidago curtisii Torr. & Gray

One of a few *Solidago* species with flowers in short axillary clusters, the upper clusters sometimes considerably longer and the inflorescence appearing more terminal than axillary. Plants 0.3–1.5 m; main stems angled by fine lines running from petiole bases, not glaucous. Leaf blades 3–10 times as long as wide, largest ones elliptic-lanceolate to oblanceolate, and pinnately veined from a single prominent midvein. Involucre (2.5)3–5(6) mm; ray flowers mostly (2)3–5(6); disc flowers mostly 5–9. Perhaps best treated as a var. of *S. caesia*. Occasional. Rich deciduous woods, ravines, slopes; n half of Ga Pied into nMiss, cWVa, and nwSC. Aug–Oct.

In the similar *S. caesia* L. plants 0.3–1 m; main stems glaucous, terete, and lacking fine lines like those in above species; involucre 3–4.5 mm; ray flowers (1)3–4(5), disc flowers mostly 5–7. Common. Similar habitats; Fla into eTex, Wisc, sQue, NS, and Va. Aug–Oct. In *S. lancifolia* Torr. & Gray inflorescence appears more terminal than axillary; involucre 4.5–7 mm; ray flowers 5–8, disc flowers 6–12. Mostly above 5,000 ft; wNC into eTenn, and swWVa. Sept.

402   Wrinkle-leaved Goldenrod
Solidago rugosa Mill.
Plants 30–150(250) cm from long creeping rhizomes. Stems with spreading hairs. Leaves nearly sessile, numerous, crowded; blades hairy at least below, with prominent teeth, and 1 main vein with several laterals. Ray flowers 6–11, disc 4–8. Achenes short-hairy. A quite variable species usually divided into ssp. and vars. Common. Thin woods, moist to wet shrubby areas, meadows, swales, freshwater marshes, bog edges; nFla into Tex, Mo, Mich, sOnt, Nfld, Mass, and NC. Aug–Oct.

403   White Goldenrod; Silverrod
Solidago bicolor L.
Easily recognized by its narrow elongate panicle and flower heads with 7–14 white or whitish rays. Plants to ca 90 cm, phyllaries whitish to straw-colored except for the light green tip; disc flowers light yellow, inconspicuous, about the same number as rays. Dry sterile soils in open or thin woods; nGa into nAla, Mo, sOnt, swQue, NS, and ceNC; sMiss; nLa. Sept–Oct.

404   Seaside Goldenrod
Solidago sempervirens L.
Plants 0.4–2 m with a coarse rootstock and without long thin rhizomes. Leaves abundant; blades fleshy, smooth, entire, tapered to base. Rays 7–17, disc flowers 10–22; pappus of prominent capillary bristles. Common. Edges of salt marshes, bay shores, swales, minidunes, overwash areas. Nfld into Fla and seTex; into tropical Amer. Aug–Nov.

405   Skunk Goldenrod
Solidago spithamaea M. A. Curtis
Plants 10–40 cm. Most leaves basal, upper one not clasping stem, glabrous or nearly so, smooth, sharply serrate; largest blades elliptic to ovate or subrhombic, 5–10 × 1.5–4 cm. Inflorescence a dense corymb to 10(15) cm across; involucre 5–6 mm; phyllaries firm, narrow, green-tipped, not striped longitudinally; disc flowers mostly 20–60; rays ca 8(13), 2–3.5 mm; achenes short-hairy, pappus 2.5–3.5 mm. Very rare. Rock crevices at upper alt in BR of NC and Tenn, notably on Grandfather and Roan Mts. July–Aug.

S. rigida L. (Stiff Goldenrod) is a more common Goldenrod with inflorescence a dense corymb 5–25 cm across but plants 0.5–1.5 m; leaves scabrous, upper ones clasping stem, largest blades 6–25 × 2–10 cm. Phyllaries 6–8 mm, longitudinally striped; disc flowers 17–35, rays 7–14; achenes glabrous; pappus 4–5 mm. Common. Dry areas; thin woods, prairies and other open areas, especially in sandy soils; cAla into NM, Alta, NY, wMass, RI, cSC, and nGa. Aug–Oct.

406   Flat-topped Goldenrod
Euthamia tenuifolia (Pursh) Nutt.
Euthamia species share many features with *Solidago* and often have been included in that genus. The distinctive flat-topped inflorescences and fine glandular dots on the leaves distinguish *Euthamia;* only *Solidago odora* has such glands. Leaves alternate, the lower usually falling early. Phyllaries many, overlapping.

   E. tenuifolia plants 0.3–1 m; leaf blades 1–4 mm across and 20–50 times as long as wide, with 1 main vein. Disc flowers 5–7(9); rays (8)10–16. Common. Brackish and freshwater marshes, roadsides, swales, slough margins, meadows, thin woods, overwash areas; NS into Me, Va, nFla, and seLa. Aug–Oct.

   E. minor (Michx.) Greene is similar in that plants are 0.3–1 m; leaf blades have 1 main vein, are 1–2(3) mm across, and 20–50 times as long as wide. It is distinct in having 3–4(5) disc flowers and 7–11(13) rays. Common; the most abundant of our species. Sandy places, especially near the coast; CP from Md into Fla and seLa. Aug–Oct. In *E. graminifolia* (L.) Nutt. (Common Flat-topped Goldenrod) plants are 0.3–1.5 m. Leaf blades 3–10 mm across and only 10–20 times as long as wide, with 3 main veins. Ray and disc flowers total 20–35 or more. Common. Similar habitats; Tex into Okla, Kan, BC, Que, Nfld, Va, wNC, cTenn. Aug–Sept. In *E. leptocephala* (Torr. & Gray) Greene (Mississippi Valley Flat-topped Goldenrod) plants are 0.3–1 m. Leaf blades 3–6 mm across and 10–20 times as long as wide. Ray and disc flowers total 13–19. Open, often moist and sandy places, and thin woods; neFla into Tex, Ark, Mo, and cTenn. Aug–Sept.

407   White-topped Aster
Sericocarpus asteroides (L.) B.S.P.
Perennial 15–60 cm. Basal leaves usually present; largest blades 10–45 mm across, broadly oblanceolate to obovate or elliptic, margin of some or all toothed, base of some tapered onto a petiole. Flower heads in corymblike clusters; ray flowers 3–8; disc flowers 9–20. Common. Dry woods, clearings, roadbanks; nwFla into cwAla, sO, sMe, Va, csSC, and cGa. June–July. *Aster paternus* Cronq.

408   White-topped Aster
Oclemena reticulare (Pursh) Nesom
Conspicuous perennial 40–80 cm with 1–several stems from a rootstock; leaves and at least the upper part of the stem are glandular and hairy. Rays over 10 mm; pappus of two distinct series, the outer of short bristles and the inner of elongate capillary bristles. Common. Low pinelands; Fla into seGa, and s tip of SC. Apr–Aug. *Aster reticulatus* Pursh; *Doellingeria reticulata* (Pursh) Greene.

409 Bushy Aster

Symphyotrichum dumosum (L.) Nesom var. *dumosum*
Perennial, usually with creeping rhizomes; stems slender and much-branched, spreading or ascending to 1.5 m. Blade of stem leaves scabrous above, not auriculate-clasping; blade of middle and upper stem leaves sessile, soft, and flexible, linear to lance-linear to narrowly elliptic, and entire or nearly so. Flowering heads abundant and panicled; involucre 4–6 mm, phyllaries acute but not sharp-pointed. Rays white to blue or lavender. A very common plant in old fields, pastures, meadows; less abundant in thin woods and marshes; Fla into eTex, csMo, Va, WVa, ePa, nO, sMich, sOnt, and swMe. Aug–Oct. *Aster dumosus* L.

  S. pilosum (Willd.) Nesom var. *pilosum* is similar but the leaves often are not scabrous above and the phyllaries are sharp-pointed. A variable species with 3 vars. Common. Similar habitats; SC into nwFla, Ark, neOkla, seKan, w and seMinn, Wisc, seOnt, NS. Aug–Nov. *Aster pilosus* Willd.

410 Swamp Aster

Symphyotrichum puniceum (L.) A. & D. Löve var. *puniceum*
This is one of several species with auriculate-clasping leaves. A perennial with a stout stem to 2.5 m from a short stout rhizome: stem with spreading hairs, at least beneath the heads, and no glands. Leaf blade widely serrated, acute, gradually narrowed to the base, 7–16 cm; phyllaries, at least the inner ones, long-acuminate to long-tapering. Common. Moist thin woods, swamps, wet meadows, and other open places; Pied of Ga into neArk, Neb, Sask, sMan, Nfld, and Pied of SC. Sept–Oct. *Aster puniceus* L.

  S. premanthoides (Muhl. ex Willd.) Nesom is similar but has long rhizomes and grows to only 1 m. The leaf blade is usually abruptly contracted into the base and the phyllaries are acute to barely obtuse. Occasional. Similar places; nw and cwNC, into eIa, seMinn, Del, eMass, and csVa. Aug–Oct. *Aster p.* Muhl. ex Willd.

411 Annual Saltmarsh Aster

Symphyotrichum subulatum (Michx.) Nesom
To 1.5 m with a short taproot, glabrous, main stem straight or nearly so. Leaf blades sessile and well spaced; linear, entire, to ca 20 × 1 cm. Flower heads few to many but uncrowded; involucre 5–8 mm, phyllaries acuminate, overlapping; pappus a single circle of capillary bristles; rays 15–50, bluish to sometimes white, ca 3 mm. Common. Salt and brackish marshes, mud flats, brackish swales and sloughs; sMe into Tex. July–Nov. *Aster subulatus* Michx.

412 New York Aster

Symphyotrichum novi-belgii (L.) Nesom var. *novi-belgii*
Slender to stout perennial 0.2–1.4 m from elongate rhizomes. Leaves sessile; blades lanceolate to elliptic, 4–17 × 0.4–2.5 cm, base eared and clasping stem, entire to sharply serrate, glabrous except margins, which are stiffly ciliate. Flower heads several to many, involucre glabrous, 5–10 mm; rays 20–50, blue or occasionally white to rose, 6–14 mm; disc corollas yellow to light red. Common. Salt and brackish marshes, swamp borders, pond shores; Nfld–csSC. July–Oct. *Aster n-b.* L.

  S. novae-angliae (L.) Nesom (New England Aster) is similar but upper leaf blade surface is scabrous or stiffly appressed-hairy and the phyllaries and/or peduncles are glandular. Leaves sessile, blade 3–12 × 0.6–2 cm. Rays 45–100, bright reddish-purple to rosy. Occasional. Moist places—meadows, pond shores, sloughs, thin woods, low shrub areas; wNC into nAla, ND, Wyo, Vt, Mass, and DC; cAla to cMiss. *Aster n-a*. L. *Symphyotrichum patens* (Ait.) Nesom is also similar but leaf blades are cordate-clasping, hairy or scabrous above, 2.5–15 × 0.8–4.5 cm; phyllaries barely glandular to glandular and/or short-hairy. Rays 15–30, blue to rarely pink. Rare. Usually dry places—thin woods, old fields, meadows; nFla into Tex, Kan, NH, and NC. Aug–Oct. *Aster p*. Ait.

413   Perennial Saltmarsh Aster

Symphyotrichum tenuifolium (L.) Nesom

Glabrous, 20–70 cm with a single main stem arising from a slender creeping rhizome, stems usually zigzag, often appearing dichotomously branched. Leaves few; blades succulent, linear or nearly so, lower ones falling early. Flower heads few to many, scattered; involucre 6–9 mm, phyllaries firm, overlapping, sharply acute to acuminate; rays 15–25, blue to pink or nearly white, 4–7 mm. Common. Salt and brackish marshes, sand-mud flats; Mass–seTex. June–Dec. *Aster tenuifolius* L.

414   Many-flowered Aster

Symphyotrichum lateriflorum (L.) A. & D. Löve

This species is a perennial 0.3–1.2 m. Blades of midstem leaves entire to serrate, sessile or nearly so, linear to lanceolate or nearly rhombic, 5–15 × 0.5–3 cm. Flower heads usually small, numerous; involucre 4.5–5.5 mm, glabrous; phyllaries acute to obtuse, overlapping in several layers, with evident fairly broad green tips. Rays 9–14, white to slightly purple, 4–6.5 mm; disc corollas lobed half to three-quarters way to base, the lobes recurved. Common. Usually in dry places—thin woods, fields, meadows, sloughs, pond shores; cFla into Tex, Mo, Minn, sQue, NS–nSC. Aug–Nov. *Aster lateriflorus* (L.) Britt.

   S. ericoides (L.) Nesom (Heath Aster) also has small flower heads but is easily separated by its phyllaries with coarsely ciliate margins. Plants 0.3–1 m, colonial by elongate creeping rhizomes. Leaf blade to 60 × 7 mm; lower to middle ones often soon fall. Rays 8–20, white (blue-pink). Occasional. Dry places—thin woods or in open; sMe into Del, nVa, Tenn, Tex, Ark, SD, sSask; nMex–seAriz. Sept–Nov. *Aster e*. L.

415   Wood Aster

Eurybia divaricata (L.) Nesom

Perennial to 1 m; a quite variable species. Leaves petioled. Blades glabrous or with some long appressed hairs; blade of lower leaves ovate with a cordate base and acuminate tip; blade of lowest leaves frequently smaller than those above and often dying early; blades of upper leaves progressively less cordate and smaller toward stem tip, petioles shorter. Flower heads in a corymblike arrangement, sometimes elongated; involucre 5–10 mm, outer phyllaries to ca 2.5 times as long as wide, tips green and very short. Rays white, (10)12–16(20), 10–20 mm. Common. Woods; chiefly in mts; Ga into Ala, O, Pa, Del, and SC. Sept–Oct. *Aster divaricatus* L.

   Symphyotrichum undulatum (L.) Nesom also has leaf blades with a cordate base. Plants 0.3–1.2 m, from a branched crown of short rhizomes. Stems densely hairy to occasionally nearly glabrous below. Leaf blade 3.5–14 × 1.5–7 cm, upper ones extremely variable in size and shape. Flower heads in an open panicle that also contains numerous small bracts; phyllaries 3 times or more as long as wide; rays 10–20, blue to lilac. Occasional. Similar places; NS–SC. Aug–Nov. *A. undulatus* L.

416   Flat-topped White Aster
Doellingeria umbellata (Mill.) Nees
Perennial (0.4)1–2 m from well-developed creeping rhizomes; stems glabrous or nearly so below inflorescence. Leaves petioled; blade entire, midstem ones 40–160 × 7–35 mm, most narrowly elliptic to lance-elliptic. Heads usually numerous in a compact corymb; phyllaries green, relatively thin, rarely any over 1 mm across, acute to obtuse, base cuneate to rounded. Flowers 23–54 per head, rays (6)7–14, white, 3–8 mm, pappus double, the outer ones less than 1 mm, the inner much longer, of different lengths, and thickened toward tips. Common. Moist low places; nGa into neAla, Minn, Nfld, Va, and wNC. Aug–Oct. *Aster umbellatus* Michx.

417   Robins-plantain
Erigeron vernus (L.) Torr. & Gray
Members of this genus have many narrow pink, violet, purple, or white ligules and yellow or yellowish disc corollas. The pappus is a single circle of capillary bristles.

This species is a perennial to 60 cm with prominent basal leaves but only a few nonclasping stem leaves. Rays 25–40, white or rarely lavender. Common. Moist places, savannas, open pinelands, bogs; Fla into sLa, cCP of Ga, and seVa. Feb–June.

E. pulchellus Michx. also has prominent basal leaves, but the stem leaves are wider and clasp the stem. Plants have stolons and the flower heads, which are solitary or few, are much larger. Rays 50–100, blue-purple or sometimes pink or white. Common. Wooded slopes, meadows, roadside banks; Ga into eTex, ceMinn, and sMe; uncommon in Pied, rare in CP. Apr–June.

418   Daisy Fleabane
Erigeron philadelphicus L.
Biennial or short-lived perennial to 1 m with a few basal leaves; stem leaves several and clasping. Phyllaries over 4 mm. Rays 150 or more, 5–10 mm, pink or light rose to whitish. Occasional. Fields, rich open woods, rich waste places; Fla into se and cnTex, eKan, eND, BC, and Nfld. Mar–June.

E. quercifolius Lam. is similar but usually shorter. Phyllaries under 4 mm. Rays 100–150, only 3–5 mm, violet to blue. Occasional. Moist open pinelands or savannas, fields, roadsides; Fla into sLa, sGa, and seVa. Mar–June, rarely in fall.

419   Daisy Fleabane
Erigeron strigosus Muhl. ex Willd.
Annual or rarely biennial 30–70(90) cm, at least midstem with short appressed hairs. Midstem leaves well separated, sessile, linear to lanceolate, entire to some-

what serrate, base not clasping, to ca 15 × 2.5 cm, basal ones larger. Involucre 2–4 mm, obscurely glandular. Rays 50–100, to 6 × 1 mm, white or rarely bluish-tinged; disc 5–10 mm across; pappus of disc flowers double. Common, especially in disturbed sites. Roadsides, old fields, waste places, thin woods; a weed over most of the US and sCan. Apr–June, with a few plants blooming in Oct.

E. annuus (L.) Pers. has a similar aspect but is more robust, to 150 cm, midstem leaves somewhat crowded, to ca 10 × 7 cm, blade coarsely toothed. Stems and leaves with conspicuous spreading hairs. Heads several to very many, involucre 3–5 mm; rays 80–125, 4–10 × 0.5–1 mm. Occasional. A weed over most of the US and sCan. Apr–June, rarely in Oct.

420   Camphorweed; Marsh-fleabane
Pluchea odorata (L.) Cass.
Members of this genus are strongly aromatic; leaves alternate; phyllaries of several lengths and overlapping; no rays, no receptacular bracts, and pappus of a single ring of fine bristles. There are 7 species in the eUS.

This species is a fibrous-rooted annual to 1.5 m. Leaves petioled or sometimes the upper ones tapered to a narrow base; blade lanceolate to elliptic or ovate, most 4–15 × 1–7 mm. Central clusters of flower heads exceeded by some of the lateral flowering branches; involucre 4–7 mm, phyllaries usually pink to purple, at least at the tips, bearing small several-celled glandular-sticky hairs. Corollas rose-purplish. Common. Salt and brackish marshes, sloughs and swales, salt flats, rarely in freshwater marshes; Mass into NC, Fla, and Tex; W. Indies; nS. Amer; also scattered inland stations W of Miss R. Aug–Oct. *P. purpurascens* (Sw.) DC.

Camphorweed; Stinkweed
Pluchea camphorata (L.) DC.
Plants to 2 m, nearly glabrous, annual or perennial. Leaves petioled. Central cluster of flower heads overtops lateral clusters; involucre 4–6 mm, sometimes purplish; phyllaries finely glandular, lacking several-celled hairs. Disc 3–6 mm across. Occasional. Freshwater habitats—marsh edges, meadows, swales, swamps; Fla into eTex, eOkla, sIll, sO, Del, and NC. Aug–Oct.

421   Stinkweed
Pluchea foetida (L.) DC.
Perennial to 1 m from a fibrous-rooted crown or short rhizome, glandular and often somewhat cobwebby with short hairs; stems not winged. Leaves sessile; blade 4–10(13) × 1–4(4.5) mm, oblong to elliptic, ovate, or ovate-lanceolate, usually broad-based, and clasping stem. Heads several to many in short broad flat-topped clusters or sometimes storied; phyllaries 5–8 mm, overlapping in several series, middle ones less than 2 mm across. Disc 6–12 mm across; corollas creamy-white. Common. Wet soils—meadows, pond edges, swampy woods, freshwater marsh edges. Mainly in CP—sNJ into Fla, eTex, and sArk. July–Oct.

P. rosea Godfrey is quite similar but corollas are rose-pink to rose-purple. Phyllaries barely to not overlapping. Disc 5–9 mm across. Common. In woods or open—wet pinelands, pond shores, intermittent ponds, swales, ditches, poorly drained woods; seNC into Fla, and eTex; Mex, W. Indies. June–July.

422   Black-root

Pterocaulon virgatum (L.) DC.

Perennial from large dark roots. Leaf undersides and stems with the feel of kid leather because of short, densely felted, and light-colored hairs. Flowers very small, these in tight heads which in turn are in tight elongate terminal clusters. Three clusters are shown in the photograph. Common. Sandy pinelands and open sandy areas; CP—seNC into swGa and Fla. May–June. *P. undulatum* (Walt.) C. Mohr; *P. pycnostachyum* (Michx.) Ell.

Individual leaves of some species of the related *Antennaria* and *Gnaphalium* are similar. In *G. purpureum* L. plants look like small forms of *Pterocaulon* but with the flower heads less densely arranged. Common. Thin woods, various open habitats such as yards and fields; Fla into Calif, BC, cIll, and Me. Mar–June.

423   Pussy-toes

Antennaria plantaginifolia (L.) Richards.

Male plants are shown in the picture; the female plants are usually taller. The stolons, mats of hairs on the stems and underside of the 3–5-nerved leaves, and the several heads of flowers identify this species. Common. Dry situations, in thin woods or open; nFla into La, eOkla, Minn, and seMe. Mar–May.

A. neglecta Greene is similar but the leaves are narrower and have only 1 vein, or rarely an additional 2 obscure veins. Occasional. Similar places; Va into eTenn, Ky, Kan, Ariz, Calif, Yukon, and Nfld. Mar–May. *A. solitaria* Rydb. is usually smaller and has only 1 head of flowers at tip of stem. Occasional. Woods, occasionally thin and sometimes alluvial. Ga into La, sInd, sePa, and Md. Mar–May.

424   Facelis

Facelis retusa (Lam.) Schultz-Bip.

Slender simple erect annual 3–30 cm, or more often with 1–several decumbent branches from near ground; stem loosely white-woolly. Leaves numerous, ascending, alternate, simple; most blades 7–20 × 1.5–4 mm, linear-spatulate, entire, summit mucronate and nearly truncate; upperside green and nearly glabrous; underside white-woolly. Heads with disc corollas only, leaves crowded around bases of several; involucre 8–11 mm, phyllaries overlapping in several series. Bisexual flowers few, central, with slender tubular 5-toothed corolla; pappus of strongly plumose capillary bristles well surpassing the corollas. Native to S. Amer. Common. Weed at roadsides, borders, yards, and other disturbed sites; seVa into SC, Fla, and Tex. Mar–June.

425   Rabbit-tobacco; Everlasting

Pseudognaphalium obtusifolium (L.) Hilliard & Burtt ssp. *obtusifolium*

Members of this genus are woolly annuals or perennials. Leaves alternate, simple, entire, commonly numerous and rather small and narrow. Phyllaries overlapping; flowers many, all disc, outer ones female; anthers tailed at base; pappus of capillary bristles.

This species is an annual or winter-annual (10)30–100 cm, plants with a fragrant balsamic odor. Stems have a tight mat of long whitish hairs. First leaves in a rosette, blade green and glabrous above, conspicuously and densely white-hairy below, narrowly lanceolate to narrowly oblanceolate. Pappus bristles fall separately or in small groups. Leaves are eaten by turkeys, the plants by deer in winter. Common. Fallow fields, pastures, roadsides, thin woods; Fla into csTex, cKan, seMan, and NS. Aug–Oct. *Gnaphalium obtusifolium* L.

Pseudognaphalium helleri (Britt.) A. Anderb. ssp. *helleri* is similar except stems have glandular hairs and are sticky to touch. Occasional. More likely in thin woods; Pied of Ga into neArk, seMo, sMe, neVa, and cNC; rare in sBR. *Gnaphalium h.* Britt.

426 Purple Cudweed

Gamochaeta purpurea (L.) Cabrera

Annual or biennial 10–40(100) cm. Basal leaves to 10 × 2 cm, forming a prominent basal cluster; blade spatulate to oblanceolate, apex rounded and usually mucronate, green above and densely matted with woolly white hairs. Heads numerous in a seldom-branched cluster; involucre 3–5 mm, somewhat leafy-bracted; pappus bristles united at base, falling as a unit. Native. Common. Widespread weed in yards, meadows, borders, and other places. Mar–Sept. *Gnaphalium purpureum* L.

Gamochaeta falcata (Lam.) Cabrera is also native and a common weed and is similar but the leaves on apical two-thirds of plant have blades linear to linear-oblanceolate and about equally hairy on both sides; basal leaf cluster either sparse or absent. Treated as a variety of *G. purpurea* by several manuals. Among the many thousands eradicated from our lawns during 45 years the 2 kinds are always quite distinct with no intermediates; therefore we choose to treat them as separate species. Common. Generally restricted to eVa into Fla and La. Mar–Sept. *Gnaphalium falcatum* Lam.

427 Yellow Leafcup; Bears-foot

Smallanthus uvedalius (L.) Mack. ex Small

Perennial to 3 m. Leaves opposite, palmately lobed or cut, the blades about as long as wide, their shape promoting the name "Bears-foot." Receptacular bracts absent, disc flowers sterile and with an undivided style, ray flowers fertile and producing rounded and faintly grooved achenes, pappus absent. Common, but only scattered in the CP. Usually in rich soil of low places, deciduous woods, pastures; Fla into cTex, seKan, ceO, sNY, and NJ. June–Oct. *Polymnia uvedalia* L.

In *Polymnia canadensis* L. the leaves are pinnately lobed; stems with small stalked glands; and rays white to pale yellow. Occasional. Rich woods, usually calcareous places; nwGa into nwArk, cIa, cMinn, sVt, Conn, and nwNC. July–Oct. In *P. laevigata* Beadle the leaves are usually smaller and pinnately lobed, stems glabrous, flower heads smaller, rays white, and achenes 5-angled. Rare. Rich woods; in nwFla, nwGa, cAla, seTenn, and seMo.

428 Starry Rosin-weed

Silphium asteriscus L. var. *laevicaule* DC.

Members of this genus are coarse erect perennials with yellow corollas, sterile disc flowers, no receptacular bracts, a pappus of 2 small awns or absent, and winged achenes.

To 2.5 m with well-developed alternate or opposite leaves to high on the stem, basal leaves usually absent. Receptacular bracts with small stalked glands on the back near the tip. Common. Dry thin woods, uncultivated fields; Fla into ceMiss, Pied and adjRV of Ga, cnNC, and SC. May–Sept. *S. dentatum* Ell.

In *S. a.* var. *asteriscus* the upper stem has coarse spreading hairs or is very rough to touch. Common. Dry woods and open places; cn and nwFla into nw and cGa; Pied of NC into neTenn and csVa. May–Sept. Other species with well-developed stem leaves but lacking glands on the bracts include: *S. trifoliatum* L. with the upper stem glabrous or nearly so, leaves rough and usually whorled. Occasional. Similar places; cnGa into sw and neInd, Pa, and neNC. June–Sept. *S. laevigatum* Pursh with essentially glabrous leaves and stems. Rare. SC into nAla, sInd, and sO. June–Sept.

429   Green-eyes
Berlandiera pumila (Michx.) Nutt.
Perennial to 1 m from a large root. Leaves alternate, coarsely crenate or only slightly lobed at the base, hairy on both surfaces. Phyllaries 4 mm wide. Pappus absent; ray flowers bearing fruits, disc flowers with stamens and pistils but sterile. When in bud the disc flowers are green, thus the common name. Occasional. Sandy soils in thin woods or open; Fla into seTex, CP of Ga, and upper CP of SC. Mar–frost.

B. subacaulis Nutt. has very similar flower heads but with almost leafless stems and the lower leaves pinnately lobed. Occasional. Dry pinelands; ne and cnFla and southward. Flowers in frost-free periods.

430   Chrysogonum; Green-and-gold
Chrysogonum virginianum L.
Perennial, at first stemless as shown in the picture, or very short-stemmed, later elongating to 5–40 cm, often trailing and rooting in open situations; in the shade becoming mostly erect. Leaves opposite, hairy, with winged petioles. Ray flowers bearing fruits, the disc flowers opening but sterile. Phyllaries overlapping. An excellent ornamental for open, sunny borders. Common. Well-drained soils in woods; SC into nwFla, seLa, seO, csPa, and Va. Mar–June; occasionally Nov. *C. australe* Alexander.

431   Wild-quinine; American Feverfew
Parthenium integrifolium L.
Coarse perennial to 1.1 m from a tuberous-thickened root. Stem single or branched near the top, glabrous to minutely hairy. Leaves alternate, firm, coarsely toothed. Rays 5, white, inconspicuous, 1–2 mm; flowers fertile; disc flowers sterile. Receptacular bracts present. Achenes conspicuously flattened. Pappus of 2–3 scales or awns. Common. Dry open woods, prairies; Pied of Ga into eOkla, seMinn, Pa, seNY, and ceNC. May–Sept.

In *P. auriculatum* Britt. the stems have coarse, spreading hairs. Rare. Dry woods, old fields; cnNC into cnVa. May–Aug.

432   Narrow-leaf Sumpweed
Iva microcephala Nutt.
Slender annual 0.4–1 m, stem simple, top wand-shaped. Basal to midstem leaves opposite, remainder alternate; blade linear, most 25–65 × 1–3 mm. Heads

numerous, phyllaries 4–5 and separate, nearly obovate, 1.5–2 mm. Slender receptacular bracts generally present; female flowers commonly 3, corolla tubular, ca 1 mm; male flowers in same head, mostly 4–6, style undivided. Occasional. Wet low places; pinewoods, disturbed sites; CP—Ala into Fla and SC. Sept–early Dec.

   I. angustifolia Nutt., another annual, is about as tall and leaf blade about as long but 1–8 mm across. Female flowers 1–2, tubular corolla ca 1.5 mm; male flowers 1–4. Occasional. Pinelands and most low places, often in disturbed soil; La into Tex, Okla, and Ark; introduced in Mo and in Wakulla Co., Fla. Sept–Nov.

433   Common Ragweed
Ambrosia artemisiifolia L.
Annual 30–100 cm, apical portion prominently branched. Leaves opposite below, alternate above, blade deeply dissected (1)2 times, broadly ovate to elliptic in outline, 4–10 cm, those on basal portion of stem petioled. Flowers all disc, male and female in separate heads; involucre of male flower heads 1.5–2 mm, female 3–5 mm. Waste places, disturbed soils, fields, cropland; most of the US and sCan. Mostly July–Oct.

434   Giant Ragweed
Ambrosia trifida L.
Annual, most 2–5 m, upper part of stems hairy, lower portion often glabrous. Leaves opposite, scabrous; blade sometimes broadly elliptic but generally ovate to suborbicular in outline, often 20 cm, serrate, palmately 3–5-lobed, or unlobed on small plants. Involucres of male heads 1.5 mm; fruiting ones 5–10 mm and several-ribbed. Common. Moist soil; waste places, cornfields; throughout the US except pen Fla. Late Sept–Oct.

435   Common Cocklebur
Xanthium strumarium L.
Coarse freely branched annual to 2 m from a taproot; highly variable. Leaf blade broadly ovate to roughly orbicular, base cordate or nearly so. Flowers unisexual. Male in many-flowered heads at end of branches, phyllaries in 1–3 series, the receptacle columnar; flowers die and usually fall before fruits drop. Female flowers 2 per head, enclosed by the involucre and forming a conspicuous 2-chambered bur bearing strong hooked bristles. Native of Europe. Common. Waste places, cultivated and old fields, gardens, pond shores, ditches, stable dune areas; Fla into Tex, Tenn, WVa, and Va; distribution N not clear, perhaps to Nfld. July–frost. *X. echinellum* Greene.

436   Eclipta; Yerbadetajo
Eclipta prostata (L.) L.
Little- to much-branched weedy annual bearing scattered stiff appressed ascending hairs. Stems to 1 m, weakly ascending to trailing and rooting at nodes. Leaves opposite, blade lanceolate to linear-lanceolate or lance-elliptic, 20–100 × 4–25 mm. Flower heads 1–3 in axil of one of a pair of leaves or on terminal peduncles 1–5 cm. Ray flowers female, rather numerous, rays white, ca 1 mm, slender; disc flowers many, bisexual, corollas dusky-white, 4-lobed. Receptacular bracts bristlelike; pappus absent or at most a conspicuous crown. Common. Pond

shores, stream banks, sloughs, ditches, freshwater marshes, depressions; Fla into Tex, Neb, Ind, sOnt, and Mass. June–frost. *E. alba* (L.) Hassk.

437   Tetragonotheca

Tetragonotheca helianthoides L.

The generic name is from Greek words meaning "four-angled case," which alludes to the 4 large phyllaries that enclose the unexposed flower buds in the head. Two such heads are seen in the picture. A perennial with 1–several erect to ascending stems, to 90 cm. Occasional. Dry soil, usually sandy; Fla into sMiss, Ga, and seVa; ceTenn. Apr–July.

   Heliopsis h. (L.) Sweet, Ox-eye, is similar but often taller, to 1.5 m, without the 4 large phyllaries, and with no hairs on the stems and leaves. Occasional. Open places and thin woods; Fla into NM, sSask, and seMe. June–Sept. Incl. *H. scabra* Dunal.

438   Black-eyed-Susan

Rudbeckia hirta L.

Members of this genus have alternate leaves, rays are yellow to orange or marked with purple or reddish-brown at the base. Disc flowers are fertile and accompanied by receptacular bracts, but the ray flowers are neither. The receptacle is strongly conic or columnar. Identification to species may be difficult. There are reports of livestock being poisoned from eating plants of some species.

   This species is a biennial or short-lived perennial with unlobed leaves and dark purple disc corollas with no pappus. The receptacular bracts are acute to sharp-pointed but not spine-tipped. Common. Old fields, pastures, roadsides, thin woods; Fla into csTex, BC, and Nfld. May–frost.

   In *R. mollis* Ell. stems are densely covered with spreading hairs, leaves are clasping, and receptacular bracts are obtuse. Dry places in thin woods or open; Fla into seAla, sGa, and swSC. June–Oct.

439   Narrow-leaved Cone-flower

Rudbeckia nitida Nutt.

Fibrous-rooted perennial to 1.3 m, essentially glabrous. Basal leaves long-petioled, the blade nearly linear to elliptic, to 25 × 6.5 cm. Heads solitary or few, peduncles long; disc light reddish-brown, to 4.5 cm at maturity; rays yellow, drooping, 2–5 cm, apical portion of receptacular bracts bearing sticky hairs. Rare. Moist low areas naturally but has been one of our hardiest ornamentals in Ga Pied for ca 35 years growing in well-drained loamy soil. Scattered in CP—Fla and Ga to eTex. Late June–Sept.

   R. maxima Nutt. is similar but 1–2.5 m, glaucous. Blade of basal leaves to ca 26 × 16 cm. Disc 4–8 cm at maturity. Occasional. Moist low areas; La into eTex, Okla, Mo, and Ark. June–Sept.

440   Cut-leaf Coneflower

Rudbeckia laciniata L.

Perennial from a woody base, most 0.5–3 m; stem glabrous, often glaucous. Leaves large, petioled; blade coarsely toothed, some or most blades from deeply cut to merely 3-lobed. Heads 1–many; rays yellow, drooping, 3–6 cm; disc yellow or grayish, 1–2 cm across, hemispheric; tips of receptacular bracts bearing short sticky hairs; pappus a short crown. Moist places; eUS W to Rocky Mts. July–Sept.

441 Purple Coneflower
Echinacea pallida (Nutt.) Nutt.
Perennial to 1.2 m. Leaves alternate, hairy above and below, lanceolate to linear-lanceolate, tapered at base, never serrate. Receptacular bracts have spine tips and are longer than the disc flowers. Rays 4–9 cm, deep pink to purplish-pink, fading with age. Pappus a toothed crown. Rare E of Miss R. Thin woods, rocky glades, prairies; cGa into eTex, eKan, nMich, and cnNY. May–July.

   In *E. purpurea* (L.) Moench the leaves are broadly to narrowly ovate, rounded at base, often serrate, and rough above. Rare E of Miss R. Similar places; swGa into neTex, eOkla, seIa, sMich, and csNC. June–Sept. In *E. laevigata* (C. E. Boynt. & Beadle) Blake the upper surface of the leaves is smooth. Rare. Fields, thin woods; neGa into sePa and cwSC. June–July.

442 Coneflower
Ratibida pinnata (Vent.) Barnh.
Perennial to 1.3 m from a strong rootstock. Leaves alternate, pinnately compound, with 3–9 mostly lanceolate leaflets, the upper ones small and sometimes simple. Rays 5–10, spreading or drooping, 3–5 cm; flowers sterile. Receptacular bracts not spine-tipped. Columnar receptacle. Achenes flattened and smooth. Pappus none. The crushed fresh receptacle has an aniselike odor. Occasional. Dry places, thin woods, in open, roadsides, fencerows, prairies; nwGa into cAla, eOkla, eSD, sOnt, and cO. May–Sept.

   In *R. columnifera* (Nutt.) Woot. & Standl. the leaves are mostly linear and the rays less than 3 cm. Achenes usually slightly winged. Rare. Prairies, roadsides, along railroads, usually calcareous soils; cAla into NM, Mont, sMan, and Ill; cGa. June–Sept.

443 Narrow-leaved Sunflower
Helianthus angustifolius L.
Members of this genus are mostly coarse erect plants with undissected leaves, flower heads 1 cm or more broad, light to dark yellow to reddish-purple corollas, sterile ray flowers, fertile disc flowers, receptacular bracts, nonwinged achenes, and a pappus of 2 awnless scales. Most species are difficult to name.

   This species is a perennial to 2 m with a single stem that is often much-branched in upper half. Leaves occur all along the stem, to 18 cm, usually less than one-tenth as wide, the few basal leaves (if any) much wider. Lobes of disc corollas usually purple. Common. Moist shady places or open depressions; Fla into sTex, seOkla, Tenn, NC, seVa, and seNY; Ohio R valley into O. July–frost.

   In *H. longifolius* Pursh, the basal as well as upper leaves are narrow and lobes of the disc corollas yellow. Rare. Dry rocky soil; cw and nwGa into ce and neAla; swNC. Aug–Oct.

444 Purple-disc Sunflower
Helianthus atrorubens L.
Perennial to 2 m, without rhizomes. Basal leaves large with petioles often as long as the blade, lower stem leaves only slightly smaller, the upper one much reduced.

Lobes of disc flower corollas red. Common. Thin woods in dry situations, uncultivated open places; SC into nwGa, seLa, Tenn, seKy, and Va. July–Oct.

Other similar species include: *H. silphioides* Nutt., with leaves gradually smaller upward and the petioles much shorter than the blades. Occasional. Thin woods or open; seTenn into nAla, cLa, Ark, and sIll. *H. occidentalis* Riddell, with disc flower corollas yellow and much-reduced stem leaves. Rare in the seUS. Dry often sandy soils, thin woods or open; nwFla into eTex, Minn, sWVa, DC, and nGa; ceNS. *H. radula* (Pursh) Torr. & Gray, with rays lacking or only 1–2 mm. Rare. Low open pine barrens; Fla into seLa and swSC.

445   Viguiera
Helianthus porteri (Gray) Heiser
Annual to 1 m, often occurring in spectacular flowering colonies. Stem much-branched. Lower or lower and middle leaves opposite; most lower blades 5–10 × 2–10 mm, short-petioled. Flower heads in great abundance, to 4 cm across from tip to tip of the rays; disc 7–15 mm across, conic; phyllaries loose and narrow but overlapping; all corollas yellow; rays ca 8, 10–18 mm; anthers dark brown; a receptacular bract clasping each achene. Common. On or about granitic outcrops. Pied of Ga and Ala; Alexander Co., NC. July–Oct. *Viguiera p.* (Gray) Blake.

446   Cucumber-leaved Sunflower
Helianthus debilis Nutt.
Annual or short-lived perennial with prostrate to erect stems to 1 m. Leaf blade scabrous or rough-hairy to nearly glabrous, 3–10 × 1.5–9 cm, base truncate to cordate; most petioles 1.5–7 cm. Heads on naked peduncles; phyllaries lanceolate, to 3 mm across; rays 11–21, 1–3 cm, central receptacular bracts inconspicuously short-hairy. Occasional. Beaches, dunes, grasslands, waste places; Fla into Tex; rare introduction inland. May–Oct.

447   Jerusalem Artichoke
Helianthus tuberosus L.
Perennial (0.7)1–3 m with well-developed commonly tuber-bearing rhizomes; stems stout but tall plants arch under weight of mature flower heads. Leaves numerous; largest blades 10–25 × 4–12 cm, scabrous above, serrate; petiole (1.5)2–8 cm, somewhat winged. Heads several to numerous; phyllaries dark green, apical portion loose but not reflexed; rays yellow, 10–20, 2–4 cm; disc yellow, 15–25 mm across. Occasional. Moist soil, waste places; widely planted for its tubers and escaping; e and cUS and adjCan. July–Sept.

448   Pale-leaf Sunflower
Helianthus strumosus L.
Rhizomatous perennial 1–2 m; stem glabrous below the inflorescence or with a few long hairs. Leaf blades entire to inconspicuously finely serrate, tapering onto the 6–30 mm petiole; upperside scabrous; underside pale, sometimes glaucous. Phyllaries nearly equal, lanceolate, somewhat loose, especially the long acuminate tips, shorter to slightly longer than disc. Disc and rays yellow, 8–15, 1.5–4 cm. Occasional. Woods, open areas; widespread in eUS, S into nFla and La. *H. saxicola* Small. Late June–Sept.

449   Woodland Sunflower
Helianthus decapetalus L.
Perennial 0.5–1.5(2) m from slender rhizomes; stem short-hairy in the inflorescence, otherwise glabrous. Leaves mostly evenly distributed along the stem; blade thin, pale and faintly to moderately scabrous beneath, serrate, broadly lanceolate to ovate, long-acuminate, most 8–20 × 3–8 cm, abruptly tapered onto the 15–60 mm petiole. Phyllaries very loose, often some spreading. Disc yellow, 1–2 cm across; rays yellow, 8–15, 1.5–3.5 cm. Common. Woods and along streams; widespread in neUS, S into NC, Pied of Ga, and cTenn. Sept–Nov.

450   Melanthera
Melanthera nivea (L.) Small
Coarse erect scabrous perennial 0.5–2 m. Leaf blades with 3 prominent main veins, narrowly to broadly ovate or triangular, often 3-lobed or hastate (sometimes with a long narrow terminal lobe and flaring lateral lobes), the better-developed ones mostly 5–15 × (2)2.5–10 cm. Heads 1–2 cm across. Achenes 2.5–3 mm, warty. Occasional. In a variety of habitats, moist to dry, natural to disturbed; CP—SC into Fla and La; widespread in tropical Amer. July–Sept; all year where freeze-free. *M. hastata* Michx.

451   Crown-beard; Wingstem
Verbesina occidentalis (L.) Walt.
Single-stemmed perennial to 2.5 m. Stems 4-winged. Phyllaries overlapping, less than 3 mm wide. Rays 2–5, more than 5 mm, unevenly spaced around the head. Fruit flattened, not winged, with 2 strong awns. Common. Woods, fields, pastures; less common in the CP; Fla into neMiss, sePa, and seVa. July–Oct.

   V. alternifolia (L.) Britt. ex Kearney has similar individual flowers but the heads are globose, the phyllaries reflexed, the rays 2–10, and the fruits 2-winged. Common, but rare in CP. Moist places, woods, swamps, marshes, pastures; CP of Ga into cLa, Mo, cKan, Ill, sOnt, seNY, and Pied of SC. July–Oct. *Actinomeris a.* (L.) DC.

452   Tickweed
Verbesina virginica L.
Perennial to 2.5 m, with a single, densely fine-hairy, winged stem. Leaves alternate, ovate to ovate-lanceolate, to light green beneath, to 25 cm, with winged petioles. Ray and disc corollas white; rays 1–5 and 5–10 mm. Fruits flattened, hairy, usually winged, with 2 short awns. Common. Fla into e and cnTex, seKan, e and ncKy, nwGa, Pied of Ga, and cNC; ceVa. Aug–Oct. *Phaethusa v.* (L.) Small.

453   Tall Coreopsis
Coreopsis tripteris L.
Members of this genus have opposite or rarely alternate leaves, ray flowers around the margin of the head, the rays usually yellow or rarely pink-purple or white. The disc flowers are fertile, the phyllaries are in 2 series and of 2 sizes, about 8 of each, the outer narrower and somewhat spreading, the inner broader and appressed. Narrow receptacular bracts are present. The achenes are flattened and usually winged. The pappus consists of 2 short rows of teeth.

This species is a perennial to 2.5 m. Leaves are distinctly petioled, compound, and have 3, rarely 5, linear to lanceolate or narrowly elliptic blades. Common. Thin woods, open places; Ga into seLa, neTex, eKan, Wisc, Mass, and NC. July–Sept.

454 Whorled-leaf Coreopsis
Coreopsis major Walt.
Perennial to 1 m. Although there appear to be 6 whorled leaves at each of the middle and upper nodes, the leaves are really opposite and deeply palmately divided into 3 segments, each segment 5–30 mm across. Stems and leaves glabrous to prominently hairy. Common. Thin woods, usually dry places; Fla into seLa, sO, sPa, cVa, and SC. May–Aug. *C. m.* var. *rigida* (Nutt.) Boynt.

C. verticillata L. also has sessile and segmented leaves that often appear whorled, but the leaf segments are under 2 mm across. Rays yellow. Occasional. Dry thin woods and pinelands; nFla into eArk, eTenn, eWVa, Md, c and eVa, and cnSC. May–Aug.

455 Swamp Coreopsis
Coreopsis nudata Nutt.
Perennial to 1.2 m from an elongate rootstock about 6 mm across. Leaves few, terete except toward base, 3 mm or less across. This is unlike our other *Coreopsis* species in having purplish-pink to reddish-purple rays. Occasional. Swamps, ditches, and other depressions; lower CP of Ga into Fla and sMiss. Apr–May.

C. auriculata L. is unusual in having stolons and usually having only the lower half of the stem with leaves, or the leaves basal only. Rays yellow. Occasional. Rich woods or openings; seSC into neLa, neMiss, Ky, sWVa, and seNC. Apr–June.

456 Large-flowered Coreopsis
Coreopsis grandiflora Hogg ex Sweet
Perennial to 1 m; stems leafy almost to summit; leaf blade deeply cut into linear-filiform to narrowly lanceolate segments, the lateral ones to ca 5 mm across, terminal segment to 10 mm across. Peduncles 5–20 cm, slender, less than half as long as the leafy part of stem. Occasional. Usually dry and sandy places; granitic rocks; CP and Pied of Ga and Ala; Tex, Kan, and Ark; spotty in NC, SC, Miss, and La. May–Aug.

C. lanceolata L. is similar in several aspects but plants 20–70 cm, stem leafy chiefly toward the base, and peduncles generally nearly as long as or longer than the leafy part of the stem. Occasional. Dry often sandy places; disturbed soils; Mich, n shore of Lake Superior S into Fla and NM. Apr–June.

457 Shepherd's-needle
Bidens pilosa L.
Members of this genus are annuals or perennials with opposite simple or compound leaves and disc flowers only or both ray and disc flowers. The head is surrounded by 2 well-separated circles of phyllaries. The achenes are flattened and without a neck or wings, in 2 species spindle-shaped and quadrangular. A pappus of 2–6 usually retrorsely barbed awns enable the achenes to cling to clothing, hence the name "Beggar's-ticks" applied to several species.

This species is an annual or short-lived perennial. The rays are white or nearly so, the leaves compound, and the achenes quadrangular and spindle-shaped. Occasional. Waste places, old fields, roadsides; Fla into seTex, sGa, and seNC. Mar–frost.

Tickseed-sunflower
Bidens aristosa (Michx.) Britt.
Annual or biennial to 1.5 m. Leaves once or twice pinnately compound, the segments lanceolate or lance-linear, acuminate and sharply serrate. Rays 10–25 mm. Achenes flat, the margins ciliate, mostly obovate to elliptic-obovate, 5–7 mm. Often forms dense colonies that are spectacular when in flower. Common. Marshes, meadows, ditches, open low ground; nSC into eTex, sMinn, Me, and NC; mostly absent from Appal Mts. Aug–frost.

    B. mitis (Michx.) Sherff is similar but the leaves are often more finely divided. The achenes are not ciliate on the margins or only slightly so and only 2.5–5 mm. Occasional. Fresh or brackish marshes or swamps; Fla into seTex, sGa, and eMd.

458   Wild-goldenglow; Bur-marigold
Bidens laevis (L.) B.S.P.
A glabrous perennial to 1 m, stems ascending or reclining and rooting at the nodes, often forming dense colonies. Leaves serrate and unlobed. Rays about 8, yellow, 15–30 mm. Recaptacular bract at base of each flower, tip reddish. Occasional. Marshes, margins of pools and streams, fresh or brackish; Fla into Tex, CP of Ga, sNH; Calif; locally inland to sInd and eWVa. Sept–frost.

    B. cernua L. also has simple leaves, which often clasp the stem and even may have opposite leaf bases united. Stems glabrous or rough-hairy. Rays seldom over 15 mm. Involucral bracts yellowish-tipped. Occasional. Low wet places, marshes, bogs; nwGa into Colo, Wash, BC, PEI, and cwNC. July–frost.

459   Annual Beggar's-ticks
Bidens frondosa L.
Annual, glabrous or nearly so, 0.2–1.2 m. Leaves pinnately compound with 3–5 lanceolate acuminate serrate leaflets to 10 × 3 cm; petiole 1–6 cm. Heads bell-shaped to half-spheres with only disc flowers, disc to 1 cm across; outer phyllaries 5–10, usually 8, green and somewhat leafy, margin ciliate at least at base. Achenes flat, narrowly cuneate, strongly 1-veined on each face, 5–10 mm, commonly dark brown to blackish. Pappus of 2 retrorsely barbed awns. Occasional. Waste places, especially wet soils; nFla into La. July–Oct.

    B. vulgata Greene is similar but may be distinguished by phyllaries 10–16(21), usually 13, and by achenes to 12 mm and commonly olive to somewhat yellowish. Occasional. Wet or occasionally dry waste places; wNC, Tenn, Kan, Wyo, Alta, sQue, Nfld, NS, and Va; Calif, Wash, Nev. July–Oct.

460   Swamp Beggar's-ticks
Bidens coronata (L.) Britt.
Glabrous annual or biennial 0.3–1.5 m. Leaves to 13 cm, blades with pinnately parted segments, mostly 3–7, lance-linear to linear, incised-dentate to entire, apex

acute to acuminate; petiole 3–15 mm. Heads with ca 8 rays 10–25 mm; disc 8–15 mm across. Achenes flat or nearly so, 5–9 mm, 2.5–4 times as long as wide, narrowly cuneate-oblong to cuneate-linear, dark; pappus with 2 short erect strong awns or awn scales. Common. Waste places; Va into Ky, Neb, Wisc, Ont, and Mass. July–Oct.

461   Spanish-needles
Bidens bipinnata L.
Glabrous or minutely hairy annual 0.3–1.7 m. Leaves 4- 20 cm including the 2.5 cm petiole, most 2–3 times pinnately dissected. Heads narrow, disc only 4.6 mm across at pollen-shedding; outer phyllaries 7–10, linear, not expanded toward tip, shorter than inner phyllaries. Achene linear, 4-sided, most 10–13 mm; pappus of 3–4 yellowish retrorsely barbed awns. Common. Moist to wet places; cFla into Tex, Kan, RI, and NC; Mex. July–Oct.

462   Annual Balduina
Balduina angustifolia (Pursh) Robins.
Members of this genus are easily recognized by their yellow ray corollas and the honeycombed receptacle surface.

In this species the stem has 0–20 branches. It resembles Bitterweed, *Helenium amarum*, but has narrow rays and fewer leaves and is less branched above. Both species are annuals and have linear leaves and yellow disc and ray corollas. Occasional. Deep well-drained sandy soils of pinelands, sandhills, and scrub oak; Fla into sMiss; lower CP of Ga. Sept–Nov. *Actinospermum angustifolium* (Pursh) Torr. & Gray.

One of the 3 species of the genus, *B. uniflora* Nutt., resembles *Helenium vernale* (see under this for characteristics) except receptacular bracts are conelike, persist on disc, and form a toothed honeycomb in which flowers are set. Occasional. Moist to dry pinelands and savannas; CP of NC into nFla and wLa. Aug–Oct. The third species, *B. atropurpurea* Harper, is also similar to *B. uniflora* but may be recognized by the dark purple disc flowers. Occasional. Pitcher plant bogs and other wet habitats; CP of SC, Ga, and nFla. Sept–Oct.

463   Galinsoga; Peruvian-daisy
Galinsoga quadriradiata Ruiz & Pavón
Freely branching somewhat hairy annual 20–70 cm. Leaves petioled; most blades 2.5–7 × 1.5–5 cm, ovate, coarsely toothed. Heads numerous on well-developed plants; peduncles slender, hairy; phyllaries 3–4 mm; rays white, mostly (3)5(6), prominently 3-toothed, 2–3 mm. Achenes black, quadrangular; slender scales of pappus taper to a short awn tip. Native of tropical Amer. Occasional. Weed in borders, gardens, yards, and waste places; Fla into Neb, Minn, Ont, Que, New Eng, and NC. Apr–Nov. *G. ciliata* (Raf.) Blake.

464   Barbara's-buttons
Marshallia graminifolia (Walt.) Small var. *cynanthera* (Ell.) Beadle & F. E. Boynt.
Members of this genus are glandular-dotted perennials with alternate leaves, peduncle over 5 cm, involucral bracts entire, no ray flowers, receptacular bracts present, achenes with 5 finely hairy angles and a pappus 5–6 short thin scales.

This species has linear to linear-lanceolate stem leaves, gradually reduced to mere bracts at top. Basal leaves are horizontally spreading, spatulate to oblong-ovate, and leave no old wiry fibrous bases. Stem usually branched about mid-way, each branch with 1 head of flowers surrounded by strongly acuminate to subulate-tipped phyllaries. Common. Thin longleaf pine areas, moist pinelands, bogs, swamps; Fla into se and ceTex and seGa. July–Sept. *M. tenuifolia* Raf.

In *M. ramosa* Beadle & Boynt. phyllaries are acute to barely obtuse and 4–6 mm. The stem is branched above the middle and bears 4–20 heads. Rare. Places dry except during rainy seasons; rocky outcrops, pine barrens; cCP of Ga. May–June.

Marshallia

Marshallia obovata (Walt.) Beadle & F. E. Boynt.
Fibrous-rooted perennials 15–60 cm, usually single-stemmed. Leaves with 3 longitudinal veins; blade 30–90 × (5)8–20 mm, oblanceolate or spatulate to elliptic. Heads solitary, most 20–35 mm across; phyllaries obtuse; receptacular bracts linear-spatulate and obtuse. Occasional. Old fields, meadows, thin woods, sandhills, and flatland pines; csVa into cs and seNC, cw and nwSC, most of Ga, seAla, and wFla. Apr–May.

465   Bitterweed

Helenium amarum (Raf.) H. Rock
Members of this genus have alternate stem leaves, usually yellow rays, no receptacular bracts, a pappus of papery scales, and truncate style tips.

An annual to 1 m, much-branched above, the leaves linear, to 7 cm, 1–4 mm wide, often with smaller axillary clusters of leaves. A serious pest in pastures. Although bitter and usually avoided by animals, it is often eaten when forage is scarce. The milk of cows that have grazed on the plant contains a bitter flavor and horses and mules have been reported poisoned by it. Common. Pastures, fields, roadsides, waste places; Fla into csTex, eKan, sInd, and Mass; rare in the mts and the northern parts of its range. May–frost. *H. tenuifolium* Nutt.

466   Sneezeweed

Helenium flexuosum Raf.
Perennial with 1–several finely hairy stems from a rootstock. Leaves alternate, base of the blades extending down the stems making them winged. Flower heads usually many and short peduncled. Pappus of scales. This and the species below are poisonous when eaten. Animals usually avoid eating the bitter plants but may do so when other forage is scarce. Common. Moist to wet meadows, pastures, waste places; nFla into eTex, seKan, sMich, O, swPa, and swMe. May–Aug. *H. nudiflorum* Nutt.

H. brevifolium (Nutt.) Wood is similar but has only 1–4 heads and blooms in the late spring. Rare. Bogs, swamps, depressions; cwGa into nwFla, seLa, and neAla; NC into seVa. May–June.

467   Spring Helenium

Helenium vernale Walt.
Perennial to 70 cm with 0–3 long glabrous branches, each with a head of flowers. Sometimes with more than one stem from the basal cluster of spatulate to linear

leaves. Leaves opposite. Achenes glabrous. Common. Wet pinelands, pond margins, open swamps, ditches; nFla into seLa, sGa, and seNC. Mar–May.

H. pinnatifidum (Nutt.) Rydb. is similar but the lower leaves are often lobed or cleft and the achenes are finely hairy on the ribs. Occasional. Similar habitats; pen, cn, and neFla into sGa and seNC. *Balduina uniflora* Nutt. is also similar but flowers July–Sept, has a honeycombed receptacle surface and thick- instead of thin-margined phyllaries. Common. Moist to wet savannas and pinelands, ditches; nFla into seLa, CP of Ga, and swSC; seSC into seNC. July–Sept. *Endorina u.* (Nutt.) Barnh.

468   Sneezeweed
Helenium autumnale L.
Fibrous-rooted perennial 0.5–1.5 m, finely hairy to nearly glabrous. Leaves many, blade-toothed, 4–15 × 1.4 cm, most elliptic to oblong or lanceolate, narrowed to a sessile base and decurrent on stem. Heads several to many; ray flowers female, rays ca 13–21, yellow, and most 15–25 mm; disc corollas yellow. Pappus of ovate to lanceolate scales tapering to a short awn, ca 1 mm overall. Common. Moist to wet meadows, pastures, waste places; nFla into eTex, seKan, sMich, O, swPa, swMe, and NC. July–Nov. *H. latifolium* Mill.; *H. parviflorum* Nutt.

469   Gaillardia; Fire-wheel
Gaillardia pulchella Foug.
Annual, or biennial in the warmer parts of our area, to 70 cm, decumbent to erect, branches few to many but the plant not compact. Leaves alternate, entire to serrate or pinnately cut. Flower heads long-peduncled, phyllaries overlapping. Rays 15–25 mm, red to purplish-red, or the tip yellow, or all yellow. Disc corollas the same color as the rays. Achenes 4-angled. Pappus a crown of 5–7 long tapering scales. Occasional. Fields, roadsides, dunes, prairies, usually sandy soils; Fla into Ariz, sNeb, La, CP of Ga, and seVa. Apr–frost. *C. picta* Sweet.

In *G. aestivalis* (Walt.) Rock the rays are 10–20 mm, yellow and tinged with red, sometimes absent. Disc corollas purplish-red or nearly maroon. Pappus of 7–10 scales. Occasional. Sandy soils in thin woods or open; Fla into eTex, sIll, La, nwGa; CP of Ga, into swCP of NC. May–frost. *G. lanceolata* Michx.

470   Common Yarrow; Milfoil
Achillea millefolium L.
Aromatic herbaceous perennial, often rhizomatous. Leaves alternate; blade finely and pinnately dissected. Heads numerous in a flat to round-topped cluster; involucre 3.5–5 mm; phyllaries overlapping in several series, dry with thin edges. Rays mostly 3–5, white (pink), 2–3 mm, flowers female; disc flowers 10–20, bisexual; receptacular bracts chaffy; pappus none. Occasional. In a variety of habitats, especially disturbed sites; nFla into La, cTex, sCan, Lab, and NC. Apr–Sept.

471   Ox-eye Daisy
Leucanthemum vulgare Lam.
Perennial to 1 m with short rhizomes. Leaves alternate, the numerous basal ones usually pinnately lobed or cleft. Flower heads 1–few, with no receptacular bracts. Rays 15–30, white, 10–25 mm. Disc flowers produce achenes, the corollas yellow.

Pappus absent. Useful as an ornamental, either in clusters or colonies, or as cut flowers, but can be a troublesome weed. Common. Fields, pastures, lawns, waste places; nearly throughout the US and from BC into Lab. Apr–July; sometimes in the fall. *L. leucanthemum* (L.) Rydb.; *Chrysanthemum l.* L.

472   Tansy
Tanacetum vulgare L.
Aromatic perennial to 1.5 m from stout rhizomes. Leaves numerous, twice–pinnately divided, glandular-dotted. Heads as many as 200 in a corymblike inflorescence; phyllaries dry and overlapping; receptacular bracts absent; rays none; disc flowers rich yellow, bisexual, and fertile. Achenes 2-ribbed; pappus a minute crown, sometimes almost absent. Occasional. Fields, roadsides, around old dwellings, waste places; Ga into Mo, Okla, BC, Nfld, Va, and nwNC. July–Oct.

473   Fireweed
Erichtites hieraciifolia (L.) Raf. ex DC.
Fibrous-rooted annual weed 0.1–2.5 m, glabrous to spreading-hairy throughout. Leaves numerous and evenly distributed, alternate, blades up to 20 × 8 cm, sharply serrate, often some auricled-clasping. Heads with swollen base, whitish; phyllaries a single series of narrow equal somewhat herbaceous bracts 10–17 mm. Receptacle flat or nearly so; ray flowers none; central flowers bisexual, corollas narrowly tubular, 5-toothed; these surrounded by numerous female flowers in several series, flowers with filiform-tubular corollas. Achenes with 10–12 ribs, 2–3 mm; pappus of numerous bright white long slender bristles that eventually drop. Common. In many wet to dry often disturbed habitats; Fla into seTex, Minn, wOnt, PEI, and NC. Sept–frost, or all year where frost-free.

474   Golden Groundsel
Senecio aureus L.
Members of the genus have all leaves alternate; phyllaries herbaceous and tender, most equal, ca 90 in a primarily single circle; both ray and disc flowers present, corollas yellow, disc flowers bisexual and fertile; receptacular bracts absent; pappus of capillary bristles.

This species is a perennial to 80 cm with stolons, lightly hairy at first, becoming glabrous. Basal leaf blades up to 11 × 11 cm, nearly circular, base of some cordate; stem leaves much reduced; petiole long. Heads several to many, disc 5–11 mm across, phyllaries strongly purple-tipped, rays 6–13 mm. Common. Moist woods and swamps; cnGa into seKy, Ill, nArk, eND, sOnt, Lab, Va, and cnSC. Apr–June.

S. obovatus Muhl. ex Willd., another perennial, has a similar inflorescence. Recognized by its well-developed stolons, obovate basal leaves, stem leaves rapidly reduced upward. Occasional. Rich woods, usually on slopes; nFla into Tex, eKan, sOnt, sNH, wNC, and cwSC. Mar–June. *S. glabellus* Poir. has similar flowers in a tighter inflorescence. A fibrous-rooted annual 15–90 cm, often forming dense stands. Stems usually single and unbranched below inflorescence. Leaves deeply pinnately lobed, the largest basal, to 20 × 7 cm, slightly but progressively reduced upward yet still conspicuous into the inflorescence. Suspected of being poisonous when eaten. Common. Wet places in open or in woods, especially in stream bottoms; Fla into eTex, eKan, swO, nw and ceGa, and seNC. Mar–Apr.

475 **Hairy Groundsel**
Senecio tomentosus Michx.
Perennial to 70 cm, spreading by basal offshoots and stolons, cottony-hairy, especially at base, this sometimes lost after flowering. Leaves mostly basal; blades
crenate to nearly entire, abruptly contracted onto the petiole. Phyllaries in 1 series
except for a few very short ones at base. Rays 5–10 mm. Pappus off-white. Occasional. Wet places in thin woods and open; chiefly on or near CP; sNJ into eNC,
cnFla, seTex, eOkla, and Ark. Apr–June. *S. smallii* Britt.

476 **Southern Ragwort**
Senecio anonymus Wood
Perennial to 70 cm, the stem densely woolly at base. Basal leaves cuneate, lanceolate, and crenate to serrate or once pinnately dissected. Common. Fields, pastures, open woods, savannas, roadsides; Fla into Ala, csKy, sePa, and sNS. Apr–
June.
    S. millefolium Torr. & Gray is similar but the leaves are twice pinnately divided
or are divided into finer divisions, less than 3 mm wide. Rare. On or near rock
outcrops in mts; neGa into swNC and nwSC.

477 **Yellow Thistle**
Cirsium horridulum Michx.
Members of the genus have no ray flowers, the leaves are spiny, and the pappus
consists of plumrose capillary bristles.
    This species is usually a biennial to 1 m. The phyllaries are spiny tipped and are
closely surrounded by a series of narrow, spiny-toothed leaves. The terminal head
may be 5 cm broad. Corollas may be yellow, purple, or rarely white. Plants at first
dense and unbranched, later often becoming branched and to 1.2 m. Common.
Roadsides, fields, and other open places; seTex into Ga, Fla, and SC; then into
sMe. Mar–June. *C. smallii* Britt.; *Carduus spinosissimus* Walt.

478 **Purple Thistle**
Cirsium carolinianum (Walt.) Fern. & Schub.
Biennial to 2.5 m and the flower heads on naked peduncles, no leafy bracts being
present beneath the heads. Heads usually less than 2.5 cm broad, the middle phyllaries tipped with spines about 3 mm. Stem leaves ca 10–30. Corollas bright red-
purple. Achenes 3–4 mm. Young stems of this and other species can be eaten after
peeling off the rind, cutting into pieces, and boiling in salted water. Occasional.
Thin woods, dry, often sandy or rocky soils; cwSC into eTex, sMo, sO, Ky, and
cNC. May–July. *Carduus carolinianus* Walt.
    C. virginianum (L.) Michx. is similar but the spines on the middle phyllaries
are only 2 mm or less and the stem leaves more abundant, ca 35–70. Occasional.
Moist to wet places, pine barrens, savannas, roadside dit hes; neFla into sNJ.
Aug–frost. *Carduus virginianum* L.

479 **Bull Thistle**
Cirsium vulgare (Sav.) Ten.
Biennial weed 0.5–1.5 m, stem with conspicuous spiny wings from the decurrent
leaf bases to next node or nearly so. Heads several, involucre 25–40 mm, phyllaries all tipped with long spreading spines. Occasional. Fields, meadows, pastures,

roadsides, and waste places; native of Eurasia, now widely distributed in N. Amer; S into Va, nSC, cMiss, La, and Ark. Aug–Oct. *C. lanceolatum* (L.) Hill misapplied.

480 Spotted Knapweed
Centaurea maculosa L.
Biennial or short-lived perennial 0.3–1.2 m. Leaves numerous, main stem ones deeply pinnately divided. Heads terminal on numerous branches; involucre 10–13 mm; phyllaries longitudinally striped with green, middle and basal ones with a short dark tip; pappus to 2 mm or rarely absent. Occasional. Native of Europe, commonly escaped in neUS, S into Va, cNC, neSC, Tenn, nArk; N into Que to BC; Calif. July–Oct.

481 Common Chicory; Blue-sailors
Cichorium intybus L.
Milky-juiced perennial to 1.7 m from a long taproot. Basal leaves numerous, the stem usually much-branched. Flower heads sessile or short-peduncled, 1–3 of them in axils of the much smaller upper leaves. All flowers with rays, these usually bright blue, sometimes pink or white. Achenes angled. Pappus of 2–3 rows of very short scales. Young leaves are used by some as a potherb. The water is usually poured off twice to remove the bitter taste. The ground roots are roasted and used as a substitute for or to flavor coffee. Occasional. Fields, roadsides, waste places; nearly throughout the US but less common in CP. May–frost.

482 Dwarf Dandelion
Krigia virginica (L.) Willd.
Members of this genus have milky juice, all flowers have yellow ligules, phyllaries are of equal size (or nearly so), and achenes are not beaked.
    This species is an annual 3–40 cm, lacking underground tubers; stems un-branched; leaves all basal. Phyllaries mostly 9–18, 4–6.5 mm at pollen-shedding, up to 9 mm and reflexed when achenes are mature; pappus of 5 short thin scales alternating with as many scabrous bristles several times as long. Common. Open places, often in sandy soils; Fla into se and cnTex, Wisc, sMe, and NC. Mar–June.
    K. occidentalis Nutt. is also an annual but only 2–20 cm, with phyllaries 5–8 and persistently erect. Rare. Prairies and other dry open places, often in sandy soil; swMo into adjArk and Kan; La into Tex. Apr–June.
    K. dandelion (L.) Nutt. (Dwarf Dandelion) is a perennial with slim rhizomes which frequently bear a globose to ovoid tuber a few cm underground. Phyllaries 7–15 mm. Occasional. Thin woods or open places; chiefly in sandy soils; Fla into se and cnTex, seKan, sInd, seTenn, cwGa, Pied of SC, and sNS. Mar–May. *Cynthia d.* (L.) DC.

483 Dwarf-dandelion
Krigia montana (Michx.) Nutt.
Perennial with leafy stems. The leaves are much alike and may be crowded near the base at first flowering. Peduncles single, arising from axils of ordinary foliage leaves. Phyllaries 4–8 times as long as wide. Occasional. Moist places on exposed rocks, stream margins; nwSC into cnGa and eNC. May–Oct. *Cynthia m.* (Michx.) Standl.
    K. biflora (Walt.) Blake is similar but the peduncles are 1–several, arising from

the axils of the upper, smaller leaves. Common. Rich woods, sometimes in fields and pastures; cwGa into neOkla, seKan, Ill, seMan, Mass, and swNC. Mar–July. *K. cespitosa* (Raf.) Chambers is also similar, but an annual with the phyllaries 1.5–3 times as long as wide. Common. Fields, pastures, disturbed ground; nwFla into se and cnTex, cKan, swInd, nw and Pied of Ga, and seVa. Mar–June. *K. oppositifolia* G. H. Weber ex Raf.; *Serinia o.* (Raf.) O. Ktze.

484   Cat's-ear

Hypochaeris radicata L.

Members of this genus recognized by having milky juice, all or most leaves basal, all flowers with rays, receptacular bracts chaffy, and pappus of plumose bristles.

This species is a hairy perennial 15–60 cm with all leaves basal, possibly some bracts on main stem, blades 3–35 × 0.5–7 cm. Heads a few to several, rays ca 4 times as long as wide. Native of Eurasia. Common. Weed in lawns, pastures, disturbed sites; widely established in US and sCan; S into NC, nFla, and La. Apr–July, sporadically until frost.

H. glabra L. is similar but is a taprooted annual to ca 40 cm and essentially glabrous; leaf blades 2.5–15 × 0.7–3.5 cm; heads solitary to several. Native of Europe. Occasional. Disturbed and waste places; NC into nFla and La. Apr–June, sporadically until Oct.

Two other species are similar but have a few conspicuous leaves at a few nodes above basal leaves. One, *H. brasiliensis* (Less.) Hook. & Arn., is a coarse perennial to 1 m, middle and lower phyllaries hairy, and rays yellow; leaf blades mostly 7–30 × 1.5–8 cm; heads several to numerous. Native of warm temperate S. Amer. Occasional. Weed of roadsides, fields, waste places, often in sandy soil; NC into nFla and La. Mar–June. *H. alata* (Wedd.) Griseb. The other, *H. microcephala* (Schultz-Bip.) Cabrera has glabrous phyllaries and white rays. Rare. Weed of roadsides and waste places; La into sTex. Mar–June.

485   Dandelion; Blowballs

Taraxicum officinale G. H. Weber ex Wiggers

Perennial from a deep taproot to 1 cm thick. Leaves all basal, barely lobed to sharply pinnately cut or divided. Flowers with rays only, many in each head, and surrounded by more than 1 set of phyllaries. Heads single, 2–5 cm broad, on hollow naked stems to 50 cm. Fruits olive green to greenish-brown, with a long neck and "parachute" at the top. Leaves are used as a salad and potherb and are best gathered young and tender. For a potherb the leaves are boiled in water for a short time; the water is changed once or twice if the bitter taste is undesirable. Common. Lawns, pastures, other open places; nearly throughout the US and sCan. Mar–Sept, occasionally in winter.

T. laevigatum (Willd.) DC. (*T. erythrospermum* Andrz. ex Bess.) is recognized by some as a species; the fruits are reddish-brown. Occasional. Similar places and distribution. Mar–Dec.

486   Spiny-leaved Sow-thistle

Sonchus asper (L.) Hill

Members of this genus are annuals or perennials with alternate leaves and milky juice. Heads composed of 80–250 yellow ray flowers. Pappus of capillary bristles only; achenes beakless, 3(4–5) mm, flattened, each face longitudinally ribbed.

This species is an annual to 2 m. Leaf blades pinnately lobed or obovate and lobeless, margin prickly toothed, auricles rounded on all leaves. Heads 15–25 mm across, involucre 9–13 mm in fruit; achenes ribbed longitudinally only. Native to Europe. Common. A weed widely distributed in the US and sCan. Apr–Nov.

S. oleraceus L. (Common Sow Thistle), another annual, is similar but can be recognized by the acute base on leaf auricles and by achenes transversely roughened with tubercles as well as longitudinally ribbed. Native to Europe. Common. A weed widely distributed in the US and sCan. Apr–Nov, all year in frost-free areas. *S. arvensis* L. (Perennial Sow Thistle), to 2 m, has heads 30–50 mm across and involucre 14–22 mm; achenes strongly roughened. Native to Europe. Common. A weed widely distributed in the US and sCan. July–Oct.

487   Wild Lettuce

Lactuca canadensis L.

Members of this genus have milky juice and all flowers have rays. Heads have 5–56 flowers, rays yellow, orange, blue, or purple; achenes flattened, the pappus a single circle of capillary bristles.

Annual or biennial to 3 m with green to reddish stems. Leaves oblanceolate to lanceolate, entire or toothed to pinnately lobed or sagittate. Heads with 13–22 flowers, rays yellow. Involucre 10–15 mm when achenes are mature. Achenes 5–6 mm including a narrow neck 2–2.5 mm, each side of the body with 1 main rib and sometimes a pair of indistinct ones. Pappus bristles 4.5–6 mm. Common. Fields, pastures, waste places, thin woods; Fla into eTex, Sask, and PEI. June–frost.

L. graminifolia Michx. has similar fruits but with mostly basal linear to oblanceolate rarely toothed leaves, blue to violet rays, and pappus bristles 7–8 mm. Occasional. Fla into Ariz, Pied of Ga, and sNJ. Apr–Sept. *L. serriola* L. has prickly margined leaves and 5 or more ribs on each side of the achenes. Throughout most of the US. Common in seUS. June–frost.

488   False-dandelion

Pyrrhopappus carolinianus (Walt.) DC.

Milky-juiced annual or biennial to 1.2 m from a taproot. Stem leaves 3–12, upper ones gradually smaller, basal ones oblanceolate to narrowly elliptic, pinnately lobed or dissected to merely toothed. All flowers with rays, these yellow or pale cream-colored. The longest phyllaries are 2-lobed or widest at the tip. Achenes not flattened, finely 5-grooved. Pappus a ring of very light tan capillary bristles. Common. Fields, pastures, roadsides; Fla into e and cnTex, seNeb swInd, Va, and Del. Mar–June.

489   White Lettuce

Prenanthes trifoliata (Cass.) Fern.

Perennial herb 45–120 cm with milky juice. Stem stout, glabrous. Leaves highly variable in shape, well over 1 cm across. Heads nodding; involucre glabrous, 10–13 mm, cylindric, with (7)8(9) phyllaries. Flowers (9)10–11(13) per head, all bisexual and with ligules. Achenes columnar or nearly so; pappus straw-colored to light brown, of numerous deciduous simple capillary bristles. Photograph of plant in shallow soil on rocks on summit of cliff overlooking ocean. Common. Woods;

Pa into Nfld and Md; S in CP into NC; S in mts into Ga, Tenn, and NC. Sept–Oct. *Nabalus trifoliatus* Cass.

490   **Poor-robins Hawkweed; Rattlesnake-weed**

Hieracium venosum L.

Members of this genus are perennials with milky juice. They have a short to elongate rhizome or rootstock and all are fibrous-rooted. All flowers have yellow, sometimes red-orange, rays. Achenes beakless, terete to prismatic; pappus of numerous whitish to more often tawny or brownish capillary bristles.

This species 20–80 mm; leaves all or most basal, 3–16 × 0.8–5 cm including short petiole; blade ovate to broadly lanceolate, upperside with conspicuous reddish-purple veins when alive. Heads few; involucre 7–10 mm, glabrous to sometimes bearing stalked glands; flowers 15–40 per head. Common. Dry thin woods; cGa into cAla, Mich, NY, Va, and cSC. Apr–Aug.

H. traillii Greene is similar but hairier and leaves not purple-veined; involucre to 12 mm and conspicuously hairy. Occasional. Dry thin woods, sometimes on shale barrens; mts of sPa into wVa and WVa. June–Sept. *H. greenii* Porter & Britt.

491   **Hairy or Beaked Hawkweed**

Hieracium gronovii L.

0.3–1.5 m. Stems mostly solitary, spreading hairs conspicuous toward base. Leaves prominent from base to midstem or higher; basal leaves 4–20 × 1.2–5 cm, obovate to broadly oblanceolate or elliptic, some dying first year, others evergreen. Inflorescence thin, columnar; involucre mostly 6–9 mm, flowers 20–40 per head; achenes 2.5–4 mm. Common. Dry thin woods; cFla into Tex, Kan, sOnt, and NC. June–frost or Apr–Dec in s part of range.

H. megacephalum Nash is similar but inflorescence broadly and openly corymblike, involucre 9–11 mm, and achenes 3.5–5 mm. Rare. Dry sandy woods, commonly with pines, sometimes with oaks or palmettos, or in hammocks; pen Fla into sGa. Sept–May.

492   **King-devil; Devil's Paintbrush**

Hieracium aurantiacum L.

Plants 10–60 cm, with slender stolons and ordinarily with a long rhizome. Leaves all basal or 1–2 also on lower portion of stem, most 5–18 × 1–3.5 cm, oblanceolate to narrowly elliptic. Inflorescence a compact corymb; heads 5–20, the rays red-orange (unique among our species), becoming deeper red in drying. Native to Europe. Rare. Fields, meadows, and roadsides; mts of neGa into wNC, w and nVa, WVa, neUS, and seCan. June–Aug.

The two following species are also native to Europe and are similar. One, *H. caespitosum* Dumort. (King-devil), is 25–90 cm, herbage not glaucous, and rays bright yellow. Occasional. Weed in fields, pastures, thin woods, roadsides; nGa into eTenn, neUS, seCan, Va, and csNC. Apr–June. *H. pratense* Tausch. *H. florantinum* All. is 20–100 cm, herbage glaucous, and rays bright yellow. Similar habitats; nGa into neUS and seCan. June–Aug.

493   **Mouse-ear; Mouse-bloodwort**
Hieracium pilosella L.
Abundantly stoloniferous, 3–25(40) cm, densely hairy. Leaves all basal, possibly a smaller one above, blade 2–13 × 0.6–2 cm, oblanceolate, underside with tawny stellate hairs. Heads solitary, seldom 2(4); involucre 7–11 mm; achenes 1.5–2 mm. Native of Europe. Occasional. Weed in pastures, meadows, thin woods; c and wNC into eTenn, WVa, neUS, seCan, and ceVa. May–Aug.

Monocotyledons

Leaves of most species parallel-veined; vascular bundles often scattered throughout the stem or in a band surrounding a pith (notably most grasses); floral parts, when of definite number, borne in sets of 3, seldom 4, and almost never fewer (carpels often fewer, occasionally many); embryos mostly with one cotyledon.

TYPHACEAE: Cattail Family

494 **Common Cattail**
Typha latifolia L.
The dense columns of tiny flowers identify Cattails. Male flowers, which drop soon, are in a column above the larger mass of female flowers.

In this species male and female flower clusters are usually adjacent and the stigmas are lanceolate to ovate. Rootstocks can be eaten raw or cooked or ground into meal. Pollen also can be used as flour; young stems and young flowers can be cooked and eaten; leaves are used in weaving chair seats. Common. Shallow water, sometimes brackish, in open; all of US and sCan. Apr–July.

Two other species occur in eUS. In *T. domingensis* Pers. (Southern Cattail) the spikes are pale brown, the masses of male and female flowers are usually separated, and the stigmas are linear. Similar habitats; Fla into coastal Ga, eVa, and Del; along the Gulf coast into Tex, inland into csKan, and W into Calif and swOre. In *T. angustifolia* L. the spikes are deep brown, male and female sections separate, and styles linear. Common. Marshes, tolerant of salt and alkali habitats: nearly throughout our area, most abundantly along the coasts.

SPARGANIACEAE: Bur-reed Family

495 **Bur-reed**
Sparganium americanum Nutt.
Leaves elongate, sometimes 2 cm wide, with rounded tips. Flowers and fruits in dense globose heads, with staminate flowers on the upper and outer parts of the inflorescence. Pistillate flower heads mature into "burs," which are usually in leaf axils. Individual sections (fruits) of the "burs" are dull and finely pitted. Underground tubers have been boiled or roasted for food. Occasional. In mud at edge of or in water; Fla into eTex, eND, and Nfld. May–Sept.

S. androcladum (Engelm.) Morong is similar but has smooth shiny fruits. Rare. Similar places; Va into Mo, eTex, Minn, sQue, and sMe. May–Sept. *S. erectum* L. ssp. *stoloniferum* (Graebn.) Hara is another species with lustrous fruits, but 1 or more of the "burs" are not in axil of a leaf or bract. Rare. Similar places; nwNC into wPa, cO, Ia, sQue, and Nfld. May–Sept. *S. chlorocarpum* Rydb.

ALISMATACEAE: Water-plantain Family

496 Arrowhead
Sagittaria latifolia Willd.
The generic name comes from the arrow-shaped leaves of some species. Others have no such leaves. Sagittarias have many small "seeds" (carpels) in dense rounded heads. Lower flowers female, or all flowers female, or all male; flowers never bisexual.

In this variable species the fruit-bearing pedicels are not curved downward. The bracts at pedicel bases are thin and obtuse at apex. Beaks on the "seeds" are at right angles to the body to erect. Common. Various wet places; Fla into Me and to the W, except for a few states. June–Nov. *S. longirostra* (Micheli) J. G. Sm.

Other aquatic species with arrow-shaped leaves are: *S. engelmanniana* J. G. Sm., with erect beak and thick pedicels. Rare. Ark into sInd, Mass, and nSC. July–Oct. *S. montevidensis* Cham. & Schlecht., with mature pedicels curved downward. Rare. Fla into La; SC into NC; Del into ePa and NJ. July–Sept.

497 Narrow-leaved Sagittaria
Sagittaria graminea Michx.
Perennial to 60 cm. Above-water leaves linear to ovate, to 30 cm. Submersed leaves narrow, to 50 × 2.5 cm. Male flowers with many stamens, the filaments finely hairy and widened at the base, equal to or shorter than the anthers. Common. Edges of or in water, especially in swamps; Fla into cs and eTex, eKan, sMinn, and sLab. Apr–Oct.

S. subulata (L.) Buch. also has narrow leaves, occasionally with some ovate or lanceolate blades on the floating ends of the leaves, but the filaments are glabrous. Occasional. Shallow water and mud, in open; CP of Fla into sMiss, sGa, and Mass. May–Sept.

POACEAE: Grass Family

Grasses are a substantial, highly visible, valuable component of practically all habitats, and a study that excludes them is incomplete. They do have a reputation, often justified, for being difficult to identify, and some members of other families occasionally are confused with them. The species described here are representative of some of the types of grasses and possibly familiar but generally ignored. Knowing them by their proper names is a satisfying experience and can be accomplished rather easily upon familiarity with the descriptive terminology.

Grasses have leaves that are 2-ranked (described later), an arrangement usually quite evident. The leaf consists of 2 major parts: the basal portion, called a *sheath,* which closely surrounds the stem and has margins with free edges that often overlap or, rarely, are fused; and the upper portion, the *blade,* which diverges from the stem and is mostly long-linear.

Grass flowers lack a perianth and, except for one microscopic part that is unimportant here, consist of 3 (rarely 1–6) stamens, or 1 pistil, or both. These are inserted between and at the base of 2 "bracts"; the outer and larger "bract" is known as the *lemma,* the inner and often inconspicuous one is the *palea.* The lemma, palea, stamen(s), and/or pistil combined are known as the *floret.*

Florets are sessile and attached singly or in alternate 2-ranked sets of 2–many. In most grasses there are 2 empty (thus sterile) "bracts" called *glumes* at the base of the single floret or of the 2–many florets. These together with the floret(s) constitute the *spikelet*. Spikelets may be sessile or pedicellate, and most are arranged in panicles. Some spikelets, however, are pedicellate and in racemes or spikelike racemes; others have sessile spikelets that are arranged in spikes, some 2-sided, some 1-sided.

The fruit of a grass is a *grain* in which the ovulary wall is completely fused with the seed or, in two genera that we have included (*Sporobolus, Eleusine*), the fruit consists of a thin wall with the enclosed seed connected only by its stalk.

In identifying grasses, much help is obtained by noting certain easily recognized characteristics such as spikelet size, whether or not the spikelets have pedicels or are sessile, the number of florets per spikelet, the kind, size, shape, and compactness of the inflorescence bearing the spikelets, and the presence, location, and character of hairs. Some grasses have a hard, shiny, smooth lemma, a few species have sharp spines associated with the spikelets, others have bristles mixed with spikelets, and many have awns on lemma tips. A 10× hand lens is often useful.

Plants vegetatively similar to grasses are predominantly members of the JUN-CACEAE (Rushes), LILIACEAE (Lilies), AMARYLLIDACEAE (Amaryllises), COMMELI-NACEAE (Spiderworts), IRIDACEAE (Irises), XYRIDACEAE (Xyrises), ASTERACEAE (Composites), CYPERACEAE (Sedges), TYPHACEAE (Cattails), and ERIOCAULACEAE (Pipeworts).

Members of the first 7 families listed above may cause some confusion but they all have a perianth that is sufficiently developed to be recognizable, usually readily, and the fruits are other than grains.

Members of the last 3 families are easily separated from grasses by features specific to the individual families. Pipeworts have a perianth that is difficult to recognize, but the tiny flowers are in conspicuous dense heads on the tip of a leafless stem. The tiny flowers of cattails are arranged along the top of the stem in huge numbers in coarse columnar clusters. Sedges are the most similar to grasses and can be separated by the characters given below. An asterisk (*) indicates there are a few exceptions.

POACEAE	CYPERACEAE
1. Leaves 2-ranked	1. Leaves 3-ranked
2. Leaf sheaths with free margins*	2. Leaf sheaths with margins fused
3. Internodes hollow*	3. Internodes solid*
4. Stem cross-section circular except for axillary bud notch	4. Stem cross-section triangular*
5. Flower between 2 "bracts"*	5. Flower in axil of 1 "bract"
6. Stigmas 2	6. Stigmas 2–3
7. Fruit a grain*	7. Fruit an achene or nutlet

In using leaf rank as a distinguishing feature, attention should be focused on the leaf blade bases, noting whether they are in 2 rows up and down the stem or in

3 rows. Seedlings with 3 or more leaves are easily separated in this manner. This is an especially useful character to determine which of the 2 families is involved in lawn and garden weeds.

498   Gamma Grass
Tripsacum dactyloides (L.) L.
Perennial from coarse rhizomes. Plants to 2.5 m. Leaves largely basal and on lower half of stem, blades to 60 × 3 cm. Flowers in spikes, unisexual, the sexes borne on different parts of the same spike with the male above the female. Spikes 1–a few, terminal on the principal stem axis and on the few erect lateral branches. Male portion of spike almost identical in appearance to sections of the "tassel" of corn plants, dropping off as a whole shortly after shedding pollen. Individual female flowers and grains developing from them are in bony beadlike joints of the spike, these breaking apart at maturity. Plants in thin clusters or colonies, or occasionally solitary. Occasional. Usually moist places—ditches, depressions, swales, thin woods, waste places; eTex into seKan, Ill, Ky, NY, NH, Va, and Fla. May–Nov.

499   Sugarcane Plumegrass
Saccharum giganteum (Walt.) Pers.
This is one of our largest native grasses, growing to 3 m. Flowers are quite small and consist mostly of a few stamens and a pistil with 2 fuzzy stigmas, all borne between 2 stiff scalelike structures, the lemma and palea. A grain, part of which is the seed, develops between the lemma and palea. The lemmas have long copious hairs fastened to their bases and a terete awn on the tip. Common. Moist open places; Fla into eTex, seOkla, Ky, Va, and seNY. Mostly absent from the Appal Mts. Sept–Oct. *Erianthus giganteus* (Walt.) P. Beauv.

    S. alopecuroides (L.) Nutt is equally as tall and also has copious hairs on the spikelet but the awn is flat and spirally coiled. Common. Various open places and thin woods; nFla into eTex, eAlta, sMo, sInd, eKy, Va, and seNJ; Cuba. Aug–Nov. *Erianthus a.* (L.) Ell.

    S. brevibarbe (Michx.) Pers. is similar but slightly smaller, to 2.5 m, panicle slimmer and less dense, and the hairs on the spikelet shorter than the spikelet. Occasional. Similar places; Del; eNC into Ga, neFla, La, and Ark. Sept–Oct. *Erianthus coarctatus* Fern. *E. brevibarbis* Michx. as used in manuals.

500   Bushy Broomsedge
Andropogon glomeratus (Walt.) B.S.P.
Andropogon species are prominent features of many open habitats in the fall. They are perennials and much alike vegetatively. Differences between species mostly involve inflorescences, all of which have conspicuous soft silky hairs in the spikelet-bearing racemes. Spikelets are in pairs in the racemes (unusual arrangement in grasses); one spikelet is sessile, the other is stalked and different in size or shape, or may be absent in which case represented by its pedicel. Identification to species is often difficult partly due to differing interpretations, so our treatment will not agree with all manuals. Those Broomsedges having only a single raceme on each peduncle are placed in the genus *Schizachyrium* in some recent manuals.

    Bushy Broomsedge grows to 1.5 m. It is easily recognized by the large dense terminal mass of inflorescences making the plant bushy-topped. The paired racemes are short-peduncled and sit between 2 clasping bractlike leaves. There are 5 varieties; the photograph is of var. *glomeratus.* Common. Moist to wet places in the open—swales, ditches, sloughs, meadows, freshwater marshes, margins of brackish marshes, depressions; cFla into eTex, seOkla, Miss, cTenn, and eKy; ePa into seNY: eMass into eNC and cGa. Aug–Oct. *A. virginicus* L. var. *abbreviatus* (Hack.) Fern. & Grisc.

501 Chalky Broomsedge

Andropogon virginicus L. var. *glaucus* Hack.
Plants to 1 m, usually in clumps; glaucous, especially toward the base. Inflorescences distributed somewhat uniformly in the upper half or more of the plant. Racemes shorter than the bractlike leaves clasping them; stem branches just below these bracts glabrous. There are 3 vars. of *A. virginicus.* Occasional. In thin woods or in the open; swales, stable dunes, ditches, margins of freshwater marshes, meadows, roadsides, depressions; sVa into Fla and La. Aug–Oct. *A. capillipes* Nash.

   A. virginicus L. var. *virginicus* (Virginia Broomsedge) has a similar aspect to the above 2 species and is otherwise similar but is not glaucous. Hairs 5–7 mm on the raceme axis and on the peduncle supporting the 2 racemes. Common. In thin woods or in the open—dunes, swales, meadows, marsh margins, roadsides, old fields, pinelands; cFla into seTex, eOkla, eKan, sIll, sMich, O, Pa, cNY, Mass, and Del. Aug–Oct.

   Two similar Broomsedges are: *A. glomeratus* (Walt.) B.S.P. var. *glaucopsis* (Ell.) C. Mohr, also glaucous, usually heavily so, and otherwise much like *A. virginicus* var. *glaucus* except there are long hairs on the stem branches just below the bracts clasping the racemes. Occasional. Similar places; wVa into Fla. Aug–Oct. *A. gyrans* Ashe var. *gyrans* (Elliott's Broomsedge) is easily recognized. It has dense clusters of racemes, but in much smaller masses than *A. glomeratus.* The raceme pairs are clasped by leaflike bracts as in several other species, but in addition groups of these clasped raceme pairs are clasped by much enlarged overlapping leaf sheaths just below; these sheaths are 5–10 mm wide. Occasional. Old fields, thin live oak and maritime woods, pinelands, between stable dunes; eTex into Mo, sIll, sO, neKy, ePa, NJ, NC, and Fla. Aug–Oct. *A. elliottii* Chapm.

502 Splitbeard

Andropogon ternarius Michx.
This species grows to 1.2 m and is vegetatively much like *A. virginicus* but the inflorescence differs significantly. The paired racemes are on peduncles 5–12 cm with only the base of the peduncles tightly clasped by the bractlike leaves, which are inconspicuous because of their small size and clasping nature. The photograph is of the inflorescence of a plant just before the racemes break apart. The racemes are copiously hairy, the hairs to 9 mm; the sessile spikelets are 7 mm. Common. Dry places—thin woods, pinelands, stable dunes, old fields, roadsides, meadows; cFla into eTex, eKan, sMo, Ky, and Tenn; Md into Del and SC. Aug–Oct.

503 Little Bluestem

Schizachyrium scoparium (Michx.) Nash
This genus is much like *Andropogon* except racemes are solitary at top of each peduncle; joints of the raceme are flat.

In this species the stems are loosely to densely tufted, to 1.2 m, and freely branched in upper portion. Blades 3–7 mm across. Joints of raceme ciliate; awns 7–14 mm, twisted near the base. There are 3 vars; the photograph is of var. *littoralis* (Nash) Gould, which occurs along the Gulf and Atlantic coasts. Common. Dry soils; roadsides, old fields, prairies, thin woods; Tex into Alta, sOnt, sQue, wNH, Mass, and Fla; Mex. Aug–Oct.

504   Drooping Woodgrass

Sorghastrum secundum (Ell.) Nash
Perennial to 2.2 m with stems growing singly or more frequently in clumps. Leaves mostly basal. Inflorescence a narrow 1-sided panicle to 45 cm. Spikelets yellowish-brown, shiny, bearing conspicuous stiff ascending hairs mostly 1–1.5 mm on one side only and a pedicel without spikelets alongside. Spikelets with an awn 25–35 mm, twice-geniculate, flat, twisted when dry except near the tip, and easily pulled free of the spikelet when mature. Occasional. Pinelands, scrub oak–pine, live oak–pine; SC into cGa, cFla, and sAla; seTex. Sept–Oct.

S. *elliottii* (C. Mohr) Nash (Elliott's Woodgrass) is much the same but spikelets are chestnut-brown and in a loose but not 1-sided panicle to 30 cm. Occasional. Similar places and old fields; cFla into eTex, cOkla, Tenn, eMd, and cSC. Sept–Oct.

505   Vasey Grass

Paspalum urvillei Steud.
Perennial to 2 m, erect, branching at the base and forming tufts. Inflorescence composed of 7–20 ascending to arched-spreading racemes on a single axis. Additional inflorescences develop throughout the growing season on ends of erect branches from leaf axils. Identity is assured if spikelets are 2–3 mm and bear conspicuous silky hairs, a rare occurrence in *Paspalum*. Plants often can be located from a distance due to movements caused by birds seeking the spikelets for food. Common. In most moist to wet places in the open, or occasionally in drier habitats; eVa into eNC, sGa, Fla, sAla, seTex, and ceArk; sCalif. May–Oct.

P. *dilatatum* Poir. (Dallis Grass) also has spikelets with long silky hairs but spikelets are larger, 3–3.5 mm. Stems are in clumps, often decumbent, leaves are mostly basal. Common. Very weedy and can be expected almost anywhere in the open—most frequently at roadsides, in lawns and fields, doing best in moist places; sNJ into eNC, sGa, Fla, Ala, eTex, cOkla, and Tenn; Colo; sAriz into sCalif. May–Oct.

506   Bahia Grass

Paspalum notatum Flugge
The V-shaped inflorescence high above the basal tufts of leaves makes Bahia Grass conspicuous. In addition, plants are usually in dense mats formed largely by means of coarse vigorous rhizomes. Plants to 80 cm. The inflorescence consists of the 2 racemes, each of which arises from the tip of the peduncle; occasionally there is another raceme or rarely 2 just below. Spikelets are ca 3.5 mm, rarely to 4 mm. Abundantly planted for forage and along highways, commonly escaping. Common. In almost any habitat except dense woods, in water, or on strongly active dunes; NJ into Fla, eTex, and Ark. June–Oct.

P. *distichum* L. (Knotgrass) also has 2 racemes at the end of a stem but is not outstanding because of its size—flowering stems are rarely over 40 cm. However, it frequently occurs in thick showy mats or carpets formed largely through extensive rhizome growth. Knotgrass may be separated from other similar species by its loose leaf sheaths, racemes at the end of a leafy stem, and spikelets 2.3–3 mm with conspicuous short soft fine hairs. Occasional. Swales, marsh and pond margins, accretion areas, and depressions; also poorly drained spots in pathways, roads, lawns, golf fairways; ePa into NJ, eNC, seGa, and cFla. June–Aug.

Axonopus fissifolius (Raddi) Kuhlm. has an inflorescence similar to that of Knotgrass. It forms extensive carpets; spikelets are about 2 mm, and their backs are turned away from the axis of the spikes. Spikes are slimmer than those of Knotgrass and to 9 cm. A third and sometimes fourth spike occur below the top 2. Most leaf tips are blunt. Common. Wet or poorly drained places—live oak woods, pond margins, ditches, thin woods; occasionally in dry habitats; sAla into sLa; sArk into seOkla, Tex, sAriz, sCalif, Nev, eWash, sIda, and Utah. June–Oct. *A. affinis* Chase.

507   **Switch Grass**
Panicum virgatum L.
Although spikelets of various species of *Panicum* vary in size (1–8 mm) and in shape, the genus is easily recognized. Spikelets are in panicles and consist of a hard shiny smooth fertile inner lemma that is rounded on the back and on the other side encloses the similar-surfaced palea, and 3 membranous veined units clasping the lemma, the inner one a sterile lemma and the outer 2 glumes. Identification beyond genus is another matter; only a few specialists can name with confidence most of the over 100 species in the eUS. *Panicum,* as described above, has been split by some botanists into 2 genera, probably correctly so, namely *Panicum* and *Dichanthelium. Panicum* species are annuals or perennials and have blades of the basal and stem leaves similar, winter rosettes not formed, primary and secondary panicles alike or nearly so, spikelets all fertile. *Dichanthelium* species are perennials, their basal and stem leaves usually different in shape, the former closely crowded and forming a winter rosette (2 species excepted); primary panicles are borne on simple stems in early summer, the branches forked at the very base, the spikelets often sterile; later these stems usually branch freely, the branches bearing numerous smaller panicles that are often concealed by the leaves, the spikelets never opening and pollination not possible but the spikelets fertile.

Switch Grass serves as a good example of *Panicum* because it is so widely distributed and parts of the spikelet are easily seen. It is a perennial to 2 m with coarse scaly rhizomes; stems erect to arching, usually growing in clumps. Spikelets 3–5 mm, on long pedicels in large thin panicles 15–50 cm and nearly as wide. Common. At margins of and in marshes and ponds, depressions, thin woods, swales, wet pinelands, prairies, dunes; Fla into Ariz, Tex, N into Nev, Wyo, ND, sOnt, sNS, and Me; Mex; S. Amer. June–Oct.

508   **Seaside Panicum**
Panicum amarum Ell.
Glaucous glabrous perennial 0.2–3 m with large extensive rhizomes. Stems solitary to more frequently in clumps. Lower stem widths vary from ca 2 to 10 mm.

Spikelets 4–7.7 mm in panicles 10–40 × 2–10 cm, the principal branches ascending. Common. Active dune areas, beaches, swales behind foredunes; along the coasts, Conn into Fla; Miss into Tex. July–Oct.

509   **Capscale**

Sacciolepis striata (L.) Nash

Glabrous perennial with much-branched stems that are often decumbent and rooting at the nodes, sometimes forming dense masses or falling over and growing more than 2 m. Spikelets 3.5–5 mm, much like those of *Paspalum, Panicum,* and *Echinochloa* species, the inner unit being hard, shiny, and smooth. Easily recognized by its inflorescence, which is a slender panicle with branches so short that it may appear spikelike, and by the unique second glume of the spikelet. This glume is about as long as the spikelet and has a cupped or bulging base appearing much like baggy knees; the first glume is short and acute at apex. Important food source for wildlife. Common. Marshes, swales, sloughs, ditches, pond margins, depressions; NJ into cFla, cGa, cMiss, and cTenn; La into eTex, and eOkla. July–frost.

S. *indica* (L.) Chase, an annual, has similar but smaller spikelets, 2.5–3 mm, and the plants are smaller. Rare. Similar places; Ga into NC. July–frost.

510   **Water Grass**

Echinochloa walteri (Pursh) Heller

Annual to 2 m with 1–several usually erect stems from the base, each bearing a terminal inflorescence; other inflorescences, usually smaller, often occur on erect lateral branches. Inflorescences are panicles that are usually nodding; the branches are spikelike with the spikelets congested on short secondary branches. Spikelets lanceolate, about 3 times as long as broad; the inner unit (the fertile lemma) hard, shiny, and smooth, as in *Paspalum, Panicum,* and *Sacciolepis* species, but the tip of the fertile lemma is abruptly pointed and does not enclose the tip of the palea. The second glume and sterile lemma are long-awned. Common. Wet places or in shallow water in thin woods or in the open—marshes, ponds, sloughs, depressions, ditches; Ark into Ia, sWisc, sMich, O, NY, Mass, eNC, Fla, and Tex. July–frost.

E. *crus-galli* (L.) Beauv. (Barnyard Grass) is similar but spikelets are ovate, about twice as long as broad, and awnless or with awns under 10 mm. Common. Dry to wet places in the open, mostly at low to medium alt; waste places, fields, roadsides, marshes, depressions; Wash into NS, S into Calif and Fla. July–frost.

511   **Green Foxtail**

Setaria viridis (L.) Beauv.

Setaria species attract attention because of their bristly spikelike panicles. These are distinctive in that 1–several bristles are attached just under the base of each spikelet; as the spikelet or most of it falls free the bristles remain. Spikelets are much like those of *Paspalum* and *Panicum* species in having a hard shiny inner unit, and the fertile lemma tightly clasped by the membranous infertile lemma and 2 glumes. Fruits are an important source of food for wildlife.

This species is an annual to only 80 cm with stems branching from or near the base. Leaves mostly erect, the sheath margins ciliate. Panicles 2–10 cm; spikelets 1.8–2.5 mm, the fertile lemma minutely roughened; bristles 1–3, green until spikelets begin falling. Occasional. Roadsides, disturbed soils, yard margins, gardens, parking areas; BC into Nfld, cFla, sGa, and sLa; Cuba. June–frost.

   S. corrugata (Ell.) J. A. Schult. is also an annual but the fertile lemmas are clearly horizontally ridged and the bristles purplish. Occasional. Roadsides, fields, waste places, parking areas; eNC into cFla, sGa, sAla, and sLa; Cuba. July–frost.

512   Knotroot Bristlegrass
Setaria parviflora (Poir.) Kerguelen
Perennial to 80 cm with knotty usually branching rhizomes. Stems generally several from the base and often with branches above. Leaves mostly erect, the sheath margins lacking cilia. Panicles 2–8 cm. Spikelets 2–2.5 mm, the fertile lemma finely glandular to very finely roughened, bristles 7–12 below each spikelet. Common. Thin woods, marsh margins, swales, wet pinelands, pond shores, roadsides, fields, yard borders, fencerows; eOkla into sKan, sIll, Mo, Tenn, WVa, sMass, eNC, Fla, and Calif. May–frost. *S. geniculata auct. non* (Willd.) Beauv.

513   Giant-millet; Giant Foxtail
Setaria magna Griseb.
Coarse annual to 4 m, stems to 2.5 cm thick at base. Inflorescence to 50 × 6 cm, arching to nodding. Spikelets 2–2.2 mm; the smooth fertile lemma falls free when mature, leaving the remainder of the spikelet and the bristles. Common. Freshwater and brackish marshes, pond margins, swales, sloughs, ditches; sNJ into eMd, eVa; near coasts into Fla, and Fla into seTex; seArk. July–frost.

   S. italica (L.) Beauv. (Italian Millet) is similar but smaller, to 1 m, inflorescence to 30 cm and spikelets 2.5–3 mm. Native of E. Indies. Cultivated. Rare. Roadsides, waste places, fencerows; eNC into seNY; neArk into eNeb, Mo, sMich, O, and nKy; Miss. July–Oct.

514   Yellow Bristlegrass
Pennisetum americanum (L.) Leeke
This species is much like those of *Setaria* but is easily recognized by the bristles falling with the spikelets.

   P. americanum is a tufted annual to 1.5 m, branching, and often forming dense colonies. Leaves mostly spreading, the sheath margins lacking cilia. Panicles 4–12 cm, spikelets 2.8–3.5 mm, the fertile lemma finely ridged horizontally; bristles 5–20 beneath each spikelet. Native of Europe. Common. Dry to wet places in the open—roadsides, fields, stable dune areas, around buildings, shrub borders; BC into NB, S into Ga; Fla into Calif. July–frost. *P. glaucum* R. Br.; *P. p., non* L.; *Setaria glauca* (L.) Beauv.; *S. lutescens* (Weigel) F. T. Hubb.

515   Dune Sandbur
Cenchrus tribuloides L.
Seaside sandburs are easy to recognize: 1–8 spikelets are enclosed in a spiny bur, the spines minutely retrorsely barbed near the tip. Plants often not noticed until burs are clinging to clothes or skin. Parts of spines can break off under the skin and cause pain.

   Dune Sandbur, an annual, is easily identified by the densely hairy cuplike burs that, including spines, are 10–15 mm across. Burs with 15–43 spines in spikelike racemes, the pedicels swollen. Leaf sheaths inflated, blades usually folded. Common. Loose sands in thin woods or in the open; along coasts; seNY into Fla; Ala into La. Aug–Sept.

516   Southern Sandbur
Cenchrus echinatus L.
Tufted annual with ascending to sprawling branched stems. Burs with a ring of terete bristles at base nearly as long as but slenderer than the larger-based flattened spines above them. Common. Roadsides, lawns, fields, stable dune areas; eSC into sGa, Fla, and eTex; NM into sCalif. June–frost.

Other tufted annuals include: *C. carolinianus* Walt. (Coastal Sandbur), a tufted annual with short rhizomes, sometimes overwintering. Stems branching and often sprawling. Leaf blades 2–6 mm wide. Burs, including spines, to 7 mm across, glabrous or with many very short hairs; spines 8–40, broad at base. Ring of basal bristles lacking. Common. Stable dune areas, minidunes, roadsides, waste places, thin woods; eVa into eNC, sGa, Fla, and eTex. July–frost. *C. pauciflorus* Benth. *C. longispinus* (Hack.) Fern. has 50 or more slender spines on the burs, those at the base most numerous, pointing downward, and shorter than those on the body of the bur. Common. Dunes, sandy meadows, roadsides, fields; Ore into Ont, S into Calif, Tex, and Fla. July–Oct.

517   St. Augustine Grass
Stenotaphrum secundatum (Walt.) Kuntze
Low mat-forming stoloniferous perennial having thick glabrous succulent flat leaf blades with an obtuse apex; leaf sheaths keeled, loosely clasping stem. Inflorescence a spike, the spikelets partially embedded in the thick flattened axis. Spikes 6–9 × 5–8 mm. Common. Roadsides, pastures, swales, meadows, brackish marshes; widely cultivated in yards; seNC into Fla and Tex. July–Oct.

518   Annual Wildrice
Zizania aquatica L.
Coarse plants to 3 m; mostly perennial southward and annual northward. Principal leaves 120 × 4–5 cm, margins finely toothed. Inflorescence a large terminal panicle. Spikelets unisexual, the sexes in the same panicle, the male ones hanging from the spreading lower branches of the panicle, the female ones erect on the stiffly erect upper branches. One of the few grasses with 6 stamens per flower. Common. Freshwater and brackish marshes; ND into Que, NC, Fla, La, and Mo. May–Oct.

Zizaniopsis miliacea (Michx.) Doell. & Asch. (Southern Wildrice) is much the same vegetatively except it is always a perennial. The spikelets are also unisexual and in large panicles but the sexes are intermixed throughout and all branches of the panicle are spreading. Common. Similar places; NC into Fla, eTex, eOkla, Ky, Va, and Md. Apr–Aug.

519   Pink Muhlenbergia
Muhlenbergia filipes M. A. Curtis
At least 15 species of this large genus occur in the eUS. Identification of most species is difficult.

This species is a tufted perennial to 80 cm. The inflorescence is a loose panicle, pinkish when mature, rarely white. When these panicles are in full color and bend with the wind they are outstandingly attractive, especially when there are many clumps over a large area. The spikelets have 1 floret and are borne on thin pedicels at the ends of the panicle branches. The stems are collected along the SC coast and used in the weaving of baskets, table mats, and the like. Occasional. Along or near the coast in moist open places, between dunes, in pinelands; NC into Fla and eTex. Sept–Oct. *M. capillaris* var. *filipes* (M. A. Curtis) Beal.

520   **Smut Grass**
Sporobolus indicus (L.) R. Br.
Sporobolus is one of many genera with 1 floret per spikelet and spikelets in panicles, but it is fairly easy to recognize when fruits are mature as it has seeds free from the thin ovulary wall that closely covers it. Rolling mature fruits between thumb and finger will remove some of the wall, revealing the shiny and reddish seed.

Smut Grass is a tufted perennial to 80 cm with mostly basal leaves and a slender panicle to 40 cm. Leaves are 2–5 mm across and quite tough, difficult to cut with mowers. Branches of the panicle are often conspicuous though closely ascending and the lower 1–few occasionally not overlapping those immediately above. Healthy panicles are greenish when mature, turning light tan upon drying. Seeds at maturity are reddish and loose, tending to stick to the mucilaginous ovary wall from which they came. Panicles are often infected with a smut (a fungus), which turns them black. Common. Meadows between dunes, swales, thin maritime or live oak woods, thin pinelands, lawns, pathways, roadsides; Va into Fla, eTex, cOkla, cn and seArk, seMo, and Tenn. Mar–Nov. *S. poiretii* (R. & S.) Hitchc.

521   **Redtop**
Agrostis stolonifera L.
Inflorescences of *Agrostis* species are loose panicles. Spikelets have 2 glumes which are equal to or longer than the single floret; glumes and lemma are thin. Awns are lacking or sometimes a short one is present on the back of the floret, not at the tip. Florets fall from the panicle at maturity, leaving the glumes.

Redtop is separated from other species of *Agrostis* by having spikelets borne from the top almost to the base of the panicle branches. Spikelets are 2–3 mm and individually pedicelled. Native of Eurasia. Common. Open usually moist to wet places—freshwater to brackish marshes, meadows, pond edges, fields, roadsides, swales; Alas into Greenl, S into cArk, cGa, and NC. June–Oct. *A. alba* L. as used in some manuals.

In 3 similar species spikelets are absent from the basal portion of the panicle branches. In *A. perennans* (Walt.) Tuck. (Autumn Bentgrass) panicle branches begin at or just below the middle. Spikelets are 2–2.5 mm and scattered along most of the secondary branches. Common. Similar places; Minn into Que, S into eTex and Fla. July–Oct. In the other 2 species the branches fork well beyond the middle and the spikelets are distributed near the ends of the ultimate branches. Spikelets are 1.5–1.8 mm long in *A. hyemalis* (Walt.) B.S.P. (Ticklegrass). Common. Roadsides, old fields, meadows, pond shores; Ark into Okla, Ia, sWisc, Ind, Ky, Va, seNY, and Mass. Mar–June. In *A. scabra* Willd. spikelets are 2–2.7 mm.

Occasional. Places similar to Ticklegrass; Alas into Greenl, Nfld, S into cCalif, Ariz, La, and nGa. June–Nov.

522   Velvet Grass

Holcus lanatus L.

Tufted annual to 1 m, the entire plant grayish velvety-hairy and soft to touch. Glumes 3.2–4.5 mm, ciliate on the keeled back. Florets 2; the lemmas 1.8–2 mm, the upper one with a hooked awn 1–2 mm. Native of Europe. Occasional. Fields, roadsides, waste places, meadows; thin woods; often in damp places; Wash and Ida into NS, S into La, Ga; Neb and Colo; Ariz into Calif. May–Sept.

523   Big Cordgrass

Spartina cynosuroides (L.) Roth

Spartina is the most important genus in seaside habitats. Only 6 species, all perennials, are involved but the number of plants is astronomical. The genus is fairly easily recognized. Spikelets are arranged in 2 rows on 1 side of the spikes, which are distributed along a central axis in numbers ranging from 3 to 75. Spikelets consist of 2 glumes and 1 floret, the palea usually evident as well as the lemma. The lemma is keeled and of the same general texture as the glumes, distinguishing it from the hard, shiny, and round-backed floret of the spikelets in the 1-sided spikes of *Paspalum* species.

Big Cordgrass grows to 3.5 m, the stems hard. Fresh leaf blades flat, the largest 10–25 mm wide, the margins scabrous. Panicle with 5–67 spreading spikes; longer of the 2 glumes one-third longer than the floret. Common. Brackish or freshwater tidal marshes, brackish sloughs; sMass into Fla and seTex. June–Oct.

S. pectinata Bosc ex Link (Prairie Cordgrass), which grows to 2.5 m, also has hard stems; fresh leaves flat, 5–15 mm wide and margin scabrous. Spikes vary from 5 to 50 and are loosely appressed to spreading. The longer of the 2 glumes is about twice as long as the floret, longer than in any other species. Common. Freshwater and saltwater habitats—marshes, shores, ditches, pond shores; Wash into Alta, Nfld; S into Ore, NM, Tex, Ky, WVa, and NJ; Miss. June–Sept.

Other important Spartinas include: *S. alterniflora* Loisel (Smooth Cordgrass), the most important species in seaside habitats, covering vast areas to the exclusion or nearly so of other species. Plants to 2.5 m with soft stems. Leaf blades are flat, not scabrous on the margins, and 3–25 mm wide. Panicles erect to arching; spikes 3–25, loosely appressed, separated or more frequently moderately overlapping. Plants in the high salt marshes, especially at edges of salt pans, may be only 40 cm including the inflorescence. Mid- to high-tide levels in salt and brackish marshes; Que into Nfld, S along coasts into Fla and seTex. June–Oct. *S. spartinae* (Trin.) Merr. ex Hitchc. (Gulf Cordgrass) grows to 2 m in clumps to 70 cm across; leaf blades are involute when fresh and only to 5 mm wide. Spikes 6–75, tightly appressed, and closely overlapping in panicles 6–70 cm. Common. Dunes, sandy beaches, roadsides, ditches, meadows, salt flats, marshes; Gulf coast, wFla into seTex. Mar–Oct. *S. patens* (Ait.) Muhl. (Marsh-hay; Saltmeadow Cordgrass; Salt-hay) grows to 1.5 m, solitary or in small clumps from widely spreading slender wiry rhizomes. Leaf blades involute or rarely flat, 0.5–5 mm wide. Spikes 2–15, alternate, often well separated, spreading to less commonly loosely appressed; longer glume blunt-pointed, about one-fifth longer than the lemma. Once abundantly cut for hay and still harvested in some northern areas. Common. High salt marsh, salt meadows, brackish flats, beaches, overwash areas, low dunes, swales; Que into Nfld; S along coasts into Fla and c and seTex; wNY; seMich; W. Indies; sFrance. *S. bakeri* Merr. (Bunch Cordgrass) is similar in many respects but rhizomes are lacking and plants are in circular colonies to 1 m across. The longer glume is sharply tapered to a fine point. Common. Swales, depressions, pond margins, at edge of brackish marshes, wet pinelands, meadows; sSC into Fla. Dec–Mar, occasionally in summer.

524 Fingergrass

Eustachys petraea (Sw.) Desv.
Tufted perennial to 60 cm, with short rhizomes, stems often decumbent. Leaves mostly basal, glabrous, apex rounded to broadly acute. Spikelets many, 1.8–2.2 mm, closely arranged on 2 sides of a 3-sided spike. Spikes 3–6, to 6 cm, all arising at about the same point at end of stem, axis extending slightly beyond the spikelets. Fertile lemma shiny brown, at maturity falling and leaving the glumes, the spikes then appearing feathery. Common. Swales, sandy flats, roadsides, grasslands, margins of freshwater and brackish marshes, thin live oak woods; seNY into eVa, sGa, Fla, and eTex. June–Oct. *Chloris p.* Sw.

   E. glauca Chapm. is larger, to 1.2 m, and has 8–16 ascending to erect spikes 8–20 cm. Rare. Margins of freshwater and brackish marshes, depressions, ditches; sNC into sGa, and Ala. June–Oct. *Chloris g.* (Chapm.) Wood.

525 Goosegrass

Eleusine indica (L.) Gaertn.
Coarse tufted annual with mostly decumbent stems and overlapping leaf sheaths; leaves and stems tough. Spikelets many, arranged in 2 rows on 1 side of spikes, 3–6-flowered, lemmas and glumes awnless and with essentially the same texture. Spikes clustered at tip of stem or occasionally 1 below, the axis not extending beyond the spikelets. So tough that goosegrass is the last to disappear in well-used footpaths or vehicle ways. Common. Footpaths, vehicle trails, lawns, parking lots, around buildings, waste places; Minn into Que; S into Calif, sMich, Mass, and Fla. June–Oct.

526 Common Reed

Phragmites australis (Cav.) Trin. ex Steud.
Coarse rhizomatous perennial to 4 m, forming extensive dense colonies. Leaf blades to 50 × 5 cm, the sheaths overlapping. Inflorescence a dense tawny to purplish panicle to 50 × 20 cm. Spikelets have thin glumes and lemmas. Glumes 2, shorter than the 3–8 loosely arranged glabrous lemmas. The conspicuous hairs in the spikelets arise from the axis. Common. Marshes, especially tidal; pond margins; ditches; disturbed habitats; BC into NS; S into Mex, Mo, Ind, Pa, and Del; Ga coast. July–Oct. *P. communis* (L.) Trin.

   Arundo donax L. (Giant Reed) is nearly identical except to 6 m, the glumes longer than the lemmas, and hairs in the spikelets arising from the base of the lemmas. Planted and persisting or spreading vegetatively in dense colonies by means of short thick rhizomes. Rare. Va into Fla, Tex, and eArk; probably in scattered localities of many other states. Sept–Oct.

527 **Purple Lovegrass**
Eragrostis spectabilis (Pursh) Steud.
A large genus with inflorescences of panicles varying from large, open, and thin to small and dense. Spikelets with glumes and lemmas of similar texture, glumes shorter than the largest lemma, florets 2–many. Most plants are difficult to name.

Purple Lovegrass is a tufted perennial to 80 cm with stiffly erect to spreading stems. Each panicle is usually about two-thirds the height of the plant and has branches spreading to ascending near the top. The whole panicle eventually breaks away and tumbles before the wind. Spikelets 4–8 mm, usually purplish, on slim but rigid pedicels longer than the spikelets. Florets 3–10. Common. Stable dune areas, loose sands, fencerows, fields, dry pinelands, thin live oak woods; Minn into Me; S into Ariz and Fla; Mex. July–Oct.

Two similar species have flexible capillary pedicels. In *E. refracta* (Muhl.) Scribn. the pedicels are longer than the spikelets, which, except for terminal ones, are pressed against the inflorescence branches. Common. Marshes, pond margins, ditches, depressions; Md and Del; S in CP into Fla, eTex, and sArk. July–Oct. In *E. campestris* Trin. the spikelets are longer than their pedicels and are divergent. Common, but rare along the Atlantic Ocean. Similar places; seTex into Fla, sGa, and eNC. Aug–Nov. *E. elliottii* S. Watson.

528 **Sea-oats**
Uniola paniculata L.
Coarse perennial to 2 m with extensive creeping rhizomes, readily rooting at nodes when covered with sand; the rhizomes, roots, leaves, and stems are important in stabilizing dunes. Inflorescence a panicle; spikelets conspicuously flattened, with 8–20 florets. In most states, as a preservation measure, it is unlawful to gather any part of the plant. Common. Dunes, beaches, loose sands near seashores; along coasts; Va into Fla and Tex. June–Sept.

529 **Quaking Grass**
Briza minor L.
The spikelets of *Briza* species are so different from those of other grasses that they readily attract attention. They are deltoid, hang down on thin curved pedicels, and are easily moved by wind. Some details of the spikelet are also interesting; the lemmas are almost horizontal and have thin broad margins and a rounded apex.

This species is an erect glabrous annual to 70 cm. Inflorescence an oblong to pyramidal panicle 2–15 cm long and nearly as wide. Spikelets are about 3 mm long and a little wider. Occasional. Swales, fields, waste places, thin woods; eVa into Ga, Fla, and eTex; s and cArk; Ore into Calif. Mar–May.

530 **Annual Bluegrass**
Poa annua L.
Plants to 40 cm but usually under 15 cm; bright green, tufted, often forming dense mats excluding other plants; dying in late spring and leaving light-colored mats that are unsightly and therefore troublesome in lawns. Inflorescence a pyramidal panicle 1–12 cm; spikelets with glumes and lemmas alike, lemmas 3–6. Common. Lawns, swales, paths, roadsides, waste places, parking lots, crevices of

sidewalks, shrub and flower plantings; Alas into Lab and Greenl; S into Calif and Fla. Mar–May.

531   Wild Ryegrass
Elymus virginicus L.

Elymus species are erect tufted perennials with short rhizomes. Spikelets are sessile, arranged in groups of usually 2 each alternately on 2 sides of a spike. This arrangement is most apparent at the base of the spike or if the spike is bent. Glumes and lemmas are awned or rarely blunt or merely acute. The glumes are not on opposite sides of the spikelet but are side by side so that if there are 2 spikelets there is an arc of 4 glumes; if 3 spikelets, an arc of 6 glumes.

In *E. virginicus* the largest glume is 1.5–2 mm broad at widest part and the awns are nearly straight. An extremely variable and widely distributed species. Common. In thin woods or in the open; depressions, meadows, shores of freshwater marshes, swales, shell mounds, fields; Alta into Nfld; S into Ore, Ariz, and nFla. June–Oct.

E. villosus Muhl. ex Willd. is similar but glumes are under 1 mm wide. Rare. Similar places; sCan; S into eWyo, eTex, and SC. June–Aug.

CYPERACEAE: Sedge Family

Sedges are common constituents of most habitats and can be abundant. Some are easy to identify, many require a specialist. We are including examples of most genera to illustrate variation in the family.

Most members of this family are much like grasses vegetatively and are often confused with them although they are easy to separate. In sedges leaves are 3-ranked, internodes are usually solid and triangular in cross-section, and edges of leaf sheaths are fused. Florets are single in the axil of each scale (also called bract). The scales are either 2-ranked or in tightly spiraled groups called spikelets, the lower 1–2 scales sometimes without florets. The fruit is an achene. Comparison of the grasses and sedges in tabular form is included on p. 139 in the introduction to grasses.

532   Yellow Nutgrass
Cyperus esculentus L.

Cyperus is one of the easiest sedges to recognize. Leaves have a midrib. Florets are in terminal inflorescences and are 2-ranked; the spikelets therefore are much like those of some grasses. They differ, however, in that all scales of *Cyperus* bear florets in contrast to the empty glumes in almost all grasses. Approximately 40 *Cyperus* species occur in the eUS. We are including 2 species to illustrate the 2 types of spikelet arrangement: one in which the individual spikelets are easy to locate, the other difficult.

Yellow Nutgrass is a perennial to 70 cm with slender rhizomes that bear hard tubers. Leaves mostly basal and well-developed. Spikelets straw-colored to golden-brown, numerous but individually easy to identify, 8–30 florets, and most attached horizontally in spikes. Spikes a few to 30 or more in terminal panicles, the spike axis readily seen. Achenes 3-angled, shedding with scales at maturity leaving the spike axis. The tubers are edible. A troublesome weed in many places. Common. Fields, yards, gardens, borders, meadows, dunes, beach sands, gravel bars of streams, swales, meadows, roadsides, parking areas, paths; Tex into Wash, sMan, sOnt, sQue, NS, NC, and Fla. June–Nov.

533 Flatsedge

Cyperus retrorsus Chapm.
Perennial to 1 m, base thickened and hard. Leaves mostly basal. Spikelets 2–6 mm, with 1–4 florets, but individual spikelets not easily recognized; they are numerous, tightly and spirally arranged in spikes, the uppermost erect or nearly so, the middle ones about at right angles, and the lower ones retrorse. Spikes are few to many, short-columnar and easily confused with the spikes of some *Scirpus* species, which also have spikelets spirally arranged. Common. Dry to wet places; swales, margins of fresh and brackish marshes, meadows, pond margins, ditches, fields, pinelands; seNY into Pa, NJ, eVa, NC, Fla, and Tex; Okla into Ky. July–Oct. *C. globosus* Aubl.

   C. croceus Vahl is similar but the spikes are globose, the spikelets with 3–6 florets and radiating outward in all directions. Common. Similar places; eTex into Okla, seMo, NY, NJ, eVa, and Fla. July–Oct.

   Dulichium is the only other genus of sedges with florets 2-ranked. Short axillary clusters of spikelets separate it from *Cyperus* with its terminal spikelets. *D. arundinaceum* (L.) Britt. is the only species of the genus. It is common, occurring in borders of ponds, sloughs, mucky soils, and other wet habitats; Minn into Nfld, Fla, and Tex; BC into Calif and Mont. July–Oct.

534 Umbrella-grass

Fuirena pumila (Torr.) Spreng.
In *Fuirena* species the triangular stems have rounded corners and the lowest leaves consist entirely or essentially of a sheath only. Floret scales are many, spirally and compactly arranged in spikelets; spikelets may be terminal and solitary or sometimes with an additional short-peduncled spikelet in each of 1–3 of the next leaf axils below. Each floret has a perianth of 3 distinctive paddlelike scales alternating with 3 bristles. There are only about 5 species in the eUS.

   F. pumila is a tufted annual to 30(60) cm, with soft stems. The upper leaf blades are spreading to ascending, to ca 20 × 1 cm, basal edges with spreading hairs. The spikelet scales are ca 3 mm and have a conspicuous terminal awn; the 3 perianth scales behind each of them are about as long and have a mostly ovate blade, and the tip bristlelike. Common. Moist to wet places; swales, pond margins, depressions, swamps; sArk into neIll, nwInd, sMich, Mass, and Fla. July–Oct.

   F. squarrosa Michx. is similar but is a perennial from short scaly rhizomes, sometimes taller, to 1 m, and with similar leaf blades. Common. Similar places; eTex into eOkla, sArk, Ky, NY, NJ, eVa, SC, Fla, and Miss. July–Oct. *F. hispida* Ell. *F. scirpoidea* Michx. (Leafless Fuirena) is an erect glabrous perennial from well-developed rhizomes beneath the soil surface. Leaves mostly a sheath only; the blade, if any, is cupped, erect, and to 4 mm. Spikelets ovoid to lanceoid or ovoid-lanceoid, 7–12 mm, 1–5 sessile in a terminal cluster, the basal bract shorter than the spikelets. Common. Wet pinelands, swales, depressions, edges of brackish marshes, ditches; Fla into sGa and Miss. Mar–Aug.

535 Saltmarsh Bulrush

Scirpus robustus Pursh

Scirpus species are erect perennials with unbranched stems. Each flower hidden by a scale; scales many, spirally arranged in compact spikelets; spikelets 1–many in terminal inflorescences, sometimes appearing lateral when there is only 1 scale and it appears to be an extension of the stem. The perianth is represented by bristles that may be tiny and missed even if observed with a hand lens or may be conspicuous when the bristles project beyond the scales.

This species is a perennial to 1.5 m from a rhizome; stem hollow. Basal leaves of acuminate sheaths only, those along stem to near inflorescence with blades. Scales at base of inflorescence 2–4, the largest similar to leaves just below. Spikelets 3–many, 10–30 × 8–12 mm, some to all pedicelled. Common. Higher portions of salt or brackish marshes, swales, and depressions; nFla into eTex, sArk, NS, Mass, and nFla; Calif; W. Indies; e S. Amer. July–Oct.

536 Marsh Bulrush

Scirpus cyperinus (L.) Kunth

Perennial to 2 m, usually in large clumps, often in extensive colonies. Leaves long, curving, and conspicuous, basal ones especially long. Inflorescence large, the branches arching and drooping. Scales at base of inflorescence 2–3, at least 1 about as long as leaves just below. Spikelets ovoid, 3–5 mm, reddish-brown; perianth bristles 6 per floret, curly, much longer than scales, causing the spikelets to appear wooly. Achene yellowish to off-white, 0.7–1 mm. We have used with pleasant results the lower stems gathered at an early stage as a flower arrangement. Common. Freshwater marshes, margins of ponds and gentle streams, swamps, wet meadows, sloughs; eTex into sOkla, Minn, nIll, O, NY, Nfld, nMd, NC, and Fla. July–Oct.

537 Swordgrass

Schenoplectus americanus (Pers.) Volk ex Schinz & R. Keller

Much like some *Scirpus* species but may be recognized as a perennial arising from coarse reddish branching rhizomes. Plants to 1.5 m; stems triangular and sharp-edged. Lower leaves with a sheath only, above these a few leaves with flat blades to 20 cm. Spikelets 1–25, sessile, 5–20 mm, scales awnless or with a minute to conspicuous terminal awn. There is 1 bract at base of spikelet(s), 3–12 cm, and appearing as an extension of the stem, thus the spikelets appear lateral. Achenes ca 3 mm. Also called Chair-maker's Rush. Common. Freshwater or brackish marshes, swales; Fla into Tex and Mex; Mo into Neb, Minn, Ky, sOnt, sQue, Nfld, and eVa; Ida and Ore; C. and S. Amer. June–Sept. *Scirpus a.* Pers.

538 Spike-rush

Eleocharis tuberculosa (Michx.) R. & S.

Eleocharis is a distinctive genus. Leaves consist of a bladeless sheath only. Flowers are in axil of scales spirally arranged in a single terminal spikelet with no bracts below it; the perianth consists of 6–9 inconspicuous bristles; fruits are achenes. Species vary from 1 cm to 1 m. Identification of most species is difficult; good light and magnification are needed to determine size, shape, surface features, and accompanying structures among other criteria used to identify the 30 or so species in the eUS.

E. tuberculosa is a densely tufted perennial to 80 cm with firm stems 0.5–1 mm across and somewhat flattened. The achene is obovoid, 1–1.5 mm, surface roughly and coarsely honeycombed; on the summit is a roughly ovoid tubercle about as long and broad as the achene; the tubercle is fitted well on achene summit but actually barely attached. Common. Moist or wet places, in sun or shade; freshwater marshes, depressions, pond margins; eTex into sArk, Tenn, seNY, NS, ceNH, neMass, RI, sConn, eVa, CP of NC and SC, and Fla. June–Sept.

539   Fimbristylis

Fimbristylis castanea (Michx.) Vahl

This genus is not very distinctive and sometimes is difficult to distinguish from some other genera; several characters need to be checked. Leaf sheaths are short-ciliate or entire; no conspicuous bracts occur below the inflorescence; scales are spirally arranged, spikelets several to many, each spikelet having 1 empty scale at base. Flowers have no perianth of any sort; base of style is enlarged; fruit with no tubercle on summit. Identification of most species requires good light and magnification and perhaps a little luck.

This species is a densely tufted perennial to 1.8 m and set deeply in the ground. Upper part of stems nearly circular in cross-section or differing at the most to elliptical. Leaves thick, ascending, mostly basal, rarely any from the upper two-thirds of the main stem, usually under 2 mm across. Any bracts at base of inflorescence shorter than it is. Spikelets 5–10 mm. Achenes obovoid to lenticular, 1.5–2 mm. Common. Salt or brackish marshes, swales, sloughs; seNY into eVa, cNC, seGa, and Fla. June–Oct.

F. caroliniana (Lam.) Fern. is similar but rhizomatous, the stems solitary or in small tufts; upper part of stems flattened instead of rounded. Common. Beaches, swales, pond shares, ditches, limestone outcrops, brackish or saline sands; along coasts; NY into Fla, sGa, and Tex; also Mo into Ill, sOnt, and Pa. Jun–Oct.

A related genus, *Bulbostylis,* is much like *Fimbristylis* except leaf sheaths are long-ciliate and achenes bear a tubercle on the summit.

540   Beaked-rush

Rhynchospora corniculata (Lam.) Gray

In this genus scales are arranged spirally in spikelets of several to many scales but only 1–2 with achenes. Spikelets are numerous in terminal and/or axillary clusters on the upper portion of the plant; single spikelets or small clusters of them have evident stalks. Fruits are achenes terminated by a tubercle 0.1–20 mm; nearly always some perianth bristles at base.

R. corniculata is a perennial to 1.5 m, usually tufted, with coarse leaves to 2 cm across. Inflorescence to 55 × 30 cm. Spikelets lanceolate, about 25 mm. Achene 3.5–5 mm, the 10–20 mm tubercle long-subulate and protruding prominently beyond scales, perianth bristles shorter than to as long as achene body. Common. Shallow water, pond and lake shores, ditches, depressions, swamps, sloughs; eTex into Mo, Ky, Del, e and csVa, csNC, SC, and Fla. June–Oct.

R. macrostachya Torr. ex Gray grows to 1.3 m and has almost identical tubercles but the perianth bristles are longer than the achenes; spikelets are in tight clusters and the inflorescence branches are more erect. Occasional. Similar places; eTex into wArk, Mo, NY, seMe, eNC, and cGa. July–Oct.

541 Whitetop Sedge

Rhynchospora latifolia (Bald. ex Ell.) Thomas

The conspicuous white leaves (bracts) at the top of this grasslike plant identify it and also Star-rush, making them visible from considerable distances. The bases of the bracts surround the inconspicuous flowers.

This species has 7–10 bracts; the plants are 30–70 cm and arise from a rhizome 2–4 mm across. Occasional. Moist to wet situations in open or thin woods; depressions, savannas; CP of seVa into Fla and cTex. Apr–Sept. *Dichromena l.* Baldw.

The similar but generally smaller *R. colorata* (L.) H. Pfeiff. (Star-rush) usually has less than 7 bracts and the rhizome only 0.5–1.5 mm across. Common. Similar habitats. CP; seVa into Fla, cTex, and sArk; cAla. Apr–Sept. *Dichromena c.* (L.) Hitchc.

542 Nut-rush; Stone-rush

Scleria triglomerata Michx.

Scleria plants are easily distinguished by the white to light cream or grayish bony achenes unique to this genus. They are annuals and perennials; stems are sharply 3-angled; the lowest leaves lack blades, the upper ones have keeled blades. Flowers unisexual, the sexes in the same cluster. Fruit without a tubercle. Perianth absent. Identification of the ca 10 species in the eUS can be difficult.

This species is a tufted perennial, generally 50–75 cm from a short thick rhizome; leaf blades to ca 40 cm × 4–8 mm; sheaths usually reddish. Spikelets are few-flowered, scales spirally arranged. Achenes are white (rarely to gray), smooth, shiny, subglobose to oblong, 2.5–3.5 mm, base surrounded by a thin finely granular disc. Common. Live oak and maritime woods; mixed deciduous woods, pinelands; eTex into Kan, Minn, sOnt, Vt, Mass, and Fla. May–Oct.

S. reticularis Michx. is similar, to 70 cm; blades only 2–5 mm across. Achenes dull white to grayish, surface with a network of small ridges, the base clasped by 3 bractlike structures. Occasional. Moist or wet places; pond shores, depressions, low pinelands, slough margins, low meadows; eTex into Ark, sMinn, sWisc, sMich, sOnt, NY, sMass, NC, and nFla. June–Oct.

543 Sedge; Caryx

Carex lupuliformis Sartwell ex Dewey

The genus requires only 2 characters to identify: leaf blades have a distinct midvein and the pistil (and fruits) are surrounded by a unique saclike structure, the *perigynium.* Flowers are unisexual, the sexes in separate spikes or in different parts of the same spike. Both types may be seen in the species illustrated; the terminal spike contains only male flowers, the basal spike only female flowers, and the 3 other spikes have male flowers toward the summit (one is hidden). Spikes of *Carex* species vary much in size and shape. They are borne singly on stalks as in the photograph, or in dense clusters. There are around 250 species in the eUS; identification of most is complicated and tedious. Little can be gained by giving the diagnostic features of any species in a book of this nature.

The species illustrated is a common one, occurring in depressions, sloughs, swales, swamps, pond shores, and bogs; nArk into Minn, Vt, Conn, Va, Ky, La, and eTex. June–Oct.

544   Fraser Sedge

Cymophyllus fraserianus (Ker-Gawl.) Kartesz & Gandhi
This genus and *Carex* are the only genera with a sac (perigynium) around the pistil; the lack of a midvein on the leaves identifies *Cymophyllus.*

This is the only species of the genus. A perennial from short thick rhizomes; stems to 40 cm, obscurely 3-sided, coarsely ribbed, and glabrous. Leaves basal; blades flat, 10–45 × 3–4 cm, both surfaces glabrous, apex acute to obtuse, margin wavy and slightly scabrous. Inflorescences single terminal spikes; flowers unisexual, each in axil of a scale, female below the male; scales faintly veined, white, apex obtuse. Perigynia 3-veined, white, ovoid to subglobose, 4–6 mm. Rare. Rich woods; sPa, WVa, wVa, wNC, eTenn, and nGa. May–June.

ARACEAE: Arum Family

545   Golden-club; Never-wet

Orontium aquaticum L.
The common name "Never-wet" alludes to the leaves which, when submerged, will come out of the water dry and water drops will run across the surface. If a drop remains on the surface it often appears jewellike when reflecting light. The flowers are many at the surface of the golden club (the spadix). Fruits blue-green, thin-walled, 1-seeded. Although the raw fruits usually cause an unbearable burning sensation, the Indians used the fruits abundantly for food, boiling them extensively in water or drying them thoroughly before eating. Roots were sliced and dried, or boiled or roasted and eaten or used for flour. Occasional. Swamps, bogs, and shallow streams; Fla into La, seKy, cNY, and Mass. Feb–June.

546   White Arum

Peltandra sagittifolia (Michx.) Morong
Perennial with arrow-shaped leaves. A large white sheath (spathe) surrounds the fleshy axis (spadix) which bears the flowers. The flowers are of 2 sexes, the male flowers on the upper part of the spadix and female flowers on the lower part. The mature fruits are red. Apparently all parts of the plant contain calcium oxalate, which may cause severe burning when eaten. Thorough and prolonged drying of the rootstocks has been reported to render them edible. Rare. Swamps; CP of NC into Fla and Miss. July–Aug. *P. glauca* (Ell.) Feay.

The vegetatively similar *P. virginica* (L.) Schott has a green spathe that is tighter around the spadix, and has greenish to amber fruits. The fruits are an important food of the wood duck. Common. Wet habitats; Fla into Tex, seOkla, eMo, sOnt, and Me. May–June.

547   Jack-in-the-pulpit; Indian-turnip

Arisaema triphyllum (L.) Schott
A variable harbaceous perennial from a corm up to 5 cm thick. Plants sometimes over 1 m. Leaves 1–2, palmately divided into 3–5 leaflets, the outer leaflets

sometimes with a lobe near the base. Flowers unisexual, sessile on the fleshy axis (the spadix or "Jack"), the male flowers above and female below, or all one sex. The spadix is sometimes curved. The spathe (the "pulpit") has a tube and a hood that arches over the spadix. The upper part of the spadix may be green to deep maroon and the spathe green to partly maroon (sometimes completely so on the inner side). Fruits fleshy, 1–few-seeded, scarlet when mature. The plant contains small needlelike crystals of calcium oxalate, particularly in the corm, which, if taken into the mouth, become imbedded in tissues and provoke intense irritation and a burning sensation. The corms usually are not edible even after cooking. Thin slices thoroughly dried for several months are reported to be palatable. Common in rich woods, often moist to wet; Fla into eTex, Minn, and PEI. Mar–June.

In *A. dracontium* (L.) Schott, the Green-dragon, the flowers are arranged similarly to those of Jack-in-the-pulpit. The spadix tapers to a long slender tip beyond the much shorter sheathing spathe. Fruits in a tight cluster, orange-red when ripe. The 1 leaf is divided into 7–15 leaflets arranged somewhat in a band parallel to the ground. The corms are similar to those of Jack-in-the-pulpit in respect to ingestion. Common. Usually moist places, rich soil of dense deciduous woods; Fla into cTex, seMinn, sQue, and seNH. May–June. *Muricauda d.* (L.) Small.

XYRIDACEAE: Yellow-eyed-grass Family

548 Yellow-eyed-grass

Xyris ambigua Bey. ex Kunth

Species of this genus are annuals or perennials with inconspicuous flowers and basal linear to terete-filiform leaves that often resemble in shape and arrangement those of Irises. Individual plants can be fairly easily located by looking for small yellow spots, the bright yellow corollas. These protrude from behind woody scales that are in a compact terminal spike on a leafless stem. Identification to species is often dependent on minute characters of the seeds and the peculiar sepals as well as on characters of the leaves and scape. About 18 species occur entirely or mostly in the eUS.

In this species the sheaths around the base of the scape are shorter than the larger foliage leaves, the spikes 1–3 cm and lance-ovate to ellipsoidal, and the lateral sepals shorter than the subtending scales and with margins rough to the touch. Common. Moist sands or sandy peats of pine flatwoods, savannas, bog margins, lakeshores, and roadside ditches; Fla into eTex, lower CP of Ga, and seVa; also scattered localities inland from eAla into cNC. Apr–Aug.

ERIOCAULACEAE: Pipewort Family

549 Pipewort; Buttonrods; Hatpins
550 *Eriocaulon decangulare* L.

Members of this genus are conspicuous when in flower because of the dense white heads of flowers, "buttons," on the tip of leafless stems. The buttons provide a strong contrast in a sea of other generally less conspicuous plants, such as grasses and sedges. They are perennials with basal linear leaves and are separated from related genera by having in the leaves air spaces visible to the naked eye, jointed essentially unbranched roots, and a jet-black gland on each petal lobe.

In this species the scape is finely 8–12-ridged, the leaves are longer than the sheath around the base of the scape, and the mature heads are 10–20 mm broad. Common. Sandy or peaty soils of pine flatwoods, cypress swamps, ditches, and lakeshores; Fla into seLa, seMiss, sAla, CP of SC, and c and ceNJ; se and ceTex and adjLa; nwSC into sw and cNC; wPied of Ga. June–Oct.

BROMELIACEAE: Pineapple Family

551   **Spanish-moss**
Tillandsia usneoides L.
This relative of the pineapple hangs from trees and sometimes from other objects such as telephone lines and fences. The stems are slender and wiry, the leaves filiform. Both bear numerous small silver-gray scales that trap water and nutrient-providing dust particles. The plants superficially resemble mosses but are flowering plants. A flower is shown in the picture. Plants have been used as forage and in upholstery and mattresses. Campers should be careful about using Spanish-moss as bedding because of the danger of Red-bugs or Chiggers. Common. Pendent on trees from swamp to upland; Fla into s and cTex, sePied of Ga, and eVa. Apr–July.

COMMELINACEAE: Spiderwort Family

552   **Dayflower**
Commelina erecta L.
A variable perennial from thickened roots; leaves with a closed basal sheath. Flowers borne in conspicuous folded bracts (spathes); spathes terminal or axillary, open across the top and down the side of the nearest peduncle; appearing singly each day or so. Petals 3: 2 blue, 1 whitish and much smaller; all ephemeral, sometimes withering before noon. Fertile stamens 3. Young stems and leaves of some species of Dayflowers have been used as potherbs outside the US; seeds are eaten by Mourning Doves, Quail, and several kinds of songbirds. Common. Dry often sandy soils in open or thin woods; Fla into Ariz, cnNeb, Tenn, and seNY. June–frost.

C. communis L. also has 1 pale petal. The species is a weedy annual with fibrous roots. Spathes open across the top but base is fused. Native of eAsia. Common. Moist places in woods or open, gardens, ditches; nFla into Tex, ND, NH, and NC.

553   **Woods Dayflower**
Commelina virginica L.
Erect perennial to 1.2 m from a creeping and branching rhizome. Flowers from within greenish spathes, which are usually in a terminal cluster. Spathes are fused at their bases. All petals blue, ephemeral, 1 slightly smaller than the others. Occasional. Moist to wet places, in woods or rarely in open; nFla into e and cTex, Ill, Ky, sPa, and sNJ. June–Oct.

C. diffusa Burm. f. also has 3 blue ephemeral petals but is smaller, an annual, the spathe bases are not fused, and the much-branched stems are decumbent and rooting at the nodes. Occasional in seUS; rare elsewhere. Fla into s and eTex, Kan, Minn, sO, and Del; mostly absent from mts. June–Oct.

554 Marsh-dayflower
Murdannia keisak (Hassk.) Hand.-Maz.

Plants resembling *Commelina* but flowers not borne in a conspicuous sheath, fertile stamens 2–3. Annual to 1 m, branching abundantly; extensively creeping, prostrate, or reclining; rooting at nodes. Flowers terminal and from upper leaf axils, usually developing into racemes; sepals 5–6 mm, petals 7–9 mm. Native to eAsia. Occasional. Freshwater marshes, stream banks, seepage places; chiefly Pied and CP; eMd into Fla and La; inland to Ky and Ark. Aug–frost. *Aneilema k.* Hassk.

In *M. nudiflora* (L.) Brenan (Dove-weed) flowers are axillary, sepals 2–3 mm. Occasional. Drier places; woods or in open; SC into Fla and Tex. *Aneilema n.* (L.) Sweet.

555 Spiderwort
Tradescantia virginiana L.

Members of the genus are perennials with ephemeral petals and 5–6 fertile stamens, each with the filaments bearded.

This species has blades of the upper leaves narrower than to about as broad as the basal sheath, which is unfused. Pedicels and sepals hairy their entire length. Petals blue to purple or purplish-pink, or rarely white. Occasional. Usually in well-drained areas in the open; nGa into neArk, Wisc, Pa, and Me. Mar–July.

T. ohiensis Raf. is similar but the pedicels and sepals, except perhaps the tips, are not hairy and the leaves are glaucous. Common. In a variety of open better-drained places; Fla into Tex, Neb, Minn, and Mass. Apr–July. *T. hirsuticaulis* Small is densely hairy and has glandular as well as nonglandular hairs on the sepals. Occasional. Dry woods and rocky places; Ga into Pied and mts of NC, and Ala. Apr–June.

556 Roseling
Callisia rosea (Vent.) D. R. Hunt

Perennial, glabrous or nearly so, to 50 cm with a tuft of narrow basal leaves. Flowers subtended by bracts no longer than 1 cm, in contrast to the prominent bracts of *Tradescantia* species. Common. Sandy soil, usually thin woods or in open; cFla into cCP of Ga, cn and coastal SC. May–Aug. *Cuthbertia r.* (Vent.) Small; *Tradescantia r.* Vent.

In *C. graminea* (Small) G. Tucker the leaves are less than 3 mm wide. Common. Dry sandy areas, scrub oak or pine barrens; pen Fla into cCP of Ga, and sCP of NC; seVa. *Cuthbertia g.* Small.

PONTEDERIACEAE: Pickerel-weed Family

557 Water-hyacinth
Eichhornia crassipes (Mart.) Solms

Often seen floating on water in such a dense mass of deep-green foliage that the water and the distinctive inflated petioles are not easily seen. Largest leaves on any one plant may be only a few centimeters long, or reach 1 m. Flower clusters are on erect peduncles and are a striking contrast to the deep-green foliage. Plant masses are a serious problem in drainage ditches, canals, lakes, ponds, etc. Its abundance diminishes greatly in the cooler parts of its range. Young leaves and flower clusters have been steamed or boiled and eaten outside the US. Common. sCP of NC into Fla and Tex; introduced and sometimes persisting in Pied localities, such as in nearly constant temperature water of springs. May–Sept.

558   Pickerel-weed
Pontederia cordata L.

A soft-stemmed perennial to 1 m. One leaf not far below the flowers, the others basal. Leaves cordate to lanceolate or rarely linear, the largest 1–14 cm wide. Perianth purplish-blue or rarely white, the upper segment with a yellow area. Mature fruits are reported by some to be a pleasant food; stems and leaves may be edible. The roots are inedible, producing a severe burning sensation. Common. Usually open places, at margins of or in water; Fla into Tex, seKan, ceMinn, and NS; rare to absent inland to the Great Lakes area. Mar–Oct.

   P. lanceolata Nutt. has been separated as a species and a variety, but intergradation of characters seems too great for taxonomic recognition at either level.

JUNCACEAE: Rush Family

559   Footpath Rush
Juncus tenuis Willd.

In members of this genus stems are solid; leaves 3-ranked, their bases sheathing but open down 1 side; flower structure similar to that of Onions but perianth chaffy; fruits 3-carpelled many-seeded capsules.

   This species is a highly variable perennial; leaf blades flat with chaffy ears at base; flowers borne singly on branches of an apparent terminal inflorescence shorter than the leaf at its base. Common but mostly scattered in CP. Weedy, most persistent in and at edges of beaten paths, in lawns, woods, etc.; borders, roadsides; nearly throughout N. Amer. June–Sept.

560   Black Rush
Juncus roemerianus Scheele

A perennial forming prominent dense stands or occasionally small dense clumps. Some of the basal leaves are terete, to over 1 m, and with stiff sharp points, the others being mere basal sheaths. Stem resembling the leaves, at its top, as shown in the picture, bearing a repeatedly forked inflorescence and a terminal sharply pointed leaf. The inflorescence thus appears lateral. Flowers are in clusters at the tips of the branches. Common. Salt marshes and brackish ditches along the coast; Fla into seTex and Md. Mar–Oct.

   J. effusus L. is similar but the flowers are borne singly at tips of the branches and the terminal leaf is soft-pointed. Common. Moist depression, edges of ponds and lakes, freshwater marshes; Fla into eTex, seKan, ceWisc, and Nfld. Apr–Sept.

LILIACEAE: Lily Family

561 False-asphodel
Tofieldia racemosa (Walt.) B.S.P.
Slender perennial to 70 cm from a short rhizome; leaves mostly basal, linear, to 40 cm. Flowers in a terminal raceme, usually 2–6 from each node, the upper ones opening first. Scape roughened with minute projections. Perianth persists and becomes rigid against the fruit. Mature fruit about equals the perianth. Common. Wet sandy soils of pinelands and savannas; Fla into seTex, swSC, and NJ; wPied of Ga. June–Sept. *Triantha r.* (Walt.) Small.

 T. glutinosa (Michx.) Pers. is similar except the perianth does not become rigid and the mature fruit is about twice as long as the perianth. Rare. Bogs, marshes; neGa into neTenn, eWVa, neO, neIll, Alas, and Nfld. June–Aug. *Triantha g.* (Michx.) Baker. *T. glabra* Nutt. is also similar but the scapes are glabrous and smooth to touch. Rare. Wet savannas and pinelands; nCP of SC into ce and seNC. Aug–Oct.

562 Swamp-pink
Helonias bullata L.
The beauty of the tight cluster of many small pink lilylike flowers on the leafless stalk is enhanced by the elongate evergreen leaves at the base. After flowering, the stalk elongates to as high as 1 m. Rare. Swamps and bogs; nwSC, nwGa, swNC, nw and eVa into mts of Pa and seNY. Apr–May.

563 Blazing-star; Devil's-bit
Chamaelirium luteum (L.) Gray
Glabrous perennial with male flowers on one plant and female on another. Female plants are the taller, to 120 cm. Basal leaves large, the upper much smaller. Male inflorescence yellow at first, erect or nearly so, as shown. When all flowers have opened the inflorescence droops. Female inflorescence less conspicuous, the small flowers greenish and white. Occasional. Rich woods; Fla into seArk, Tenn, sIll, sInd, sOnt, and wMass. Mar–June.

564 Featherbells
Stemanthium gramineum (Ker-Gawl.) Morong
Glabrous perennial to 1.5 in from a slender bulb, often in small colonies; main stem leafy. Flowers in a terminal panicle of racemes; sepals and petals widest at base, united slightly at base, and united with base of ovulary, lobes narrowly lanceolate. Occasional. Thin woods, thickets; meadows; nwFla into eTex, Mo, Ill, Pa, Md, ceNC, nwSC, and nAla. July–Sept.

565 Crow-poison; Black-snakeroot
Zigadenus densus (Desr.) Fern.
Perennial to 1 m from a bulb, with 1–3 basal leaves, 2–7 mm wide, those above reduced to widely separated bracts. Flowers in a raceme. Bracts at base of each flower stalk yellowish and nearly straight at the tip. Tepals 6, nearly white to pink, each with 2 small glands at their base. Stamens 6. Probably not poisonous.

Common. Wet places in thin pinelands or open; CP—eTex into Fla and seVa. Mar–May.

The above species is often confused with *Amianthium muscaetoxicum* (Walt.) Gray, Fly-poison, but the basal leaves of the latter are more abundant and usually wider, 4–23 mm; the bracts at base of flower stalks are brownish and slightly cupped at the tip; and glands are absent. Leaves and bulbs are quite poisonous. Occasional. In various situations, wooded or open, usually moist; Fla into Okla, sMo, Ky, and sNY. Apr–July.

566 Bellwort

Uvularia perfoliata L.

Glabrous perennial to 40 cm. Most leaves perfoliate, the blade-bearing leaves below lowest branch usually 3–4. The 6 tepals conspicuously granular-papillate within. Capsule lobes deeply 2-lobed or 2-horned. Occasional. Usually in deciduous woods on well-drained areas; occasionally elsewhere; nwFla into La, sOnt, and Mass. Feb–May.

Another species, *U. grandiflora* Sm., also has perfoliate leaves. There is usually only 1 leaf below the lowest branch, the tepals are smooth within, and the main capsule lobes are neither 2-lobed nor 2-horned. SC into eOkla, ND, Que, and Me. Apr–May.

567 Bellwort

Uvularia puberula Michx.

Perennial to 45 cm. Upper stems 3-angled and usually finely hairy. Leaves sessile. Tepals and stamens 6 each. Capsule pointed. Occasional. Usually in rich well-drained woods; SC into Ga, eTenn, sPa, and seNY. Mar–May. *U. pudica* sensu Fern.

Two other species have sessile leaves. Both have glabrous stems. One, *U. floridana* Chapm., may be recognized by a sessile capsule and a bract 1–3 cm near the tip of the short flowering branchlet. Rare. Bottomland and floodplain woods; nFla into eAla, and SC. Mar–Apr. The other, *U. sessilifolia* L., has no bract and the capsule has a short stipe. Common. Hardwood forested slopes; nwFla into La, ND, Que, and NS. Mar–May.

568 Day-lily

Hemerocallis fulva (L.) L.

Perennial to 1 m. The basal tube of the flower tightly encloses the ovulary but is not fused with it. Fully grown flower buds and flowers have been used for food. These may be cooked in oil and butter in a batter of eggs, flour, and milk. They may be added for flavor in final stages of cooking of soups, meats, and other foods. Occasional. Widely domesticated, and escapes occur along roadsides, abandoned homesites, and borders of fields; Fla into eTex, eNeb, Minn, and NB. May–July.

H. lilioasphodelus L., which has yellow flowers and is smaller, is natzd. locally, as are a few of the many other horticultural types that have been introduced more recently than the 2 above species. *H. flava* (L.) L.

569 **Wild Onion; Canada-garlic**
Allium canadense L.
The odor identifies this species as one of the onions. Smooth flat mostly basal leaves 2–6 mm wide, netted fiber bulb-coats, and bulblets usually replacing some or all flowers characterize this species. The bulbs and young leaves can be boiled and eaten, the liquid used in soup. The top bulbs are used for pickling. Large amounts may be harmful. Livestock eating tops or bulbs of Onions have developed anemia and jaundice. Common. Roadsides, fields, thin woods, most frequent in moist places; nFla into eTex, ND, and NB. Apr–July.

Other Onions with flat leaves include: Ramps, *A. tricoccum* Ait., which flowers as or after the broad (2–6 cm) leaves shrivel; nGa into Minn, NB, Del, and eNC. June–July.

570 **Wild Onion**
Allium cuthbertii Small
Perennial to 65 cm. Leaves long, flat, to 6 mm wide, usually withered or withering at flowering time. Bulb covered with a mat of fibers. Inflorescence of flowers only. Perianth segments acuminate. Ovulary and fruit crested. Occasional. Sandhills, scrub oak, open woods, granitic outcrops; neFla into eAla and cnSC; nw and cNC. Apr–June.

A. cernuum Roth also has flat leaves and no bulbs in the inflorescence. Bulb covered with membranes. Inflorescence nodding on a bent peduncle. Perianth segments obtuse or rounded at tips. Occasional. Rocky places, thin woods or in open; nSC into Ariz, BC, and NY. July–Sept.

571 **False-garlic**
Nothoscordum bivalve (L.) Britt.
Perennial to 45 cm from a small bulb that has a faint odor of onions when fresh. Leaves linear, less than 5 mm wide. Flowers 3–10 in terminal umbel. Perianth segments reflexed when fruits are developed. Onions have a strong odor and the perianth is not reflexed. Petals 1 cm or longer. Common. Fields, open woods, pastures, roadsides, thin soil around granitic outcrops; Fla into Tex, seNeb, cwInd, sO, nwGa, Pied of Ga, and seVa. Mar–May; Sept–Nov. *Allium bivalve* (L.) Kuntze.

N. gracile (Ait.) Stearn is similar but the plants are taller, to 80 cm, the leaves are more than 5 mm wide, and there are usually more than 15 flowers per umbel. Rare. Sandy soils, roadsides, waste ground, fields; Fla into seLa; central coast of SC. *N. fragrans* (Vent.) Kunth; *Allium inodorum* Ait.

572 **Carolina Lily**
Lilium michauxii Poir.
Perennial to 1.5 m. Leaves chiefly whorled, the blades with smooth margins and mostly broadest above the middle. Flowers nodding, usually 1–3, or to 15 when cultivated under favorable conditions, tepals strongly recurved. Capsules nearly erect. Attractive hardy plants worthy of cultivation. Named for Andre Michaux, French botanist who traveled widely in the seUS. Rare. Dry to rich woods; nFla into seTex, eTenn, sVa, and NC. May–Aug.

Other species with nodding to horizontal flowers are: *L. superbum* L., shown next. *L. canadense* L., with flared, usually lighter-colored tepals. Rare. Wet habitats; nwSC into nAla, Ky, Minn, and NS. June–Aug.

573   Turk's-cap Lily
Lilium superbum L.
Perennial to 3 m. Leaves chiefly whorled, the blades with smooth margins and usually widest at the middle. Flowers nodding, as many as 65 but more often 25 or fewer. Tepals are strongly recurved and their inside bases are green. Rare. Moist habitats; nwFla into Mo, Minn, NB, and nGa; cCP of NC. June–Aug.

Two tall cultivated species with all leaves alternate have become natzd. locally. *L. bulbiferum* L., the Orange Lily, is 0.6–1.2 m and has erect lightly spotted flowers. *L. lancifolium* Thunb. is about as tall and has nodding, heavily spotted flowers. *L. tigrinum* Ker.

574   Pine Lily
Lilium catesbaei Walt.
Perennial to 60 cm, the leaves alternate, narrow, and ascending. The single flower and fruit erect. Petals wider than the similar sepals; the tips of both, and especially the sepals, curved backward; the blades of both orange toward the tip, yellow and purple-spotted toward the base, which tapers into a slender claw. Common. Wet pinelands and savannas, bogs; CP—seVa into Fla and sMiss. June–Sept.

Our other native species with erect flowers is *L. philadelphicum* L., with 1–5 similarly colored flowers but whorled leaves. Rare. Thin woods, meadows and balds; nwGa, mts of NC, Pied of Va, Del, and N into Ky, sOnt, and Me. June–Aug.

575   Bell Lily
Lilium grayi Watson
Plants to 2 m, stems smooth and glabrous. Leaves in 3–8 whorls of (4)5–8(11); blades 4–13 × 0.8–2.5 cm, elliptic to lanceolate, margin scabrous. Flowers 1–8, narrowly bell-shaped, most nearly horizontal; perianth red to deep reddish-orange, heavily spotted almost to apex, segments 4–5.5 × 1–1.5 cm, summits slightly outcurved, petals widest beyond middle, base without a claw; stamens 2–3.5 mm. Rare. Balds; openings in forests; meadows, swamps, and low woods among the mts; cwNC into adjTenn and cwVa. June–July.

576   Dog-tooth-violet; Trout-lily
Erythronium americanum Ker
Perennial from a bulb. The yellow tepals are washed with light to dark reddish-brown. Anthers yellow or sometimes brown to lavender. The tip of the ovulary and capsule may be truncate, rounded, or pointed. Mature capsules are held well off the ground. The leaves are variously reported as emetic or edible as a potherb. Common locally. Rich woods; nAla into Okla, Minn, and NS; csNC. Feb–May.

E. rostratum Wolf has a capsule with a prominent beak at the apex. Rare. Moist rich woods; Ala into eTex, eKan, and cTenn. Feb–Apr. *E. umbilicatum* Parks & Hardin has a distinctly indented apex to the ovary and capsule, and mature capsules are reclining on or just above the ground. Common. Rich woods; nFla into eAla, sO, eWVa, and Va. Mar–May. *E. albidum* Nutt. has white tepals. Occasional. Woods and prairies; ne and cAla into neTex, Minn, Ont, Pa, and cTenn. Mar–May.

577   **Star-of-Bethlehem**
Ornithogalum umbellatum L.
A bulbous perennial, easily recognized by the conspicuous broad white filaments below the yellowish anthers. Bulbs are reported to be edible when cooked, but all parts of the plant are also reported to be poisonous to grazing animals. Caution should be taken accordingly. This introduced native of Europe has become natzd. in lawns, fields, pastures, and low places, including woods. Occasional. Ga into La, Neb, and Nfld. Mar–June.

    O. nutans L., a native of sAsia and natzd. locally, is similar but the lower flower stalks are about as long as the upper. *O. thyrsoides* Jacq., a native of South Africa, with a dense cluster of flowers, is widely sold by florists. Both species are poisonous when eaten and should be kept from children, especially.

578   **Wood-lily**
Clintonia umbellulata (Michx.) Morong
Perennial from a knotty rhizome. The 2–6 leaves are basal. Flowers mostly erect in an umbel on a single stem that may reach 50 cm. Tepals greenish-white. Fruit a berry, usually with 2 seeds and black when mature. The very young leaves have been used as a salad. Common locally only. Rich woods; nwSC into nGa, eO, NY, and NJ. May–June.

    C. borealis (Ait.) Raf., Corn-lily or Clinton-lily, is similar but the flowers are mostly drooping, the tepals are greenish-yellow and larger, and fruit has more than 2 seeds and is blue when mature. Common locally only. Rich moist woods; mts of Ga into Pa, neO, Minn, Man, and Lab. May–June.

579   **False-solomon's-seal; Solomon's-plume**
Maianthemum racemosum (L.) Link ssp. *racemosum*
Herbaceous perennial from a fleshy knotty brown rhizome. Plants finely hairy, to 1 m, most often little over half that tall. Leaves in 2 ranks, usually on the top half to two-thirds of the stem. Flowers small; tepals 6, white to greenish. Fruit a globose 1–3-seeded berry, reddish when mature, 4–6 mm across. The berries are palatable but cathartic. Common. Rich woods; Pied of Ga into Ariz, BC, and NS. Mar–June. *Smilacina r.* (L.) Desf.

580   **Two-leaved Solomon's-seal; Canada-mayflower**
Maianthemum canadense Desf.
The small white flowers and fruits are borne in a small raceme above the (1)2–3 leaves, suggestive of a diminutive False Solomon's-seal. The individual flowers are similar to those of False Solomon's-seal, the most noticeable difference being the reflexed perianth segments. Plants are rarely over 20 cm, from branched, fine, and extensive rhizomes. The fruits are a favorite food of the ruffed grouse. Common. Woods and thickets; mts of Ga, NC, and eTenn into Mich, Man, Lab, and Del. May–June.

581   Solomon's-seal

Polygonatum biflorum (Walt.) Ell.
Perennial from an elongated knotted white rhizome. Stems to 1.5 m, arching.
Leaves sessile, glabrous beneath. Flowers, axillary, 1–12 per peduncle. Fruit a
bluish-black berry 8–13 mm across. The young shoots have been prepared and
eaten like asparagus, and the white rhozomes have been eaten as a salad. Common. Rich woods; Fla into e and cnTex, Minn, sOnt, and Conn. Apr–June.
P. canaliculatum (Muhl.) Pursh.

   P. pubescens (Willd.) Pursh is similar but the leaves are short and hairy beneath,
especially on the veins. Usually not as large. Occasional; nGa into eTenn, nIll,
sMan, and NS. May–July.

582   Lily-of-the-valley

Convularia majuscula Greene
Perennial with 2–3 leaves. Flowering stem sheathed on the lower part of the leaf
bases. Inflorescence to half as high above the ground as the leaves. Plants not in
dense colonies, often scattered. Rare. Wooded slopes and coves; mts of nGa into
WVa and Va. Apr–June. *C. majalis* var. *montana* (Raf.) Ahles; *C. montana* Raf.

   C. majalis L. is similar but the plants form dense colonies and the
inflorescences are usually over half as high as the leaves. Abundantly reported in
Europe as poisonous when eaten. *C. majuscula* may therefore be poisonous. Rare.
Spreading from cultivation locally; NC and Tenn N into Me. Apr–June.

583   Purple Toadshade; Sweet Betsy

Trillium cuneatum Raf.
Trilliums are perennials with a single stem (technically a peduncle) arising from
a rhizome, the stem summit bearing a single whorl of 3 leaves (actually bracts as
the true leaves are the scales on the rhizome) and a sessile or stalked flower with
3 sepals, 3 petals, and 6 stamens. Some species are easily recognized on sight,
while others may be baffling to identify, often depending on combinations of
characters some of which may be difficult to determine. This is especially true of
pressed and dried specimens. Some descriptions below are necessarily abbreviated
and may give a questionable identification.

   T. cuneatum is one of the more abundant and widely distributed sessile-
flowered species. Leaves mottled with 2 colors, edges overlapping. Flowers with a
sweet and spicy odor; petals purple to brown or green, rarely yellow, more than
twice as long as wide, base wedge-shaped; stamens of uniform size and shape, less
than 1.5 times as long as ovulary, filaments much shorter than the erect anthers,
anthers dehiscent on side toward pistil, prolonged part of connective wider than
long; ovulary ovoid, not angled, stigmas divergent to spreading. Common. Rich
broadleaf woods, bottomlands, rocky slopes, bluffs; cw into eMiss, cs and ceKy, e
half of NC, nSC, and cwGa. Mar–Apr.

   There are several similar species including: *T. sessile* L. (Sessile Toadshade; Sessile Trillium), with filaments much shorter than anthers, connective prominently
prolonged beyond pollen sacs; stigmas 1.5–2 times as long as ovulary. Common.
Rich woods, floodplains, river terraces; mostly in limestone areas; sometimes on
dry uplands; cTenn; nArk into neOkla, e edge Kan, and sIll; cKy into eIll, sMich,
ePa, WVa, Va, and neNC; cnAla into middle third of Tenn. Mar–mid-May. *T.*

maculatum Harbison (Mottled Trillium) has petals 4.5 times or more as long as wide, narrowly spatulate; in a side view the inside of flower is visible between petal bases; anthers on the outer whorl of stamens broader than those of inner whorl; ovulary ovoid, 3-angled at base of stigmas. Rare. Rich woods, banks and bluffs of rivers; cnFla into s and ceAla, and sw corner of SC. Feb–early Apr. *T. underwoodii* Small (Underwood Trillium), the only erect Trillium with stem less than twice as long as the sessile leaves, leaves drooping as flower bud develops, often touching ground when flower fully opens; petals 4 times as long as across. Rare. Dry to rich deciduous woods; cnFla into s and ceAla, and adjGa. Late Feb–Mar.

584   **Narrowleaf Trillium**
Trillium lancefolium Raf.
Plants to ca 30 cm. Easily recognized by sessile lanceolate to narrowly elliptic leaves; petals erect, 4 times or more as long as across, clawed, upper portion of stamens strongly incurved. Common. Usually in neutral to basic soils; floodplains; rocky upland woods and thickets, rich wooded areas; scattered; cnFla into adj Ga and Ala; swAla into seTenn; cGa and cnSC. Feb–early May.

  T. recurvatum Beck (Prairie Trillium) is also easily recognized. The leaves are petioled, sepals downturned against stem; petals 2–3 times as long as across, clawed, and arched inward. Common. Rich floodplains, rich moist woods; bluffs, most often in limestone areas; e into nwMiss; seTenn into nLa and ceTex: ce and nArk, eMo, sWisc, swMich, eO, w two-thirds of Tenn, and nwAla. Late Mar–May.

585   **Yellow Trillium**
Trillium luteum (Muhl.) Harbison
Stems 14–40 cm. Flowers strongly lemon-scented; petals pure yellow, 34–66 × 10–21 mm; stamens 11–18 mm, filament 1.5–2 mm. One of our favorite trilliums for ornamental purposes. Common. Rich deciduous woods; cnGa into e third Tenn, seKy, and wNC to Va. Mar–Apr.

586   **Decumbent Trillium**
Trillium decumbens Harbison
Stems firm but decumbent with leaves on the ground litter; anther with a beak, dehiscent essentially toward the perianth. Rare. Rocky slopes, rich woods, thin woods; floodplains; loose shale slopes; nwGa into cAla, nwGa, and extreme seTenn. Late Mar–Apr.

  T. reliquum Freeman also is decumbent with leaves on the ground litter; stems firm but gently S-shaped, usually less than twice as long as leaves; leaves with 3 distinct shades of green as in *T. decipiens* (next entry); petals narrowly elliptic-oblanceolate; stamens about half as long as petals, anthers beaked and dehiscing toward pistil. Rare. Rich mesophytic woods; sPied of Ga into swGa, CP of seAla, near Fall Line of cw and cGa, above Fall Line along Savannah R in Ga and SC. Mid-Mar–late Apr.

587   **Stately Trillium; Deceptive Trillium**
Trillium decipiens Freeman
Stems more than 2.5 times as long as leaves; leaves with 3 distinct shades of green; anthers dehiscent on edges; ovulary ellipsoid, sharply 6-angled. A magnificent

Trillium for ornamental use. Rare. Rich woods of uplands, ravines, bluffs, and bottomlands. Pied of Ga into swGa; nwFla and along Ala and Ga state line on either side of the Chattahoochee R; ceGa. Feb–early Apr.

588   **Twisted Trillium**
Trillium stamineum Harbison
No others of our sessile-flowered Trilliums have narrow, horizontal, and usually spiral-twisted petals. The flower has a strong odor of carrion. Petals 15–38 mm, very dark red-purple or brown-purple and often streaked with purple; stamens erect, 16–24 mm, anthers dark purple. Occasional. Usually associated with limestone areas; dry uplands or slopes, floodplains; deciduous or pine-deciduous woods; slightly W of cTenn into n three-quarters of wAla and eMiss. Mar–May.

589   **Pale Yellow Trillium**
Trillium discolor Wray ex Hook.
No other sessile-flowered Trillium has clawed spatulate pale yellow to ivory-yellow petals. Petals 22–50 mm, with a longer life than most Trilliums; stamens dark purple, about twice as long as the pistil. A most attractive Trillium for ornamental usage. Rare but locally abundant. Rich wooded slopes, upper river terraces; in Savannah R system; Pied and BR; Ga, SC, and barely into NC. Apr–May.

   T. viride Bech. (Green Trillium) is another distinctive species; no other species has leaf upperside with scattered stomates appearing as light dots (magnification may be necessary). Petals 35–68 mm, 5.5 times longer than across, somewhat spatulate, greenish with a purplish claw or purplish throughout. Pistil two-thirds the height of stamens, ovulary deeply angled. Occasional. Rich woods, bluffs, rocky hillsides; midsection of eMo and into adjIll. Late Apr–May. *T. viridescens* Nutt. (Ozark Green Trillium) possibly can be confused with Green Trillium but can be separated by absence of stomates on leaf upperside and pistil only half the height of stamens. Petals 75–80 mm; stamens twice as long as petals or longer. Occasional. Deciduous woods, usually quite rich; nwArk into neTex, seNeb, csMo, and cArk. Apr–May.

590   **Painted Trillium**
Trillium undulatum Willd.
One of the easily recognized and widely distributed species with pedicelled flowers. Leaves are clearly petioled and acuminate. Pedicel 2–5 cm and erect with flower and fruit; petals smooth with wavy margin, white with an inverted V-shaped red blaze near base, the red radiating outward along the major veins, rarely all white; anthers smaller than filament. Plants 20–50(60) cm. Fruits scarlet, erect, fleshy, 1–2 cm. Common. In strongly acidic humus-rich soils; s part of range in pine woods, spruce forests, deciduous-conifer forests; in hemlock stands in n part of range; nGa into wNC, swVa, seKy, ne into sOnt, NS, nNJ, and wMd. Apr–June.

   T. nivale Riddell also has clearly petioled leaves, but petals are all white and margin not wavy, the 1–2 cm pedicels erect with flower, strongly downcurved beneath leaves with fruit. Plants 3–5 cm at flowering, growing to 4.5–9 cm with mature fruit, fruits white tinged with green. Common. Limestone areas, alkaline glacial drift, loess; in rock crevices, loose soil and rocks around bluffs, floodplains; distribution spotty, mostly nMo into seNeb, Ia, swMinn, sMich, swPa, and nKy. Mid-Mar–early Apr.

591 Catesby's Trillium; Rose Trillium

Trillium catesbaei Ell.

Sepals, petals, and stamens recurved, the first 2 strongly, identify this species; plants 20–45 cm. Leaves 6.5–1.5 × 4–8 cm, sides of new ones upcurved; base tapered to a petiolelike portion 4–15 mm. Flowers below or level with leaves, rarely above; pedicel declined to occasionally above leaves, 2–4(5) cm, angled; corolla opening white, pink, or rose, 35–50 × 10–20 mm, petal bases overlap, forming a cup; stamens 16–25 mm, anthers bright yellow, irregularly twisted outward; much of styles fused. Sometimes called Rose Wake-robin. Common. Usually acid soils in sun or shade; dry thin woods to deep forest coves of mts, woods and thickets of the Pied; cwGa into ceAla, seTenn, n and cSC, and csVa. Late Mar–June.

592 Vasey's Trillium

Trillium vaseyi Harbison

This species is often spectacular because of its large size, to 65 cm, and nodding flowers with maroon (rarely yellow) petals that may reach 7 cm and about as wide; flowers the largest of all pedicellate species (much smaller plants are occasional). Flowers faintly fragrant. Pedicel 4–8 cm; stamens 15–25 mm, far exceeding the pistil at pollen-shedding, anther sacs yellow to maroon, filament as long as ovulary or longer; ovulary maroon to dark red-purple, 6-ridged, 3–12 mm; styles discrete. Occasional. Steep wooded slopes, ravines, coves, rich moist woods; neAla into mts of nGa, nSC, wNC, and eTenn. Late Apr–June.

593 Red Trillium; Stinking Willie

Trillium erectum L.

A variable species; most of the many variants difficult to identify. Flowers generally with somewhat of a wet dog odor. The common type of plant has stiffly erect pedicels but some may lean or a few even decline to below leaves; sepals 1–5 cm, green or mixed colors, or entirely dark maroon; petals outcurved, generally dark red-maroon fading to dull purple-brown, 1.5–5 × 1–3 cm, coarsely textured; stamens about as long as pistil or shorter; ovulary dark purple to maroon, visible from side view. Fruit dark maroon. Also called Stinking Benjamin. Common. Humus-rich somewhat acid soils; deciduous hardwood and spruce-fir forests; under Hemlocks, Rhododendrons, and Mt. Laurel; nGa into eTenn, eO, Mich (scattered), sQue, NS, NJ, Md, nVa, and nSC; small populations in sInd and seIll. Late Apr–early June.

594 Great White Trillium

Trillium grandiflorum (Michx.) Salisb.

Over a vast range growing often in spectacular colonies, their beauty enhanced by the upright flowers often facing mostly in one direction. Leaves sessile. Pedicel 2–8(10) mm, erect-ascending to strongly erect; sepals green, 20–55 × 12–23 mm, acuminate; petals white fading to a dull pink-purple with age, 4–8 × 2–4 cm, bases erect, flaring and recurving somewhat above the middle, producing a strongly funnel-shaped corolla; anthers pale, much longer than filament, slender, base erect, outcurved toward end; stamens shrivel and persist with ripe fruits. Ovulary 6-angled, pale green. Fruit pale green. Common. Rich woods; eTenn into eKy, O, Ill (scattered), eMinn, Wisc, sOnt, sQue, w and seNH, NJ (scattered), WVa, Va mts, and nSC; cwIa. Late Apr–early June.

T. simile Gleason (Sweet White Trillium) also has large white-petalled flowers held above the sessile leaves, from a distance very much like *T. grandiflorum;* separated from the latter by a dark purple-black very strongly angled ovulary. Flower fragrance sweet, like green apples. Pedicel 4–9 cm; petals 4–7 × 1.5–4 cm, more than twice as broad as sepals; fruit dark purple-black. Plants often in clumps rather than in colonies. Occasional, but locally common. Rich coves, Rhododendron thickets, stream terraces, forest edges, roadside rock outcrops; neGa into seTenn, swNC, and neSC. Apr–May. *T. flexipes* Raf. (Bent Trillium) plants from a distance easily confused with some forms of the Great White Trillium but may be recognized by corolla not funnel-shaped; petals coarse, 2–5 × 1–4 cm; stamens at about 45°, anther cream and nearly straight; ovulary white to light pink, strongly 6-angled. Flowers with a faint sweet to musky odor. Fruit very large. Common. Wooded limestone areas; rich slopes, terraces along streams, floodplains; neTenn into wMo, adjIll and Ky, nIa, s into se and ceMinn, wWisc into Ill, w and seMich, sOnt, swNY, wPa, wWVa, and cKy; nAla into eTenn. Apr–early June.

595 Southern Nodding Trillium
Trillium rugelii Rendle
The declined flowers with white (rarely otherwise) reflexed petals 25–50 × 8–35 mm and dark purple anthers 12–16 mm and 3–5 times longer than filaments identify most plants of this species. Variation makes identification of some others a problem, e.g., corolla may be rose, maroon, dark rose-red, or rose-red with a white center. Plants 15–40 cm, leaves 6–15 × 6–11 cm. Occasional. Rich deciduous woods, steep wooded hillsides, stream terraces; mostly ceAla into nGa, nw and cnSC, adjNC, nwNC, and seTenn. Apr–early May.

T. cernuum L. (Nodding Trillium) shares several features with the above species and has been mistaken for it. Separated by narrowly elliptic to obovate petals 15–25 × 9–15 mm, scarcely longer than the sepals; anthers lavender-pink or gray, 2–7.5 mm, about equal to the slender white filament or shorter. Common. Low moist to swampy woods, stream banks, deep deciduous woods and damp peaty woods; to N also in upland conifers mixed with deciduous trees and along streams; nVa into sePa, nO, nIll, nIa, sSask, s tip of James Bay, sQue, Nfld, and NJ. Apr–June.

596 Persistent Trillium
Trillium persistens W. H. Duncan
Stems 10–26 cm, somewhat angled. Leaves acuminate to rarely acute. Pedicel 1–3 cm; sepals weakly divergent, 11–22 × 5–6 mm; petals 20–35 × 5–10 mm, undulating, apex acute; stamens erect to slightly divergent, straight, 9–14 mm; anther and filament about equal, anther splitting open toward pistil; ovulary obovate, very sharply 6-angled, 2.5–6 mm, stigmas united, 2–6 mm. Fruit green to greenish-white. Rare. Deciduous or conifer-deciduous woods, under or near evergreen *Rhododendron* spp.; restricted to the Tallulah-Tugaloo R system of neGa and nwSC. Early Mar–mid-Apr.

T. pusillum Michx. (Least Trillium) is the nearest relative of Persistent Trillium. The Least Trillium may be separated by stems 7–20 cm; leaf tip rounded; pedicel 5–20 mm; sepals 15–30 × 5–10 mm, spreading and conspicuous; petals 15–30 × 5–15 mm; stamens 8–10 mm, anther pale lavender; ovulary ovoid, obscurely

6-angled, white. Fruit off-white. Plants in many usually widely separated small populations, some sufficiently different to have been named separate ssp., sp., var., and/or forms; 6 varieties seem generally to be accepted by botanists; however, there is evidence that differences in some populations can be the result of the varying environmental conditions. Rare. Habitats differ over a wide range, possibly the most extreme the mt crests (ca 1190 m) in WVa. Populations present in eTex, eOkla, sw and nwArk, sMo, Tenn, Ky, WVa, Va, NC, SC, and Ga. Mar–May, June in WVa.

HAEMODORACEAE: Bloodwort Family

597   **Redroot**
Lachnanthes caroliana (Lam.) Dandy
Perennial with a prominent rhizome and fibrous roots, both with red juice. A rhizome exposed by removing the soil over it may be seen in the picture. Flowering plants to 120 cm. Leaves mostly basal, resembling those of Iris. Inflorescence yellowish to brownish, very hairy. Perianth segments 3, united at the very base. Stamens 3. Ovulary 3-carpelled, inferior. Reported as poisonous when eaten but doubtfully so. Sandhill cranes may consume great quantities. Common. Wet habitats, swamps, thin pinelands, ditches, open places; Fla into sLa, lower CP of Ga, and seVa; csTenn; Del into Mass; NS. May–Aug. *L. tinctoria* (Walt.) Ell.; *Gyrotheca t.* (Walt.) Salisb.

598   **Lophiola**
Lophiola aurea Ker-Gawl.
An almost complete covering of fine white hairs on the upper parts of the plant make it conspicuous, especially when in flower. The small flowers are also striking because the insides of the tepals are maroon and bearded with long yellow hairs. Plants to 70 cm. This species probably is best placed in the Liliaceae. Occasional. Moist savannas and pine barrens, bogs; Fla into sMiss, cCP of Ga and seNC; Del into NJ; wNS. Apr–Jun. *L. americana* (Pursh) Wood.

AMARYLLIDACEAE: Amaryllis Family

599   **Atamasco-lily; Rain-lily**
Zephyranthes atamasca (L.) Herb.
Plant with several flat linear sharp-edged leaves from a covered bulb. Flowers like those of true Lilies except with an inferior ovulary, on a leafless stalk. Perianth funnel-shaped. Pistil longer than the stamens, with 3 stigmas. Leaves and especially bulbs of this species are poisonous when eaten; consumption of less than 1% of animal's body weight is fatal. Common. Low areas, usually in woods; Fla into cMiss, Pied of Ga, nwSC, and seVa. Jan–Apr.

   Z. simpsonii Chapm. is similar but with the pistil shorter than the stamens. Occasional. Low areas; Fla into sGa and seSC. Feb–Apr. In *Z. candida* (Lindl.) Herb. the perianth is widely spreading and the stigma 1 with 3 lobes. Rare. Low places; Fla into seTex; coastal NC. Sept–Oct.

600   Spider-lily

Hymenocallis caroliniana (L.) Herb.
Perennial from a large bulb. Leaves linear, all basal, 60 × 2–5 cm. Flowers 3–9, sessile on a leafless stalk. Sepals and petals alike, long and narrow, white, attached to a long floral tube. Lower half or more of stamens united with a white crown shaped somewhat like a morning-glory corolla. A sharp line marks differences in colors and tissues between the floral tube (whitish) and the ovulary (greenish). Matured ovulary beaked. Rare. Swamps, riverbanks, and terraces, adj wooded slopes, rocky shoals; Fla into cs and eTex, seMo, swInd, and nwSC. May–Aug. *H. occidentalis* (Le Conte) Kunth.

In *H. floridana* (Raf.) Morton the color and tissue changes between the floral tube and the ovulary are gradual and the tip of the matured ovulary is cuneate. Swamps, marshes, wet riverbanks; Fla into seLa, Ala, swSC, and seNC. Apr–June. *H. crassifolia* Herb.

601   Rattlesnake-master

Manfreda virginica (L.) Salisb. ex Rose
Succulent perennial to 2 m. Leaves basal, at first erect and then spreading, gradually tapered to a narrow base, abruptly tapered to the weak-pointed apex, sometimes purple-blotched. Flowers scattered in a spikelike raceme. Perianth tubular at base. Fruit a subglobose 3-carpelled capsule, 1.5–2 cm across. Common. Dry, rocky, sandy places, thin woods; Fla into eTex, swMo, sInd, sO, swWVa, Ga, and cNC; absent from most of the BR. May–Aug. *Agave v.* L.; *Polianthes v.* (L.) Shinners.

Plants with purple-blotched leaves have been improperly separated as a species, *M. tigrina* (Engelm.) Small.

602   Yellow Star-grass

Hypoxis hirsuta (L.) Coville
Perennial with hairy grasslike leaves, to 30 cm × 2–8 mm. They come from a corm that has a few thin pale to brownish sheaths. Scape to 35 cm. Petals obtuse. Ovulary inferior. Common. Dry to moist thin woods and meadows; Fla into eTex, Man, and Me. Mar–Sept.

In *H. micrantha* Pollard the petals are acute and the seeds brown. Occasional. Savannas, pine flatwoods; Fla into neTex, CP of Ga, and seVa. In *H. juncea* Sm. the leaves are under 1 mm wide. Rare. Savannas, open pinelands; Fla, seAla, into sGa and seNC. Mar–May, occasionally later.

IRIDACEAE: Iris Family

603   Blue Iris

Iris virginica L.
Perennial to 1 m. Leaves 1–3 cm wide and, as in all Irises, flattened into 1 plane, at least at their bases. Petallike structures of the flower are 9. The broadest and lowest 3 are the sepals, the next 3 are the petals, which are less than 2 cm wide, and the upper 3, which are narrowest, are the stigmas. Under these are hidden the 3 stamens. Occasional. In wet places or shallow water, thin woods or open; Fla into eTex, Tenn, Pied of Ga, and seVa. Apr–May.

I. prismatica Ker has leaves less than 1 cm wide. Rare. Similar places; neGa into cTenn, NC and along the coast into NS. May. In *l. tridentata* Pursh the petals are about half as long as the sepals. Occasional. Depressions and low places in open; nFla into CP of sNC. May–June.

604 Crested Iris
Iris cristata Ait.
Perennial from rhizomes with tuberous sections separated by long thin sections bearing distinct widely distributed scales; leaves arched downward toward the tip, the upright stems to 15 cm. Sepals and petals pale lavender, the yellow to orange band on the sepals with crinkly ridges (the crest). Common. Rich woods; Pied of Ga into eOkla, seMo, neO, and DC. Apr–May.

I. verna L. (Dwarf Iris) is from a densely scaly rhizome; leaves essentially straight; the yellow to orange band on the sepals smooth. Occasional. Sandy or rocky and thin woods; seGa into nwFla, eMiss, sO, sPa, DC, and wNC. Mar–May.

605 Blackberry-lily
Belamcanda chinensis (L.) DC.
Flowers much like those of lilies but with only 3 stamens and an inferior ovulary. When not in flower the entire plant greatly resembles Iris except that the roots and rhizomes are orange. Very hardy and drought-resistant. The covering on the mature fruits splits and curls downward exposing the many black seeds on the central column, at this stage imitating a Blackberry (whence the popular name). Occasional. Roadsides, rocky areas, thin woods and old homesites; Pied of Ga into eTex, Neb, and Conn. June–Aug.

606 Blue-eyed-grass
Sisyrinchium albidum Raf.
Members of this genus have fibrous roots, similar flattened and tufted leaves and stems, linear leaves, 3 united stamens, and inferior ovularies.

This species is a perennial with matted fibers at the base, flowers from between 2 sessile bracts called spathes, perianth white to blue, and ovulary finely glandular-hairy. Occasional. Open places, dry woods, often in sandy soils; Fla into eTex, sWisc, sOnt, and wNY. Mar–May.

S. rosulatum Bickn. (*S. brownei* Small; *S. exile* Bickn.) appears somewhat like a miniature *S. albidum* but is an annual, grows only to 15 cm, and has a yellowish perianth; the stems and leaves are radially spreading, and the fruits are nearly globose. Rare. Wet places, roadsides, fields, swamps; Fla into seTex. Apr–May. *S. minus* Engelm. & Gray is another annual of a growth form similar to *S. rosulatum,* but to 20 cm, the perianth white to purple-rose, and ovulary and fruits 1.5 times as long as broad. Rare. Dry sandy or silty soils in open; Miss into s and cTex. Apr–May.

607 Blue-eyed-grass
Sisyrinchium atlanticum Bickn.
Perennial to 50 cm without persistent matted fibers at the base. Scapes narrow winged, 1–3 mm wide, terminating in a spathe from which arise 2–3 peduncled flower-bearing spathes. Fruits 3–4.5 mm. Plant pale green, mostly retaining color

upon drying. Common. Usually low places, edges of marshes, thin pinelands, and deciduous woods; Fla into seTex, csMo, sMich, and swNS. Mar–June.

S. angustifolium Mill. is similar but the stems are broadly winged and 3–5 mm wide, the fruits 4–6 mm. The plants blacken upon drying. Common. Moist places, fields, meadows, woods, and roadsides; Fla into eTex, eKan, sOnt, and seNfld. Mar–June.

CANNACEAE: Canna Family

608   **Golden Canna**
Canna flaccida Salisb.
Perennial to 130 cm, from coarse rhizomes. Leaves to 55 × 15 cm, with sheathing petioles. Upper leaves the smaller. Sepals 3, greenish, 25–30 mm. Petals 3, yellowish-green, their bases united into a tube 50–65 mm, the lobes about the same length. The showy part of the flower consists of modified stamens that look like petals. One petaloid stamen bears the only pollen-bearing anther. The style is also petallike. The inferior ovulary and fruit are covered with small elongated warts. Rare. Freshwater swamps and marshes, ditches, in thin woods or open; Fla into sMiss, csGa, and csSC. May–Aug.

A considerable variety of cultivated plants, which are hybrids of other species, occasionally persist at abandoned sites or more rarely escape into the wild. This variety of individuals in the eUS is probably best considered as *C.* × *generalis* Bailey.

ORCHIDACEAE: Orchid Family

In Orchids, 2 petals are similar; the other (the lip) is different, often radically so, and is an important character in identification. Another unique character of Orchids is the union of the 1–3 stamens and the pistil, forming much of the column, the central structure of the flower. All have inferior ovularies. None of our Orchid species has been reported to be poisonous.

609   **Pink Lady's-slipper; Moccasin-flower**
Cypripedium acaule Ait.
Members of this genus get their common names from the inflated moccasin-like lip. All are perennials and have 2 fertile stamens.

Pink Lady's-slipper is distinguished by having only 2 leaves, both basal. Occasional. In a variety of habitats, wet to dry, thin to dense woods; nGa into nAla, neIll, Alta, Nfld, and NC. Apr–June.

610   **Yellow Lady's-slipper**
Cypripedium calceolus L.
The Latin name *calceolus* means "a little shoe." The "shoes," or moccasins, in this species are golden-yellow or rarely almost white and 15–60 mm. The 1–2 flowers are above the 3–5 leaves. This species is circumboreal, occurring in Europe and Asia. The N. Amer plants are var. *pubescens* (Willd.) Correll. They are perennials

and may grow to 70 cm. Occasional. In a variety of habitats—wet to dry, thin to dense woods, and 500–9000 feet; nGa into Ariz, BC, Yukon, Que, Nfld, and cNC. Apr–June. *C. parviflorum* Salisb.

611   Showy Lady's-slipper

Cypripedium reginae Walt.
The moccasins of this species are reddish to pink or less commonly waxy white. Plants are often larger, to 85 cm, than those of the preceding species and have 3–7 leaves. Rare. Moist situations; mts of neGa, swNC, seTenn, RV of Va; Pa into nwNJ, nIll, sMo, neND, Sask, and Nfld. May–June.

612   Showy Orchid

Galearis spectabilis (L.) Raf.
A genus of ca 100 spp.: 2 in the eUS, 1 in Alas, China, and Japan; the remainder in Europe, Asia, and Africa.

   This species is a glabrous perennial to 35 cm. Leaves 2, glossy, and basal. Flowers 2–15 in a terminal raceme, very fragrant, the hood pink to mauve or rarely white. Lip without spots, lobes, or notches. Occasional. Rich hardwood forests; mts of Ga into nAla, nArk, neKan, eNeb, neMinn, sQue, and NB. Apr–June. *Orchis s.* L.

   Amerorchis rotundifolia (Banks ex Pursh) Hulten has some similarities but only 1 leaf, the lip white spotted with pale purple, 3-lobed, and notched at summit. Occasional. Wet woods and swamps; Alas into Greenl, S into BC, Minn, Mich, nNY, and Que. June–July.

613   Green Woodland Orchid

Platanthera clavellata (Michx.) Luer
Glabrous perennial to 8–45 cm, with 2–4 leaves along the stem and no basal ones, the upper leaves much the smaller. Flowers 3–16 in a terminal raceme, the pedicel plus ovulary about 1 cm, the spurs some longer, the lip oblong and with 3 short rounded teeth. Common in or at edges of water, usually in woods; Fla into eTex, Minn, Ont, and Nfld. June–Aug. *Habenaria c.* (Michx.) Spreng.

   P. integra (Nutt.) Gray ex Beck is in some ways similar but is 30–60 cm and has longer and narrower leaves. The flowers may number 70 and are light to dull orange, the spurs shorter (about 5 mm), and the lip obtuse to acute but without the 3 distinct teeth. Swamps, bogs, low pine barrens and flatwoods; chiefly CP— nFla into seTex, eTenn, O, and NJ. *Habenaria i.* (Nutt.) Spreng.

614   Ragged Orchid

Platanthera lacera (Michx.) G. Don var. *lacera*
Perennial to 80 cm, with fleshy roots and erect leaves, the upper ones smaller. The 2 upper petals oblong, their upper margins entire. The lip is 15–20 mm and deeply lacerated, hence the name "ragged." Flowers pale yellow to almost white. Occasional, northward common. Moist or rarely dry open places or thin woods; cPied of Ga into Ark, neTex, Minn, seMan, and Nfld. May–Aug. *Habenaria l.* (Michx.) R. Br.

   P. leucophaea (Nutt.) Lindl. is similar. It may be recognized by smaller flowers, lip 7–15 mm, and the 2 upper petals being wedge-shaped and their upper margins only eroded or finely toothed. Common. Wet places, in prairies and coniferous forests; La into Mo, Kan, ND, eQue, and NS. *Habenaria l.* (Nutt.) Gray.

615 Purple-fringed Orchid
Platanthera psycodes (L.) Lindl.
Glabrous perennial to 120 cm. Roots tuberous. Leaves cauline, up to 5, the upper ones gradually smaller. Lip deeply 3-parted, the 3 divisions both coarsely and finely toothed along their ends. Corolla lilac-lavender to pinkish-purple. Spur slender and curved. Occasional. Usually in moist places, thin woods or open, sometimes along small streams; mts of nGa, Va, NC, and Tenn into eKy, neIll, eNeb, sOnt, and Nfld. May–Aug. *P. grandiflora* (Bigelow) Lindl.; *Habenaria p.* (L.) Spreng.

    P. peramoena (Gray) Gray is similar, having purplish flowers, but the 3 divisions of the lip are entire or slightly uneven. The middle division is notched. Occasional. Moist places; nwSC into cMiss, cArk, ceInd, sO, sePa, and wNJ. *Habenaria p.* Gray.

616 Yellow-fringed Orchid
Platanthera ciliaris (L.) Lindl.
Some plants of this beautiful perennial grow to 1 m. The several leaves are largest near the base and gradually become smaller above. Flowers are in dense racemes to 20 cm. Lips of the flowers are oblong, 8–12 mm, and prominently fringed. The spurs are 20–33 mm. Occasional. In almost any type of habitat; Fla into eTex, Ill, Ont, and Vt. June–Sept. *Habenaria c.* (L.) R. Br. ex Ait. f.

    P. cristata (Michx.) Lindl. has orange flowers also but they and the racemes are smaller than in the above species. The spurs are much shorter and the lip is shorter than 6 mm. Occasional in sCP, rare elsewhere; cFla into eTex, CP of Ga, and seMass; cArk; c and eTenn into nwSC and swNC. June–Sept. *Habenaria c.* R. Br. ex Ait. f.

617 White-fringed Orchid
Platanthera blephariglottis (Willd.) Lindl. var. *blephariglottis*
This orchid is much like *P. ciliaris* except the flowers are white. Plants to 1 m. Lower 1–3 leaves linear or lance-linear, to 20 × 2 cm; inflorescence compact, ovoid, 5–15 × 4–5 cm, spur long and slender. Occasional. Moist habitats, usually in open. Fla into seTex, CP of Ga, eVa, nMich, and Nfld; scattered places in Pied of Ga into mts of Tenn and NC. July–Sept.

618 Snowy Orchid
Platanthera nivea (Nutt.) Luer
Perennial to 90 cm. Leaves 2–3 near the base and nearly erect, grading abruptly into as many as 10 slender erect bracts. Raceme 3–15 × 1.3–3 cm. Flowers snowy white, rarely pink-tinged. Lip turned upward, linear-oblong to linear-elliptic. Occasional. Pine barrens and flatwoods, bogs, savannas; Fla into seTex, CP of Ga, neNC, and seNJ. May–July. *Habenaria n.* (Nutt.) Spreng.

619 Rose Pogonia

Pogonia ophioglossoides (L.) Ker
Perennial 10–70 cm with an ascending leaf about midway up the stem, occasionally with 1–2 long-petioled leaves arising from base of stem. Flowers 1–3, rose to white, fragrant. The lip 25 mm or shorter, lacerate-toothed along its lower margins, and bearded along the middle of the upper side. Occasional in the CP, rare in the Pied and mts. Swamps, bogs, thin flatwoods, and other wet habitats; Fla into eTex, Minn, and Nfld.

Large Whorled Pogonia

Isotria verticillata (Muhl. ex Willd.) Raf.
Recognized by the single whorl of 5–6 leaves at right angles on top of stem, the single (rarely 2) flower with 3 narrowly lanceolate spreading sepals 3.4–6 cm × ca 3 mm, and the peduncle 2–5.5 cm. To 35 cm. Rare. Moist hardwood slopes, stream margins; n and ceGa, swGa, cnFla, eTex, sMo, Mich, Ont, cNH, swMe, neNC, and nSC. Apr–July.

 I. medeoloides (Pursh) Raf. (Small Whorled Pogonia) is similar but to 25 cm, peduncle 1–1.5 cm, and sepals 1.5–2.5 cm. Rare. Thin hardwoods or pine-hardwoods; neGa into seMo, Mich, Ont, sMe, and cNC. May–June, early July in N.

620 Rose Orchid; Spreading-pogonia

Cleistes divaricata (L.) Ames
Glabrous perennial to 75 cm, with fibrous roots. Leaves 1, sometimes 2, above the middle of the stem. Flowers usually 1, sometimes 2–3. Sepals 3, similar, linear-lanceolate, spreading, brownish to purplish or rarely almost green. Lip and other 2 petals partly united, forming a cylinder. Lip crested and with finely wavy margins near the tip. Rare. Usually moist places, pine barrens, savannas, thin woods, swamps, bogs; rarely on dry grassy slopes and mt tops, usually in acid soils; cFla into seTex, CP of Ga, and NJ; mts of nGa into eTenn, eKy, and wVa. Apr–June.

621 Autumn-tresses

Spiranthes cernua (L.) Rich.
A variable genus, with plants often difficult to name to species. The generic name means "coil-flower" in allusion to the spiral arrangement of the flowers of many species.
 This species has several spirals forming a dense spike with flowers usually in 3 longitudinal rows. The lip is 6–14 mm and has 2 small rounded projections at its base. The largest leaves are near the base of the stem and at most are 25 cm. Common. In a variety of moist habitats; Fla into NM, cKan, ND, and NS. May–frost.
 In the similar *S. ovalis* Lindl. the lip is only 4–5 mm. Rare. Moist to well-drained woods, palmetto swamplands, hammocks; cnFla into eTex, Mo, cwWVa, and eVa. Aug–frost.

622 Spring Ladies'-tresses

Spiranthes vernalis Engelm. & Gray
Perennial to 110 cm from several thick roots. Upper parts with fine dense pointed hairs. Leaves basal, erect, up to 30 × 1 cm, their bases sheathing the stem. Flowers

strongly spiraled, rarely as little as in those of the photograph. Stem, ovulary, and bracts densely and finely hairy, the hairs glandless. Lip 4.5–8 mm, widest near the base. Occasional in CP, rare farther inland. Moist places, meadows, coastal salt marshes, thin pinelands, savannas, floodplains; Fla into eTex, eKan, swLa, sO, and seNY. Feb–July.

S. longilabris Lindl. has similar leaves and the lip widest near the base, but the upper part of the plant is glabrous or nearly so, the flowers are not spiraled or only slightly so. Rare. Wet places in thin woods or open, swamps, savannas, pine barrens; Fla into seTex, CP of Ga, and seVa. Oct–Dec.

623   Kidney-leaf Twayblade
Listera smallii Wiegand
Plants 6–35 cm with 2 leaves that are somewhat kidney-shaped, thus the common name. Flowers usually 4–10, the lip broadened toward the tip, which is split less than halfway to the base. Named for George K. Small, who studied the flora of the seUS extensively for many years. Rare. Bogs or often under Hemlocks or Rhododendrons; up to 4700 ft in mts of neSC into Ga, eWVa, se and cPa, and ceVa. June–July.

L. australis Lindl. is similar but the leaves are ovate to elliptic and the lip linear and split over halfway to the base. Rare. In rich humus of moist woods, low pine barrens, marshes, and spagnum bogs; Fla into seTex, sGa, and NJ; sePied of Ga; scattered localities into Que, Ont, and Vt. Feb–July.

624   Downy Rattlesnake-plantain
Goodyera pubescens (Willd.) R. Br.
Plants of this genus are often in colonies, conspicuous because of basal rosettes of variegated leaves. The resemblance of the leaf patterns to rattlesnake skins is the basis for the common name. The leaves are prominent and attractive throughout the winter.

This species grows to 45 cm and has a dense many-flowered raceme, with flowers on all sides. Common. Dry or moist coniferous or deciduous forests; nSC into swAla, cnTenn, csMo, ceMinn, sQue, cwMe, and NC. June–Aug. *Peramium p.* (Willd.) MacM.

G. repens (L.) R. Br. ex Ait. f. is smaller, rarely over 30 cm. The raceme is 1-sided and only 3–5 cm. Occasional. Dry cool woods, dense mats of moss, usually beneath conifers; swNC into seTenn, s and eWVa, cPa, nO, neMinn, BC, Alas, Nfld, and nNJ; also in swUS and Eurasia. June–Sept.

625   Grass-pink
Calopogon barbatus (Walt.) Ames
Glabrous herbaceous perennial to 45 cm. Leaves narrow, 1–2 from near the base. Floral bracts 2–5 mm. Flowers 3–6 in a raceme, rose-pink or rarely white, and inverted, making the bearded lip erect. Petals widest below middle. Column 7–8 mm. Common. Low pinelands, grassy swamps, wet savannas; Fla into seTex, CP of Ga, and ceNC. Feb–May.

C. multiflorus Lindl. is similar but floral bracts 5–10 mm and petals widest above the middle. Occasional. Similar places; Fla into sMiss and sGa. Feb–July.

In *C. tuberosus* (L.) B.S.P. var. *tuberosus* the flowers are pink to nearly crimson; the column is 10–20 mm, upper portion winged, base of wings gradually tapered. Occasional. Low pinelands, bogs, sphagnum swamps, depressions; eTex into Minn, Nfld, and Fla. Mar–Sept. *C. pulchellus* R. Br. ex Ait. f. In *C. pallidus* Chapm. the flowers are usually lighter-colored; the column is 8–10 mm, upper portion winged, base of wings somewhat truncate. Occasional. Similar places; Fla into seLa, sGa, and seVa. Feb–July.

626   Coral-root
Corallorhiza wisteriana Conrad

These nongreen plants obtain their food from dead plant remains, probably with the aid of fungi. They grow to 40 cm but are inconspicuous, varying from tan (usually only at the base) to dark reddish-purple. The corolla lip is white, pendent, 5–7 mm, and conspicuously spotted with magenta-purple. Rare. Light to rich, dry to moist soils in various kinds of woods; Fla into e half Tex, SD and several western states, sIll, and sePa. Feb–July.

The similar, usually lighter-colored *C. odontorhiza* (Willd.) Poir. has a lip 3–4.5 mm. Occasional. Similar places; nwSC into swAla, Miss, Mo, Wisc, and Me. June–Oct. *C. maculata* (Raf.) grows to 75 cm, and the lip is 5–8 mm and unequally 3-lobed. Occasional. Rich decaying humus of upland woods or along stream banks; nGa into eTenn, O, BC, Nfld, and c and wNC; other western states. June–Aug.

627   Adder's-mouth
Malaxis unifolia Michx.

Plant glabrous 6–55 cm, from a bulbous corm. Leaf blade solitary, bright green, enlarging during growth of the fruit. Lip hanging down, 2-lobed at the end but with a small tooth between. Occasional. Rich humus soils of dry to moist woods or less frequently in the open; cnFla into eTex, Minn, Man, and Nfld. Mostly Mar–Aug.

Another species, *M. spicata* Sw., has 2 conspicuous bright green leaves. The lip is ascending and not lobed at end. Rare. Swamps, hammocks, moist woods, rich wooded slopes; pen Fla into seVa. Mostly Aug–frost.

628   Cranefly Orchid
Tipularia discolor (Pursh) Nutt.

Perennial to 65 cm, from a corm. The single leaf is green above and purplish beneath. It appears in the fall, remains green during winter, but withers in the late spring. In the summer a leafless flowering stalk develops. It usually bears 20–30 pale flowers that blend so well with the leaf litter that the plant often goes unnoticed. Each flower has a slender spur 15–22 mm. The fruits mature in the fall. They are ovoid, about 10 × 5 mm, and droop parallel to the stem. The dead main stem, occasionally with some frayed fruits, can often be found the following spring. Common. Wet to dry places in hardwood or occasionally pine forests; nFla into eTex, swArk, seInd, wNY, and NJ. June–Sept. *T. uniflora* (Muhl.) B.S.P.

629   Twayblade

Liparis liliifolia (L.) Rich. ex Ker.

A perennial from a bulbous corm. Flowers 5–40. The lip of the corolla is translucent, madder-purple, and over 10 mm. The 2 upper petals are narrowly linear. Rare. Rich woods; nGa into nAla, Mo, seMinn, and Me. May–July.

In *L. loeselii* (L.) Rich. the lip is opaque, yellowish green and long. Rare. Moist places; ceAla; mts of NC into eWVa, neO, Minn, seMo, and NS; eVa, also in Wash and W in Can into Sask. May–Aug.

630   Green-fly Orchid

Epidendrum conopseum R. Br.

An epiphyte with slender stems and thick spongy fibrous matted roots. Leaves 1–3 per stem, coriaceous; blade 30–90 × 4–14 mm, oblong to linear-lanceolate, base sheathing stem. Flowers fragrant, grayish-green. Fruit a pendent capsule 15–23 × 7–9 mm. Rare. On various species of trees in swamps and on slopes; rare on vertical faces of deep rock crevice; seNC into Fla, sGa and sMiss. July–Sept.

631   Crested Coral-root

Hexalectris spicata (Walt.) Barnh.

Saprophyte 16–80 cm; stem flesh-colored to light reddish-purple. Leaves all scalelike, sheathing stem. Flowers to ca 20 mm, larger than those of other saprophytes; on stout pedicels 4–6 mm in a raceme to 30 cm. Sepals and petals yellowish with purplish-brown striations. Fruit a strongly 3-ribbed pendent capsule to ca 25 mm. Rare. Dry woods, rarely seen in same area year after year; ordinarily in basic to neutral soils of deciduous woods; Fla into Miss, s and cMo, sInd, sO, WVa, and Va; mostly in Pied of NC, SC, and Ga; Ariz, NM. June–Aug.

1 *Saururus cernuus* × 2/3

2 *Urtica dioica* × 1/5

4 *Asarum canadense* × 1 3/8

3 *Boehmeria cylindrica* × 3/10

5 *Hexastylis arifolia* × 3/10

6 *Hexastylis shuttleworthii* × 1/3

7 *Erigonum tomentosum* × 1/6

8 *Rumex hastatulus* × 1/7

9 *Rumex crispus* × 2/5

10 *Polygonum sagittatum* × 4/5

11 *Polygonum lapthifolium* × 1/2

13 *Polygonum hydropiperoides* × 3/5

12 *Polygonum pensylvanicum* × 1/5

14 *Polygonum cuspidatum* × 3/8

15 *Polygonella polygama* × 1/6

16 *Chenopodium ambrosioides* × 1/2

17 *Chenopodium album* × 1/2

18 *Froelichia floridana* × 3/10

19 *Iresine rhizomatosa* × 3/10

20 *Phytolacca rigida* × 1/6

21 *Mollugo verticillata* × 1 1/2

22 *Sesuvium portulacastrum* × 2

23 *Talinum teretifolium* × 1 4/5

24 *Claytonia virginica* × 2/3

25 *Portulaca pilosa* × 1 1/4

26 *Portulaca oleracea* × 1 3/10

27 *Stellaria pubera* × 2/3

28 *Stellaria media* × 2 3/20

29 *Stellaria graminea* × 13/14

30   *Cerastium viscosum* × 2

31   *Minuartia caroliniana* × 2/7

32   *Silene polypetala* × 1/2

33   *Silene virginica* × 3/8

34 *Silene caroliniana* × 4/5

35 *Silene dichotoma* × 1/2

36 *Saponaria officinalis* × 1/2

37   *Nelumbo lutea* × 1/2

38   *Brasenia schreberi* × 1 3/10

39   *Nymphaea odorata* × 2/5

40   *Nuphar lutea* × 1/5

41 *Hydrastis canadensis* × 9/10   **42** *Caltha palustris* × 1/10

43 *Actaea pachypoda* × 2/5

44 *Cimicifuga racemosa* × 1/10

45 *Aquilegia canadensis* × 1 1/3

46 *Delphinium tricorne* × 2/5

47 *Anemone lancifolia* × 7/10

48 *Hepatica nobolis* var. *obtusa* × 1

49 *Ranunculus bulbosus* × 1 3/4

50 *Ranunculus fascicularis* × 1/2

51 *Ranunculus abortivus* × 1 1/3

52 *Thalictrum thalictroides* × 1/2

53 *Thalictrum pubescens* × 1/3

54 *Thalictrum clavatum* × 9/10

55 *Podophyllum peltatum* × 1/8

56 *Jeffersonia diphylla* × 2/5

57 *Diphylleia cymosa* × 3/20

58 *Caulophyllum thalictroides* × 3/5

59 *Sanguinaria canadensis* × 7/10

60 *Stylophorum diphyllum* × 1/3

61 *Dicentra cucullaria* × 1 1/10

62 *Dicentra canadensis* × 1

63 *Dicentra eximia* × 3/5

64 *Corydalis sempervirens* × 1 3/5

65 *Warea cuneifolia* × 1/2

67 *Coronopus didymus* × 1 2/5

66 *Lepidium virginicum* × 9/10

68 *Cakile edentula* × 2/3   **69** *Barbarea vulgaris* × 1/5

70 *Rorippa nasturtium-aquaticum* × 1/5

71 *Cardamine parviflora* × 2/5

72 *Cardamine diphylla* × 2/7

73 *Sarracenia leucophylla* × 1/2

74 *Sarracenia psittacina* × 1/4

75 *Sarracenia minor* × 1/6

77 *Drosera brevifolia* × 2/3

76 *Sarracenia alata* × 1/5

78 *Sedum ternatum* × 1/2

80   *Sedum rosea* × 2/5

79   *Sedum pulchellum* × 11/20

81   *Diamorpha smallii* × 1 4/5

82   *Astilbe biternata* × 1/4

83 *Saxifraga michauxii* × 1/3

84 *Saxifraga virginiensis* × 9/20

85 *Saxifraga aizoides* × 1 7/10

86 *Tiarella cordifolia* × 1/3

87 *Heuchera villosa* × 1/6

88 *Parnassia asarifolia* × 1/2   **89** *Aruncus dioicus* × 1/5

90 *Porteranthus trifoliatus* × 1/6

91 *Rubus chamaemorus* × 7/8

92 *Fragaria virginica* × 3/4

93 *Potentilla canadensis* × 1 9/20

94 *Potentilla erecta* × 1/4

95 *Potentilla anserina* × 1/2

96 *Duchesnia indica* × 1

97 *Waldsteinia fragarioides* ssp. *doniana* × 3/4

99 *Mimosa microphylla* × 1

98 *Alchemilla microcarpa* × 1

100 *Senna marilandica* × 1/5

101 *Senna obtusifolia* × 1/3

102 *Senna occidentalis* × 2/7

103 *Chamaecrista fasciculata* var. *fasciculata* × 7/10

104 *Chamaecrista nictitans* × 5/8

105 *Thermopsis villosa* × 2/7

106 *Thermopsis fraxinifolia* × 1/2

107 *Baptisia tinctoria* × 9/10

108 *Baptisia bracteata* × 1/12

109 *Baptisia alba* × 1/8

110 *Baptisia australis* × 3/5

111 *Baptisia lanceolata* var. *lanceolata* × 1/3

112 *Baptisia arachnifera* × 1/2

113 *Crotalaria spectabilis* × 1/6

114 *Crotalaria rotundifolia* × 1/3

115 *Lupinus villosus* × 1/8

116 *Lupinus perennis* ssp. *perennis* × 1/3

117 *Melilotus indicus* × 1/2

118 *Trifolium arvense* × 1 1/2

119 *Trifolium campestre* × 1/2

120 *Trifolium aureum* × 4/5

121 *Trifolium repens* × 1/6

123 *Trifolium pratense* × 9/10

122 *Trifolium hybridum* × 7/10

124 *Lotus corniculatus* × 3/4

125 *Indigofera hirsuta* × 3/5

126 *Dalea pinnata* × 4/5

127 *Dalea gattingeri* × 2 1/10

128 *Tephrosia virginiana* × 4/5

129 *Glottidium vesicarium* × 1/14

130 *Sesbania exaltata* × 3/8

131 *Coronilla varia* × 2/3

132 *Aeschynomene indica* × 3/5

133 *Stylosanthes biflora* × 1 1/6

134 *Desmodium paniculatum* × 2/5

135 *Desmodium canescens* × 1/3

136 *Lespedeza hirta* × 1/3

137 *Lespedeza repens* × 7/10

138 *Vicia sativa* ssp. *nigra* × 2/5

139 *Vicia villosa* ssp. *nigra* × 1/3

140 *Vicia caroliniana* × 3/5

141 *Lathyrus latifolius* × 2/5

142   *Clitoria mariana* × 2/5

143   *Centrosema virginianum* × 1

145   *Apios americana* × 3/10

144   *Erythrina herbacea* × 1/3

146 *Galactia elliottii* × 11/20

147 *Galactia volubilis* × 2/3

148 *Rhynchosia tomentosa* × 3/10

149 *Strophostyles umbellata* × 9/10

150 *Vigna luteola* × 1/2

151 *Geranium maculatum* × 1/2

152 *Geranium carolinanum* × 1 1/10

153 *Geranium robertianum* × 1 1/4

154 *Erodium cicutarium* × 3/5

155 *Oxalis stricta* × 1 1/3

156 *Oxalis violacea* × 3/10

157 *Oxalis montana* × 2/7

158 *Tribulus terrestris* × 11/20

159 *Polygala nana* × 2/3

160 *Polygala lutea* × 2/5

161 *Polygala grandiflora* × 1 2/3

162 *Polygala cymosa* × 4/5

163 *Polygala ramosa* × 2/5

164 *Polygala cruciata* × 1

165 *Polygala curtissii* × 1/2

166 *Polygala incarnata* × 2

167 *Polygala paucifolia* × 7/10

168 *Croton glandulosus* × 1/5

169   *Acalypha gracilens* × 1/2

170   *Tragia urticifolia* × 9/10

171   *Cnidoscolus stimulosus* × 1/2

172   *Stillingia sylvatica* × 2/3

173 *Euphorbia corollata* × 2 1/5

174 *Euphorbia cyathophora* × 3/5

175 *Chamaesyce maculata* × 1 4/5

176 *Chamaesyce hirta* × 1/2

177 *Impatiens capensis* × 9/10

178 *Modiola caroliniana* × 1 1/8

179 *Sida rhombifolia* × 9/10

180 *Hibiscus laevis* × 2/7

181 *Hibiscus moscheutos* ssp. *moscheutos* × 1/3

182 *Hibiscus palustris* × 1/5

183 *Kosteletzkya virginica* × 2/3

184 *Melochia corchorifolia* × 4/5

185 *Hypericum perforatum* × 1 1/5

186 *Hypericum gentianoides* × 11/20

187 *Hybanthus concolor* × 1/6

188 *Viola bicolor* × 1 2/5   **189** *Viola hastata* × 1/4

190 *Viola rostrata* × 1 1/10

191 *Viola lanceolata* × 1

192 *Viola floridana* × 9/20

193 *Viola pedata* × 4/5

194 *Viola nephrophylla* × 3/5

195 *Viola macloskeyi* ssp. *pallens* × 3/4

196 *Piriqueta cistoides* ssp. *caroliniana* × 2/7

197 *Passiflora incarnata* × 5/6

198 *Rotala ramosior* × 2/3

199 *Lythrum salicaria* × 1/15

200 *Rhexia alifanus* × 9/10

201 *Rhexia nashii* × 1/2

202 *Rhexia lutea* × 9/10

203 *Ludwigia decurrens* × 1/2

204 *Ludwigia linearis* × 1/2

205 *Ludwigia alternifolia* × 2/3

206 *Oenothera speciosa* × 1/4

207   *Oenothera fruticosa* ssp. *glauca* × 2/5   **208**   *Oenothera biennis* × 2/7

209   *Oenothera parviflora* × 1/3

211 *Circaea alpina* × 1

210 *Gaura angustifolia* × 1 3/7

213 *Hydrocotyle bonariensis* × 1/2

212 *Panax trifolius* × 1 3/20

214 *Centella asiatica* × 3/5

216 *Zizia aptera* × 1 7/10

215 *Eryngium yuccifolium* × 1/17

217 *Carum carvi* × 1/5

218 *Ligusticum canadense* × 1/6

219 *Angelica venenosa* × 1/14

220 *Daucus carota* × 1/10

221 *Moneses uniflora* × 1

222 *Monotropa hypopithys* × 3/5

223 *Diapensia lapponica* × 1 2/5

224 *Galax urceolata* × 1/4

225 *Samolus parviflorus* × 1 1/3

226 *Lysimachia quadrifolia* × 3/10

227 *Lysimachia lanceolata* × 1/2

229 *Glaux maritima* × 1

228 *Lysimachia terrestris* × 2/5

230 *Dodecatheon meadia* × 1/5

231 *Spigelia marilandica* × 2/3

232 *Polypremum procumbens* × 1 2/5

233 *Sabatia dodecandra* × 2/3

234 *Sabatia angularis* × 3/10

235 *Sabatia capitata* × 3/7

236 *Sabatia difformis* × 9/20

237 *Obolaria virginica* × 2

238 *Gentiana villosa* × 1/3

239 *Gentiana saponaria* × 11/20

240 *Gentienella quinquefolia* var. *quinquefolia* × 9/20

241 *Frasera caroliniensis* × 1 1/10

242 *Menyanthes trifoliata* × 3/5

243 *Nymphoides aquatica* × 3/4

244 *Amsonia tabernaemontana* × 1 2/5

245 *Apocynum androsaemifolium* × 1/2

246 *Asclepias lanceolata* × 1 1/3

247 *Asclepias variegata* × 1/5

248 *Asclepias amplexicaulis* × 1/4

249 *Asclepias perennis* × 2/3

250 *Asclepias verticillata* × 2/7

251 *Asclepias syriaca* × 1/3

252 *Asclepias tuberosa* × 1/3

253 *Asclepias connivens* × 2/3

254 *Asclepias incarnata* × 1/3

255 *Asclepias quadrifolia* × 7/10

256 *Asclepias obovata* × 2/3

257 *Asclepias humistrata* × 1/2

258 *Matelea carolinensis* × 3/5

259 *Matelea gonocarpa* × 1/2

260 *Cuscuta gronovii* × 9/10

261 *Dichondra carolinensis* × 1 4/5

263 *Jacquemontia tamnifolia* × 3/10

262 *Stylisma patens* × 1

264 *Calystegia spithamaea* × 1/3

265 *Ipomoea purpurea* × 1/6

266 *Ipomoea imperati* × 9/20

267 *Ipomoea pes-caprae* × 1/3

268 *Ipomoea coccinea* × 3/8

269 *Ipomoea hederacea* × 3/7

270   *Ipomoea pandurata* × 3/20

271   *Ipomoea cordatotriloba* var. *cordatotrilobata* × 3/10

272   *Phlox divaricata* × 1/3

273 *Phlox carolina* × 9/20

274 *Phlox paniculata* × 2/7

275 *Phlox drummondii* × 1/2

276 *Polemonium reptans* × 9/20

277 *Hydrophyllum canadense* × 1/2

278 *Phacelia bipinnatifida* × 4/5

279 *Phacelia purshii* × 1 4/5

280 *Phacelia dubia* × 7/10

281 *Hydrolea corymbosa* × 9/10

282 *Hydrolea quadrivalvis* × 3/5

283 *Mertensia virginica* × 3/8

285   *Glandularia pulchella* × 1

284   *Lithospermum carolinense* × 3/5

286   *Verbena rigida* × 2 3/5

287   *Phyla nodiflora* × 9/10

288 *Trichostema dichotomum* × 1 9/10

289 *Teucrium canadense* × 1/3

290 *Scutellaria elliptica* × 1 2/3

291 *Scutellaria integrifolia* × 2/3

292 *Scutellaria montana* × 9/10

293 *Prunella vulgaris* × 9/10

294 *Physostegia virginiana* × 1

295 *Galeopsis tetrahit* × 1/2

296 *Lamium amplexicaule* × 1 1/2

297 *Stachys tenuifolia* × 1 1/6

298 *Salvia azurea* × 11/20

299 *Salvia lyrata* × 1/4

300 *Monarda didyma* × 9/20

301 *Monarda punctata* × 7/10

302 *Blephilia ciliata* × 1

303 *Dicerandra odoratissima* × 7/10

304 *Dicerandra linearifolia* × 7/10

305 *Pycnanthemum incanum* × 1/2

306 *Pycnanthemum montanum* × 3/8

307 *Cunila origanoides* × 3/5

308 *Lycopus americanus* × 3/4

309 *Mentha arvensis* × 9/20

310 *Collinsonia canadensis* × 1/5

311 *Hyptis alata* × 2/5

312 *Physalis heterophylla* × 3/8

313 *Solanum carolinense* × 3/4

314 *Solanum sisymbriifolium* × 11/20

315 *Solanum dulcamara* × 3/5

316 *Verbascum thapsus* × 1/33

317 *Verbascum blattaria* × 7/10

318 *Nuttallanthus canadensis* × 1 1/7

319 *Linaria vulgaris* × 9/10

320 *Scrophularia lanceolata* × 1/3

321 *Chelone glabra* × 1/3

322 *Chelone obliqua* × 7/10

323 *Penstemon australis* × 3/8

324 *Penstemon canescens* × 3/20

325 *Penstemon laevigatus* × 2/3

326 *Mimulus ringens* × 5/6

327 *Gratiola ramosa* × 1

328 *Gratiola floridana* × 9/10

329 *Gratiola neglecta* × 1 1/20

330 *Bacopa caroliniana* × 1 1/6

331 *Bacopa monnieri* × 1 3/4

332 *Amphianthus pusillus* × 1 1/12

333 *Lindernia monticola* × 1 1/3

334 *Veronica persica* × 1

335 *Veronica serpyllifolia* × 9/10

336 *Agalinis fasciculata* × 5/6

337 *Agalinis tenuifolia* × 1 1/5

338 *Aureolaria pectinata* × 7/10

339 *Aureolaria virginica* × 5/8

340 *Aureolaria flava* × 1/3

341 *Buchnera americana* × 1 1/5

343 *Melampyrum lineare* × 1 1/6

342 *Castilleja coccinea* × 3/5

344 *Pedicularis canadensis* × 9/20

345 *Rhinanthus minor* ssp. *minor* × 3/4

346 *Conopholis americana* × 9/20

347 *Orobanche uniflora* × 1/2

348 *Pinguicula lutea* × 7/10

349   *Pinguicula pumila* × 1 7/8

350   *Utricularia juncea* × 1 1/2

351   *Utricularia inflata* × 1/3

352 *Utricularia inflata* × 4

353 *Utricularia purpurea* × 1/12

354 *Utricularia subulata* × 9/10

355 *Dyschoriste oblongifolia* × 1 2/7

357 *Justicia americana* × 3/4

356 *Ruellia caroliniensis* × 2/5

358 *Plantago major* × 1/5

359 *Plantago virginica* × 1/2

360 *Plantago lanceolata* × 1/8

361 *Houstonia caerulea* × 1 1/10

362 *Houstonia procumbens* × 1 4/5

363 *Houstonia purpurea* × 11/20

364 *Richardia scabra* × 9/10

366 *Diodia teres* × 1/2

365 *Diodia virginica* × 1

367 *Sherardia arvensis* × 1

368 *Galium aparine* × 1/2

369 *Galium tinctorium* × 9/20

370 *Galium hispidulum* × 4/5

371 *Triosteum perfoliatum* × 3/5

372 *Valerianella radiata* × 1/3

373 *Melothria pendula* × 4/5

374 *Echinocystis lobata* × 2/7

375   *Campanula divaricata* × 2/5

376   *Campanula rotundifolia* × 9/10

377   *Triodanis perfoliata* × 9/10

378   *Lobelia elongata* × 2/5

379 *Lobelia cardinalis* × 2/7

380 *Vernonia gigantea* × 3/20

381 *Vernonia glauca* × 1/5

382 *Elephantopus tomentosus* × 2/3

383   *Elephantopus carolinianus* × 1/6

384   *Eupatorium capillifolium* × 7/100

385   *Eupatorium rotundifolium* × 1/2

386   *Eupatorium dubium* × 3/10

387 *Eupatorium perfoliatum* × 2/7

388 *Eupatorium coelestinum* × 1/4

389 *Eupatorium fistulosum* × 1/50

390 *Ageratina altissima* × 1/3

391 *Ageratina aromatica* × 1/4

392 *Mikania scandens* × 3/8

393 *Liatris pilosa* var. *pilosa* × 2/5

394 *Liatris aspera* × 1/3

395   *Liatris helleri* × 1/2

396   *Carphephorus odoratissimus* × 1/4

397   *Chrysopsis mariana* × 3/10

398 *Pityopsis pinifolia* × 3/20

399 *Heterotheca subaxillaris* × 1/8

400 *Solidago canadensis* × 1/4

401 *Solidago curtisii* × 1/4

402 *Solidago rugosa* × 1/2

403 *Solidago bicolor* × 3/5

404 *Solidago sempervirens* × 1/10

405 *Solidago spithamaea* × 1/6

406 *Euthamia tenuifolia* × 1/3

407 *Sericocarpus asteroides* × 1

408 *Oclemena reticulare* × 1/8

409 *Symphyotrichum dumosum* var. *dumosum* × 4/5

411 *Symphyotrichum subulatum* × 3/10

410 *Symphyotrichum puniceum* var. *puniceum* × 1/8

412 *Symphyotrichum novi-belgii* var. *novi-belgii* × 2/7

413 *Symphyotrichum tenuifolium* × 2/3

414 *Symphyotrichum lateriflorum* × 1/3

415 *Eurybia divaricata* × 3/20

416 *Doellingeria umbellata* × 3/5

417 *Erigeron vernus* × 1/4

418 *Erigeron philadelphicus* × 9/10

419 *Erigeron strigosus* × 1/10

420 *Pluchea odorata* × 3/10

421 *Pluchea foetida* × 9/20

422 *Pterocaulon virgatum* × 1/3

423 *Antennaria plantaginifolia* × 1/3

424 *Facelis retusa* × 1

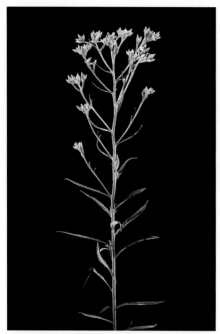

425 *Pseudognaphalium obtusifolium* ssp. *obtusifolium* × 7/20

426 *Gamochaeta purpurea* × 1/2

427 *Smallanthus uvedalius* × 1/4

428 *Silphium asteriscus* var. *laevicaule* × 1/3

429 *Berlandiera pumila* × 1/2

430 *Chrysogonum virginianum* × 7/10

431 *Parthenium integrifolium* × 9/10

432 *Iva microcephala* × 3/10

433 *Ambrosia artemisiifolia* × 1/2

434 *Ambrosia trifida* × 1/8

435 *Xanthium strumarium* × 1 1/2

436 *Eclipta prostata* × 1 1/2

437 *Tetragonotheca helianthoides* × 2/5

438 *Rudbeckia hirta* × 3/10

439 *Rudbeckia nitida* × 1/3

440 *Rudbeckia laciniata* × 1/9

441 *Echinacea pallida* × 1/5

442 *Ratibida pinnata* × 1/3

443 *Helianthus angustifolius* × 1/4

444 *Helianthus atrorubens* × 1/12

445 *Helianthus porteri* × 1/18

446 *Helianthus debilis* × 1/2

447 *Helianthus tuberosus* × 1/2

448 *Helianthus strumosus* × 3/10

449 *Helianthus decapetalus* × 1/4

451 *Verbesina occidentalis* × 1/4

450 *Melanthera nivea* × 2/5

452 *Verbesina virginica* × 1/5

453 *Coreopsis tripteris* × 1/8

454 *Coreopsis major* × 3/10

455 *Coreopsis nudata* × 1/2

456   *Coreopsis grandiflora* × 1/5

457   *Bidens pilosa* × 2/3

458   *Bidens laevis* × 2/3

459   *Bidens frondosa* × 3/4

460   *Bidens coronata* × 7/10

461   *Bidens bipinnata* × 2/5

462   *Balduina angustifolia* × 3/20

463 *Galinsoga quadriradiata* × 1

464 *Marshallia graminifolia* var. *cynanthera* × 9/10

465 *Helenium amarum* × 1/4

466 *Helenium flexuosum* × 2/5

467 *Helenium vernale* × 9/10

468 *Helenium autumnale* × 7/10

469 *Gaillardia pulchella* × 3/5

470 *Achillea millefolium* × 1/5

471 *Leucanthemum vulgare* × 1/4

472 *Tanacetum vulgare* × 3/7

473 *Erichtites hieraciifolia* × 3/5

474 *Senecio aureus* × 1/3

476 *Senecio anonymus* × 1/4

475 *Senecio tomentosus* × 1/3

477 *Cirsium horridulum* × 1/3

478 *Cirsium carolinianum* × 1

479 *Cirsium vulgare* × 3/5

480 *Centaurea maculosa* × 1 1/2

481 *Cichorium intybus* × 1

482 *Krigia virginica* × 3/4

483 *Krigia montana* × 1/8

484   *Hypochaeris radicata* × 1/2

485   *Taraxicum officinale* × 1/4

486   *Sonchus asper* × 1/5

487   *Lactuca canadensis* × 1/3

488   *Pyrrhopappus carolinianus* × 2/5

489   *Prenanthes trifoliata* × 3/5

490   *Hieracium venosum* × 2/5

491 *Hieracium gronovii* × 1/10

492 *Hieracium aurantiacum* × 1

493 *Hieracium pilosella* × 3/5

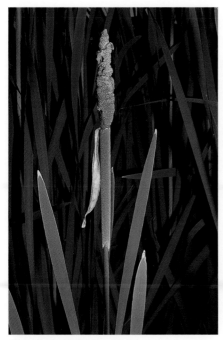

494 *Typha latifolia* × 2/7

495 *Sparganium americanum* × 1/2

496 *Sagittaria latifolia* × 2/7

497 *Sagittaria graminea* × 1/3

498 *Tripsacum dactyloides* × 2/7

499 *Saccharum giganteum* × 1/16

500 *Andropogon glomeratus* × 1/20

501 *Andropogon virginicus* var. *glaucus*
× 1/7

502 *Andropogon ternarius* × 9/20

503 *Schizachyrium scoparium* × 1/10

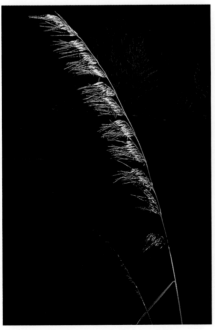

504 *Sorghastrum secundum* × 1/5

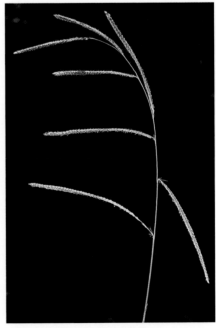

505 *Paspalum urvillei* × 1/3

506 *Paspalum notatum* × 1/10

507 *Panicum virgatum* × 1/33

508 *Panicum amarum* × 1/5

509 *Sacciolepis striata* × 7/10

510 *Echinochloa walteri* × 2/7

511 *Setaria viridis* × 3/10

512 *Setaria parviflora* × 1

513 *Setaria magna* × 1/8

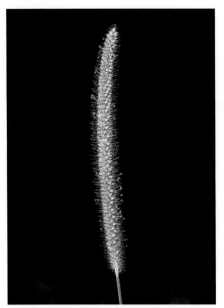

514 *Pennisetum americanum* × 1/3

515 *Cenchrus tribuloides* × 1

516 *Cenchrus echinatus* × 1 1/2

517 *Stenotaphrum secundatum* × 5/8

518 *Zizania aquatica* × 1/7

519 *Muhlenbergia filipes* × 1/7

520 *Sporobolus indicus* × 1/2

521 *Agrostis stolonifera* × 2/7

522 *Holcus lanatus* × 9/20

523 *Spartina cynosuroides* × 1/4

524 *Eustachys petraea* × 1 1/12

525 *Eleusine indica* × 3/10

526 *Phragmites australis* × 3/10

527 *Eragrostis spectabilis* × 1/4

528 *Uniola paniculata* × 1/25

529 *Briza minor* × 3/4

530 *Poa annua* × 1 4/7

531 *Elymus virginicus* × 1/3

532 *Cyperus esculentus* × 3/10

533 *Cyperus retrorsus* × 1/3

534 *Fuirena pumila* × 3/5

535 *Scirpus robustus* × 2

536 *Scirpus cyperinus* × 3/20

537 *Schenoplectus americanus* × 1 1/2

538 *Eleocharis tuberculosa* × 3/20

539 *Fimbristylis castanea* × 4/5

540 *Rhynchospora corniculata* × 7/10

541 *Rhynchospora latifolia* × 3/4

542 *Scleria triglomerata* × 3/5

543 *Carex lupuliformis* × 1

544 *Cymophyllus fraserianus* × 3/20

545 *Orontium aquaticum* × 1/5

546 *Peltandra sagittifolia* × 3/10

547 *Arisaema triphyllum* × 3/7

548 *Xyris ambigua* × 1 2/5

549 *Eriocaulon decangulare* × 1/10

550 *Eriocaulon decangulare* × 1 2/3

551 *Tillandsia usneoides* × 1 9/20

552 *Commelina erecta* × 3/5

553 *Commelina virginica* × 1/3

554 *Murdannia keisak* × 1 1/4

555 *Tradescantia virginiana* × 3/7

556 *Callisia rosea* × 2 1/8

557 *Eichhornia crassipes* × 2/5

558 *Pontederia cordata* × 2/7

559 *Juncus tenuis* × 1

560 *Juncus roemerianus* × 1/2

561 *Tofieldia racemosa* × 9/10

562 *Helonias bullata* × 1/4

563 *Chamaelirium luteum* × 1/7

564 *Stemanthium gramineum* × 3/20

565 *Zigadenus densus* × 1 3/20

566 *Uvularia perfoliata* × 2/7

567 *Uvularia puberula* × 1/4

568 *Hemerocallis fulva* × 1/15

569 *Allium canadense* × 1 1/3

570 *Allium cuthbertii* × 9/10

571 *Nothoscordum bivalve* × 1 1/4

572 *Lilium michauxii* × 3/4

573 *Lilium superbum* × 1/10

574 *Lilium catesbaei* × 2/5

575 *Lilium grayi* × 1/8

576 *Erythronium americanum* × 7/10

577 *Ornithogalum umbellatum* × 5/6

578 *Clintonia umbellulata* × 1 1/6

579 *Maianthemum racemosum* × 1/5

580 *Maianthemum canadense* × 1 1/6

581 *Polygonatum biflorum* × 1 1/2

582 *Convularia majuscula* × 1 1/2

584 *Trillium lancefolium* × 2/3

583 *Trillium cuneatum* × 9/20

585 *Trillium luteum* × 3/10

586 *Trillium decumbens* × 1/3

587 *Trillium decipiens* × 1/12

588 *Trillium stamineum* × 3/7

589 *Trillium discolor* × 3/8

590 *Trillium undulatum* × 1/5

591 *Trillium catesbaei* × 2/7

592 *Trillium vaseyi* × 3/8

593 *Trillium erectum* × 1/6

594 *Trillium grandiflorum* × 3/20

595 *Trillium rugelii* × 1/4

596 *Trillium persistens* × 2/5

597 *Lachnanthes caroliana* × 1/7

598 *Lophiola aurea* × 1 9/10

599 *Zephyranthes atamasca* × 1/2

600 *Hymenocallis caroliniana* × 3/20

601 *Manfreda virginica* × 3/10

602 *Hypoxis hirsuta* × 3/8

603 *Iris virginica* × 2/5

604 *Iris cristata* × 1/5

605 *Belamcanda chinensis* × 1/2

606 *Sisyrinchium albidum* × 3/10

607 *Sisyrinchium atlanticum* × 2 1/5

608 *Canna flaccida* × 3/10

609 *Cypripedium acaule* × 1/6

610 *Cypripedium calceolus* × 4/7

611 *Cypripedium reginae* × 9/10

612 *Galearis spectabilis* × 1

613 *Platanthera clavellata* × 1/3

614 *Platanthera lacera* var. *lacera* × 4/7

615 *Platanthera psycodes* × 3/10

616 *Platanthera ciliaris* × 9/10

617 *Platanthera blephariglottis* var. *blephariglottis* × 11/20

618 *Platanthera nivea* × 9/10

619 *Pogonia ophioglossoides* × 2/5

620 *Cleistes divaricata* × 3/5

621   *Spiranthes cernua* × 1/2

622   *Spiranthes vernalis* × 1/2

623   *Listera smallii* × 3/5

624   *Goodyera pubescens* × 1/5

625   *Calopogon barbatus* × 1 2/5

626   *Corallorhiza wisteriana* × 1/3

627   *Malaxis unifolia* × 11/20

628   *Tipularia discolor* × 2/5

629   *Liparis liliifolia* × 2/5

630   *Epidendrum conopseum* × 1 1/10

631   *Hexalectris spicata* × 3/5

Index

Photograph numbers are in **boldface.**

Acalypha
  gracilens: 46; **169**
  rhomboidea: 46
  virginica: 46
    var. *rhomboidea:* 46
    var. *virginica:* 46
ACANTHACEAE: 96
Acanthus Family: 96
Achillea millefolium: 129; **470**
Actaea
  alba: 13
  pachypoda: 13; **43**
  rubra: 13
Actinospermum angustifolium: 127
Adder's-mouth: 179
Aeschynomene
  indica: 36; **132**
  virginica: 36
  viscidula: 36
Agalinis
  fasciculata: 91; **336**
  setacea: 92
  tenuifolia: 91–92; **337**
Agave virginica: 172
Ageratina
  altissima: 107; **390**
  aromatica: 107; **391**
Agrostis
  alba: 147
  hyemalis: 147
  perennans: 147
  scabra: 147
  stolonifera: 147; **521**
Alchemilla: 27
Alchemilla microcarpa: 27; **98**
ALISMATACEAE: 138
Allium
  bivalve: 163
  canadense: 163; **569**
  cernuum: 163

  cuthbertii: 163; **570**
  inodorum: 163
  tricoccum: 163
Alumroot: 24
AMARANTHACEAE: 6
Amaranth Family: 6
AMARYLLIDACEAE: 171
Amaryllis Family: 171
Ambrosia
  artemisiifolia: 120; **433**
  trifida: 120; **434**
Amerorchis rotundifolia: 175
Amianthium muscaetoxicum: 162
Ammannia
  coccinea: 54
  latifolia: 54
  teres: 54
Amphianthus: 90
Amphianthus pusillus: 90; **332**
Amsonia
  ciliata: 66–67
  rigida: 66–67
  tabernaemontana: 66; **244**
Anatherix connivens: 69
Andropogon
  capillipes: 141
  elliottii: 141
  glomeratus: 140–41; **500**
    var. *glaucopsis:* 141
    var. *glomeratus:* 141
  gyrans var. *gyrans:* 141
  ternarius: 141; **502**
  virginicus
    var. *abbreviatus:* 141
    var. *glaucus:* 141; **501**
    var. *virginicus:* 141
Aneilema
  keisak: 159
  nudiflorum: 159

Anemone
  lancifolia: 14; **47**
  quinquefolia: 14
Anemonella thalictroides: 15
Angelica: 60
Angelica venenosa: 60; **219**
Antennaria: 117
  neglecta: 117
  plantaginifolia: 117; **423**
  solitaria: 117
APIACEAE: 58
Apios
  americana: 39; **145**
  tuberosa: 39
APOCYNACEAE: 66
Apocynum
  androsaemifolium: 67; **245**
  cannabinum: 67
Aquilegia canadensis: 13; **45**
Arabidopsis thaliana: 20
Arabis virginica: 20
ARACEAE: 156
ARALIACEAE: 58
Arenaria
  caroliniana: 9
  uniflora: 9–10
Arisaema
  dracontium: 157
  triphyllum: 156; **547**
ARISTOLOCHIACEAE: 2
Arrowhead: 138
Arrow-vine: 3–4
Artichoke, Jerusalem: 123
Arum, White: 156
Arum Family: 156
Aruncus dioicus: 25; **89**
Arundo donax: 149
Asarum
  canadense: 2; **4**
    var. *acuminatum:* 2
    var. *reflexum:* 2
ASCLEPIADACEAE: 67
Asclepias
  amplexicaulis: 68; **248**
  cinera: 68
  connivens: 69; **253**
  exaltata: 68–69
  humistrata: 70; **257**

  incarnata: 69; **254**
    ssp. *incarnata:* 69
    ssp. *pulchra:* 69
  lanceolata: 67; **246**
  obovata: 70; **256**
  perennis: 67, 68; **249**
  phytolacioides: 68–69
  purpurascens: 69
  quadrifolia: 70; **255**
  rubra: 67
  syriaca: 68; **251**
  tomentosa: 70
  tuberosa: 69; **252**
    ssp. *interior:* 69
    ssp. *rolfsii:* 69
    ssp. *tuberosa:* 69
  variegata: 67–68; **247**
  verticillata: 68; **250**
  viridiflora: 70
  viridis: 69
Asclepiodora viridis: 69
Aster: 114
  Annual Saltmarsh: 113
  Bushy: 113
  Flat-topped White: 115
  Many-flowered: 114
  New England: 113–14
  New York: 113
  Perennial Saltmarsh: 114
  Swamp: 113
  White-topped: 112
  Wood: 114–15
Aster
  divaricatus: 114
  dumosus: 113
  ericoides: 114
  lateriflorus: 114
  novae-angliae: 113–14
  novi-belgii: 113
  patens: 114
  paternus: 112
  pilosus: 113
  premanthoides: 113
  puniceus: 113
  reticulatus: 112
  subulatus: 113
  tenuifolius: 114
  umbellatus: 114

undulatus: 114–15
ASTERACEAE: 104
Astilbe biternata: 23, 25; **82**
Atamasco-lily: 171
Aureolaria
  flava: 92; **340**
  laevigata: 92
  pectinata: 92; **338**
  pedicularia: 92
  virginica: 92; **339**
Autumn-tresses: 177
Axonopus
  affinis: 143
  fissifolius: 143

Bachelor's-button: 43
Bacopa
  caroliniana: 89–90; **330**
  cyclophylla: 90
  innominata: 90
  monnieri: 90; **331**
  rotundifolia: 90
Baked-apple-berry: 25
Balduina, Annual: 127
Balduina
  angustifolia: 127; **462**
  atropurpurea: 127
  uniflora: 127, 129
Balsam-apple: 101
Baneberry
  Red: 13
  White: 13
Baptisia
  alba: 30; **109**
    var. *macrophylla:* 30
  albescens: 30
  arachnifera: 31; **112**
  australis: 30; **110**
    var. *australis:* 30
    var. *minor:* 30
  bracteata: 29–30; **108**
    var. *leucophaea:* 30
  cinera: 30
  elliptica: 30
  lanceolata
    var. *lanceolata:* 30; **111**
    var. *tomentosa:* 30
  lecontei: 29

  leucophaea: 30
  minor: 30
  nuttalliana: 30
  tinctoria: 29; **107**
Barbara's-buttons: 127–28
Barbarea vulgaris: 19; **69**
Barberry Family: 16
Barren-strawberry: 27
Batschia caroliniensis: 77
Beaked-rush: 154
Bean, Coral: 39
Bean Family: 27
Beard-tongue: 88
  Appalachian: 88
  Southeastern: 87–88
Bears-foot: 118
Bedstraw: 100
  Dye: 100
Bee-balm: 82
Beggar-ticks: 36
Beggar's-ticks
  Annual: 126
  Swamp: 126–27
Belamcanda chinensis: 173; **605**
Bellflower: 102
Bellwort: 162
Bentgrass, Autumn: 147
BERBERIDACEAE: 16
Berlandiera
  pumila: 119; **429**
  subacaulis: 119
Betsy, Sweet: 166
Bidens
  aristosa: 126
  bipinnata: 127; **461**
  cernua: 126
  coronata: 126–27; **460**
  frondosa: 126; **459**
  laevis: 126; **458**
  mitis: 126
  pilosa: 125–26; **457**
  vulgata: 126
Bindweed, Low: 72
Biovularia olivacea: 95
Birdsfoot-trefoil: 34
Birthwort Family: 2
Bittersweet: 85
Bitterweed: 128

Bivonea stimulosa: 46
Blackberry-lily: 173
Black-eyed-Susan: 121
Black-root: 117
Black-snakeroot: 161–62
Bladder-pod: 35
Bladderwort
  Floating: 95
  Horned: 95
  Purple: 95
  Slender: 95–96
  Wiry: 95–96
Blazing-star: 107–8, 161
  Plains: 108
  Smooth: 108
Bleeding-heart: 18
Blephilia: 82
Blephilia
  ciliata: 82; **302**
  hirsuta: 82
Bloodroot: 17
Blowballs: 133
Bluebells: 77
  Northern: 77
  Seaside: 77
Blue-curls: 79
Blue-eyed-grass: 173–74
Bluegrass, Annual: 150–51
Blue-hearts: 92–93
Blue-sailors: 132
Blue-star: 66–67
Bluestem, Little: 141–42
Bluet
  Summer: 98–99
  Trailing: 98
Bluets: 98
Boehmeria cylindrica: 1; **3**
Boneset: 106
Borage Family: 77
BORAGINACEAE: 77
Bouncing-bet: 11
Bradburya virginiana: 38
Bramia monnieri: 90
Brasenia schreberi: 11; **38**
BRASSICACEAE: 18
Bristlegrass
  Knotroot: 145
  Yellow: 145
Briza minor: 150; **529**

BROMELIACEAE: 158
Broomsedge
  Bushy: 140–41
  Chalky: 141
  Elliott's: 141
  Virginia: 141
Buchnera
  americana: 92–93; **341**
  floridana: 92–93
Buck-bean: 66
Buck-bean Family: 66
Buckwheat Family: 2
BUDDLEJACEAE: 63
Bugleweed, American: 83–84
Bulbostylis: 154
Bull-nettle: 85
Bulrush
  Marsh: 153
  Saltmarsh: 153
Bur-marigold: 126
Bur-reed: 137
Bur-reed Family: 137
Bush-pea: 29
Butter-and-eggs: 86–87
Buttercup
  Bulbous: 14
  Early: 14–15
  Kidney-leaf: 15
Butterfly-bush Family: 63
Butterfly-pea: 38
  Climbing: 38
Butterfly-weed: 69
Butterwort
  Dwarf: 94
  Yellow: 94
Buttonrods: 157–58
Buttonweed: 99
  Rough: 99

Cacao Family: 50
Cakile
  constricta: 19
  edentula: 19; **68**
  geniculata: 19
Callisia
  graminea: 159
  rosea: 159; **556**
Calopogon
  barbatus: 178; **625**

multiflorus: 178
pallidus: 179
pulchellus: 179
tuberosus
   var. *tuberosus:* 179
Caltha
  natans: 12
  palustris: 12; **42**
Caltrope Family: 43
Calystegia
  catesbiana: 72
  sepium: 72
  spithamaea: 72; **264**
Campanula
  americana: 102
  divaricata: 102; **375**
  flexuosa: 102
  rotundifolia: 102; **376**
CAMPANULACEAE: 102
Campanulastrum americanum:
102
Campe barbarea: 19
Camphorweed: 110, 116
Campion, Fringed: 10
Canada-garlic: 163
Canada-mayflower: 165
Cancer-root: 94
Candyweed: 44
Canna, Golden: 174
CANNACEAE: 174
Canna Family: 174
Canna flaccida: 174; **608**
Canna X *generalis:* 174
CAPRIFOLIACEAE: 100
Capscale: 144
Caraway: 59
Cardamine
  angustata: 20
  diphylla: 20; **72**
  hirsuta: 20
  parviflora: 20; **71**
  pensylvanica: 20
Cardinal-flower: 103
Cardinal-spear: 39
Carduus
  carolinianus: 131
  spinosissimus: 131
  virginianum: 131
Carex: 155

Carex lupuliformis: 155; **543**
Carpet-cress: 19
Carpetweed: 7
Carpetweed Family: 7
Carphephorus: 109
Carphephorus
  bellidifolius: 109
  corymbosus: 109
  odoratissimus: 108–9; **396**
  paniculatus: 109
  tomentosus: 109
Carrot, Wild: 60
Carrot Family: 58
Carum carvi: 59; **217**
CARYOPHYLLACEAE: 8
Cassia
  fasciculata: 28
  hebecarpa: 28
  marilandica: 28
  nictitans: 29
  obtusifolia: 28
  occidentalis: 28
  tora: 28
Castalia
  flava: 12
  odorata: 12
Castilleja
  coccinea: 93; **342**
  septentrionalis: 93
  sessiliflora: 93
Catchfly: 10–11
  Forking: 10
Catchweed: 100
Cat-clover: 34
Cat's-ear: 133
Cattail
  Common: 137
  Southern: 137
Cattail Family: 137
Caulophyllum thalictroides: 17; **58**
Celadine-poppy: 17
Cenchrus
  carolinianus: 146
  echinatus: 146; **516**
  longispinus: 146
  pauciflorus: 146
  tribuloides: 145–46; **515**
Centaurea maculosa: 132; **480**
Centella asiatica: 58; **214**

Centrosema virginianum: 38; **143**
Cerastium: 9
  glomeratum: 9
  nutans: 9
  viscosum: 9; **30**
Chamaecrista
  aspera: 29
  deeringiana: 28
  fasciculata var. *fasciculata:* 28; **103**
  nictitans: 29; **104**
    ssp. *nictitans* var. *aspera:* 29
Chamaelirium luteum: 161; **563**
Chamaesyce
  bombensis: 47
  cordifolia: 47
  hirta: 47–48; **176**
  maculata: 47; **175**
  polygonifolia: 47
  prostata: 47
Chelone
  chlorantha: 87
  cuthbertii: 87
  glabra: 87; **321**
  lyonii: 87
  montana: 87
  obliqua: 87; **322**
CHENOPODIACEAE: 5
Chenopodium
  album: 5; **17**
  ambrosioides: 5; **16**
  rubrum: 5–6
Chickweed
  Common: 9
  Giant: 8
  Mouse-eared: 9
Chicory, Common: 132
Chigger-weed: 69
Chloris
  glauca: 149
  petraea: 149
Chocolate-weed: 50
Chrysanthemum leucanthemum: 129–30
Chrysogonum: 119
Chrysogonum
  australe: 119
  virginianum: 119; **430**
Chrysopsis
  gossypina: 109

  mariana: 109; **397**
  nervosa: 109–10
  pinifolia: 109
Cichorium intybus: 132; **481**
Cicuta: 60
Cimicifuga
  americana: 13
  racemosa: 13; **44**
Cinquefoil: 26
  Rough-fruited: 26
Circaea alpina: 57–58; **211**
Cirsium
  carolinianum: 131; **478**
  horridulum: 131; **477**
  lanceolatum: 131–32
  smallii: 131
  virginianum: 131
  vulgare: 131–32; **479**
Claytonia
  caroliniana: 8
  virginica: 7–8; **24**
Cleistes divaricata: 177; **620**
Cleome
  hassleriana: 18
  spinosa: 18
Climbing-milkweed: 71
Clintonia
  borealis: 165
  umbellulata: 165; **578**
Clitoria mariana: 38; **142**
Cloudberry: 25
Clover
  Alsike: 33
  Buffalo: 34
  Hop: 33
  Low Hop: 32–33
  Rabbit-foot: 32
  Red: 33
  Sour: 32
  White: 33
  Yellow: 33
  Yellow Sweet: 32
CLUSIACEAE: 50
Cnidoscolus stimulosus: 46; **171**
Cocklebur, Common: 120
Coffee-weed: 28
Cohosh, Blue: 17
Collinsonia
  anisata: 84

canadensis: 84; **310**
punctata: 84
serotina: 84
tuberosa: 84
verticillata: 84
Columbine, Wild: 13
Columbo: 65–66
Comfort-root: 49
Commelina
  communis: 158
  diffusa: 158–59
  erecta: 158; **552**
  virginica: 158; **553**
COMMELINACEAE: 158
Composite Family: 103
Coneflower: 122
  Cut-leaf: 121–22
  Narrow-leaved: 121
  Purple: 122
Conoclinium coelestinum: 106
Conopholis americana: 94; **346**
CONVOLVULACEAE: 71
Convolvulus
  americanus: 72
  repens: 72
  sepium: 72
  sericatus: 72
  spithamaeus: 72
Convularia
  majalis: 166
    var. *montana:* 166
  majuscula: 166; **582**
  montana: 166
Copperleaf, Short-stalk: 46
Corallorhiza
  maculata: 179
  odontorhiza: 179
  wisteriana: 179; **626**
Coral-root: 179
  Crested: 180
Cordgrass
  Big: 148
  Bunch: 149
  Gulf: 148
  Prairie: 148
  Saltmeadow: 148–49
  Smooth: 148
Coreopsis
  Large-flowered: 125

Swamp: 125
Tall: 124–25
Whirled-leaf: 125
Coreopsis
  auriculata: 125
  grandiflora: 125; **456**
  lanceolata: 125
  major: 125; **454**
    var. *rigida:* 125
  nudata: 125; **455**
  tripteris: 124–25; **453**
  verticillata: 125
Corn-lily: 165
Corn-salad: 101
Coronilla varia: 35; **131**
Coronopus didymus: 19; **67**
Corydalis, Pale: 18
Corydalis
  flavula: 18
  sempervirens: 18; **64**
Cottonweed: 6
Cowslip: 12
Cow-wheat: 93
Cracca
  mohrii: 35
  virginiana: 35
Cranesbill: 41
  Carolina: 41–42
CRASSULACEAE: 22
Cress
  Bitter: 20
  Field: 19
Crotalaria, Showy: 31
Crotalaria
  purshii: 31
  retusa: 31
  rotundifolia: 31; **114**
  sagittalis: 31
  spectabilis: 31; **113**
Croton, Tooth-leaved: 45–46
Croton glandulosus: 45–46; **168**
Crowfoot Family: 12
Crown-beard: 124
Crown-vetch: 35
Crow-poison: 161–62
Cucumber, Creeping: 101
CUCURBITACEAE: 101
Cudweed, Purple: 118
Cunila origanoides: 83; **307**

Cuscuta gronovii: 71; **260**
Cuthbertia
  graminea: 159
  rosea: 159
Cymophyllus fraserianus: 156;
  544
Cynthia
  dandelion: 132
  montana: 132
CYPERACEAE: 151
Cyperus
  croceus: 152
  esculentus: 151–52; **532**
  globosus: 152
  retrorsus: 152; **533**
Cypripedium
  acaule: 174; **609**
  calceolus: 174–75; **610**
    var. *pubescens:* 174–75
  parviflorum: 174–75
  reginae: 175; **611**

Daisy, Ox-eye: 129–30
Dalea
  gattingeri: 34; **127**
  pinnata: 34; **126**
  purpurea: 34–35
Dandelion: 133
  Dwarf: 132–33
Daucus
  carota: 60; **220**
  pusillus: 60
Dayflower: 158
  Woods: 158–59
Day-lily: 162
Deer-tongue: 108–9
Delphinium
  carolinianum: 13–14
  tricorne: 13; **46**
Dentaria
  diphylla: 20
  heterophylla: 20
Desmanthus illinoensis: 27
Desmodium
  canescens: 36; **135**
  paniculatum: 36; **134**
Devil's-bit: 161
Diamorpha: 23
Diamorpha
  cymosa: 23

  smallii: 23; **81**
Dianthera americana: 96
Diapensia: 61
DIAPENSIACEAE: 61
Diapensia lapponica: 61; **223**
Dicentra
  canadensis: 18; **62**
  cucullaria: 17; **61**
  eximia: 18; **63**
Dicerandra
  Rose: 82
  White: 83
Dicerandra
  densiflora: 82
  linearifolia: 83; **304**
  odoratissima: 82; **303**
Dichondra carolinensis: 71; **261**
Dichromena
  colorata: 155
  latifolia: 155
Dicotyledons: 1
Diodia
  hirsuta: 99
  teres: 99; **366**
  tetragona: 99
  virginica: 99; **365**
Diphylleia cymosa: 16; **57**
Ditremexa occidentalis: 28
Dittany: 83
Dock, Curly: 3
Dodder: 71
Dodecatheon meadia: 63; **230**
Doellingeria
  reticulata: 112
  umbellata: 115; **416**
Dogbane: 67
Dogbane Family: 66
Dog-fennel: 105
Dog-tongue: 2–3
Dog-tooth-violet: 164
Dracocephalum
  denticulatum: 80
  virginianum: 80
Dragonhead, False: 80
Drosera
  brevifolia: 22; **77**
  capillaris: 22
  leucantha: 22
  rotundifolia: 22
DROSERACEAE: 22

Drum-heads: 44–45
Duchesnia indica: 26–27; **96**
Dulichium arundinaceum: 152
Dutchman's-breeches: 17
Dwarf-dandelion: 132–33
Dyschoriste: 96
Dyschoriste
  humistrata: 96
  oblongifolia: 96; **355**

Echinacea
  laevigata: 122
  pallida: 122; **441**
  purpurea: 122
Echinochloa
  crus-galli: 144
  walteri: 144; **510**
Echinocystis lobata: 101; **374**
Eclipta: 120–21
Eclipta
  alba: 120–21
  prostata: 120–21; **436**
Eichhornia crassipes: 159–60; **557**
Eleocharis tuberculosa: 153–54; **538**
Elephantopus
  carolinianus: 105; **383**
  elatus: 105
  nudatus: 105
  tomentosus: 104; **382**
Elephant's-foot: 104–5
  Leafy-stemmed: 105
Eleusine indica: 149; **525**
Elf-orpine: 23
Elymus
  villosus: 151
  virginicus: 151; **531**
Enchanter's-nightshade, Alpine: 57–58
Endorina uniflora: 129
Epidendrum conopseum: 180; **630**
Epifagus virginiana: 94
Eragrostis
  campestris: 150
  elliottii: 150
  refracta: 150
  spectabilis: 150; **527**
Erianthus
  alopecuroides: 140
  brevibarbis: 140
  coarctatus: 140
  giganteus: 140

Erichtites hieraciifolia: 130; **473**
Erigeron: 103
  annuus: 116
  philadelphicus: 115; **418**
  pulchellus: 115
  quercifolius: 115
  strigosus: 115–16; **419**
  vernus: 115; **417**
Erigonum tomentosum: 2–3; **7**
ERIOCAULACEAE: 157
Eriocaulon decangulare: 157; **549, 550**
Erodium cicutarium: 42; **154**
Eryngium yuccifolium: 59; **215**
Erythrina
  arborea: 39
  herbacea: 39; **144**
Erythronium
  albidum: 165
  americanum: 164; **576**
  rostratum: 164
  umbilicatum: 164–65
Eupatorium, Pink: 106
Eupatorium
  album: 105
  aromaticum: 107
  capillifolium: 105; **384**
  coelestinum: 106; **388**
  compositifolium: 105
  dubium: 106; **386**
  fistulosum: 106; **389**
  incarnatum: 106
  leptophyllum: 105
  maculatum: 106
  perfoliatum: 106; **387**
  purpureum: 106
  rotundifolium: 105; **385**
  rugosum: 107
  urticaefolium: 107
Euphorbia
  ammannioides: 47
  cordifolia: 47
  corollata: 47; **173**
  cyathophora: 47; **174**
  dentata: 47
  heterophylla: 47
  polygonifolia: 47
  prostata: 47
  supina: 47
EUPHORBIACEAE: 45

Eurybia divaricata: 114; **415**
Eustachys
  glauca: 149
  petraea: 149; **524**
Euthamia
  graminifolia: 112
  leptocephala: 112
  minor: 112
  tenuifolia: 112; **406**
Evening-primrose: 57
  Showy: 56
  Small-flowered: 57
Evening-primrose Family: 55
Everlasting: 117–18

FABACEAE: 27
Facelis: 117
Facelis retusa: 117; **424**
False-asphodel: 161
False-dandelion: 134
False-foxglove: 91–92
  Downy: 92
  Hairy: 92
  Smooth: 92
False-garlic: 163
False-hoarhound: 105
False-indigo, Yellow: 29
False-loosestrife: 56
False-miterwort: 24
False-nettle: 1
False-pimpernel: 90–91
False-solomon's-seal: 165
Featherbells: 161
Fernleaf: 93
Feverfew, American: 119
Field-pansy: 51
Figwort, American: 87
Figwort Family: 86
Fimbristylis: 154
Fimbristylis
  caroliniana: 154
  castanea: 154; **539**
Fingergrass: 149
Fireweed: 130
Fire-wheel: 129
Fivefingers: 26
Flannel-plant: 86
Flatsedge: 152
Flat-topped Goldenrod: 112
Fleabane, Daisy: 115–16

Floating-heart: 66
Flowering-wintergreen: 45
Fly-poison: 162
Foamflower: 24
Foxtail
  Giant: 145
  Green: 144–45
Fragaria
  vesca: 26
  virginica: 25–26; **92**
Frasera caroliniensis: 65–66; **241**
Froelichia
  floridana: 6; **18**
  gracilis: 6
Fuirena, Leafless: 152
Fuirena
  hispida: 152
  pumila: 152; **534**
  scirpoidea: 152
  squarrosa: 152
FUMARIACEAE: 17
Fumitory Family: 17

Gaillardia: 129
Gaillardia
  aestivalis: 129
  lanceolata: 129
  picta: 129
  pulchella: 129; **469**
Galactia: 39
Galactia
  elliottii: 39; **146**
  floridana: 40
  glabella: 40
  minor: 40
  mollis: 40
  regularis: 39–40
  volubilis: 39–40; **147**
Galax: 61
Galax
  aphylla: 61
  urceolata: 61; **224**
Galearis spectabilis: 175; **612**
Galeopsis tetrahit: 80; **295**
Galinsoga: 127
Galinsoga
  ciliata: 127
  quadriradiata: 127; **463**
Galium, Purple: 100

Galium
  aparine: 100; **368**
  hispidulum: 100; **370**
  obtusum: 100
  tinctorium: 100; **369**
  virgatum: 100
Gamochaeta
  falcata: 118
  purpurea: 118; **426**
Gaura: 57
Gaura
  angustifolia: 57; **210**
  filiformis: 57
  longiflora: 57
Gentian
  Pale: 65
  Soapwort: 65
  Stiff: 65
Gentiana
  alba: 65
  catesbaei: 65
  decora: 65
  flavida: 65
  quinquefolia: 65
  saponaria: 65; **239**
  villosa: 65; **238**
GENTIANACEAE: 63
Gentian Family: 63
Gentienella
  quinquefolia var. *quinquefolia:* 65;
   240
GERANIACEAE: 41
Geranium, Wild: 41
Geranium
  carolinanum: 41; **152**
  columbinum: 41–42
  dissectum: 41
  maculatum: 41; **151**
  molle: 42
  pusillum: 42
  robertianum: 42; **153**
Geranium Family: 41
Gerardia: 91–92
Gerardia
  fasciculata: 91
  flava: 92
  laevigata: 92
  pectinata: 92
  pedicularia: 92
  setacea: 92

  tenuifolia: 91–92
  virginica: 92
Germander: 79
Giant-millet: 145
Gillenia
  stipulata: 25
  trifoliata: 25
Ginseng, Dwarf: 58
Ginseng Family: 58
Glandularia
  pulchella: 78; **285**
  tenuisecta: 78
Glaux maritima: 62–63; **229**
Glottidium vesicarium: 35; **129**
Glycine apios: 39
Gnaphalium: 117
  falcatum: 118
  helleri: 118
  obtusifolium: 117–18
  purpureum: 117, 118
Goat's-beard: 25
  False: 23
Goat's-rue: 35
Golden-aster: 109
  Grass-leaved: 109–10
Golden-club: 156
Goldenrod
  Field: 110
  Flat-topped: 112
  Seaside: 111
  Skunk: 111
  White: 111
  Wreath: 110–11
  Wrinkle-leaved: 111
Golden-seal: 12
Goodyera
  pubescens: 178; **624**
  repens: 178
Goosefoot: 5
Goosefoot Family: 5
Goosegrass: 149
Grass
  Bahia: 142
  Barnyard: 144
  Dallis: 142
  Gamma: 140
  Quaking: 150
  Smut: 147
  St. Augustine: 146
  Switch: 143

Grass (*cont'd.*)
  Vasey: 142
  Velvet: 148
  Water: 144
Grass Family: 138
Grass-of-Parnassus: 24–25
Grass-pink: 178–79
Gratiola: 89
  Creeping: 89
  Florida: 89
Gratiola
  aurea: 89
  brevifolia: 89
  floridana: 89; **328**
  neglecta: 89; **329**
  ramosa: 89; **327**
  virginiana: 89
  viscidula: 89
Green-and-gold: 119
Green-dragon: 157
Green-eyes: 119
Ground-cherry: 85
Groundnut: 39
Groundsel
  Golden: 130
  Hairy: 131
Gyrotheca tinctoria: 171

Habenaria
  ciliaris: 176
  clavellata: 175
  cristata: 176
  integra: 175
  lacera: 175
  leucophaea: 175–76
  nivea: 176
  peramoena: 176
  psychodes: 176
Harebell: 102
Hatpins: 157–58
Hawkweed
  Beaked: 135
  Hairy: 135
  Poor-robins: 135
Heal-all: 80
Heart-leaf: 2
  Large-flowered: 2
Hedge-nettle: 81
Hedyotis
  caerulea: 98

  canadensis: 98–99
  crassifolia: 98
  longifolia: 99
  michauxii: 98
  procumbens: 98
  purpurea: 98
Helenium, Spring: 128–29
Helenium
  amarum: 127, 128; **465**
  autumnale: 129; **468**
  brevifolium: 128
  flexuosum: 128; **466**
  latifolium: 129
  nudiflorum: 128
  parviflorum: 129
  pinnatifidum: 129
  tenuifolium: 128
  vernale: 127, 128–29; **467**
Helianthus
  angustifolius: 122; **443**
  atrorubens: 122–23; **444**
  debilis: 123; **446**
  decapetalus: 124; **449**
  longifolius: 122
  occidentalis: 123
  porteri: 123; **445**
  radula: 123
  saxicola: 123
  silphioides: 123
  strumosus: 123; **448**
  tuberosus: 123; **447**
Heliopsis
  helianthoides: 121
  scabra: 121
Helonias bullata: 161; **562**
Hemerocallis
  flava: 162
  fulva: 162; **568**
  lilioasphodelus: 162
Hemlock, Water: 60
Hemp-nettle: 80
Hempweed, Climbing: 107
Henbit: 81
Hepatica: 14
Hepatica
  acutiloba: 14
  americana: 14
  nobolis
    var. *acuta:* 14
    var. *obtusa:* 14; **48**

Herb-Robert: 42
Heron's-bill: 42
Herpestis rotundifolia: 90
Heterotheca
  adenolepsis: 110
  gossypina: 109
  graminifolia: 109–10
  latifolia: 110
  mariana: 109
  microcephala: 110
  pinifolia: 109
  subaxillaris: 110; **399**
Heuchera
  americana: 24
  longiflora: 24
  parviflora: 24
  villosa: 24; **87**
Hexalectris spicata: 180; **631**
Hexastylis
  arifolia: 2; **5**
  lewisii: 2
  naniflora: 2
  shuttleworthii: 2; **6**
Hibiscus, Pineland: 49
Hibiscus
  aculeatus: 49
  coccineus: 49
  grandiflorus: 49
  incanus: 49
  laevis: 49; **180**
  moscheutos: 49
    ssp. *lasiocarpos:* 49
    ssp. *moscheutos:* 49; **181**
  palustris: 49; **182**
Hieracium
  aurantiacum: 135; **492**
  caespitosum: 135
  florantinum: 135
  greenii: 135
  gronovii: 135; **491**
  megacephalum: 135
  pilosella: 136; **493**
  pratense: 135
  traillii: 135
  venosum: 135; **490**
Holcus lanatus: 148; **522**
HOLLUGINACEAE: 7
Honeysuckle Family: 100
Horse-balm: 84
Horse-gentian, Perfoliate: 100–101

Horse-mint
  Mountain: 83
  White: 83
Horse-nettle: 85
Houstonia
  caerulea: 98; **361**
  canadensis: 98–99
  longifolia: 99
  procumbens: 98; **362**
  purpurea: 98; **363**
  pusilla: 98
  serpyllifolia: 98
Husk-tomato: 85
Hybanthus concolor: 51; **187**
Hydrastis canadensis: 12; **41**
Hydrocotyle bonariensis: 58; **213**
Hydrolea, Tall: 76–77
Hydrolea
  affinis: 77
  corymbosa: 76–77; **281**
  ovata: 77
  quadrivalvis: 77; **282**
  uniflora: 77
HYDROPHYLLACEAE: 75
Hydrophyllum
  appendiculatum: 75
  canadense: 75; **277**
  macrophyllum: 76
  virginianum: 75–76
Hydrotrida caroliniana: 89–90
Hymenocallis
  caroliniana: 172; **600**
  crassifolia: 172
  floridana: 172
  occidentalis: 172
Hypericum
  drummondii: 51
  gentianoides: 51; **186**
  perforatum: 50; **185**
  punctatum: 51
Hypochaeris
  alata: 133
  brasiliensis: 133
  glabra: 133
  microcephala: 133
  radicata: 133; **484**
Hypoxis
  hirsuta: 172; **602**
  juncea: 172
  micrantha: 172

Hyptis: 84
Hyptis
  alata: 84; **311**
  mutabilis: 84
  radiata: 84

Ilysanthes monticola: 90
Impatiens
  biflora: 48
  capensis: 48; **177**
  pallida: 48
Indian-chickweed: 7
Indian-physic: 25
Indian-pink: 63
Indian-potato: 39
Indian-strawberry: 26–27
Indian-turnip: 156–57
Indigo, Hairy: 34
Indigofera hirsuta: 34; **125**
Ipomoea
  brasiliensis: 73
  coccinea: 73; **268**
  cordatotriloba var. *cordatotrilobata:*
    73–74; **271**
  hederacea: 73; **269**
  hederifolia: 73
  imperati: 72; **266**
  lacunosa: 74
  macrorhiza: 73
  pandurata: 73; **270**
  pes-caprae: 73; **267**
  purpurea: 72; **265**
  quamoclit: 73
  sagittata: 74
  stolonifera: 72
  trichocarpa: 73–74
Iresine: 6
Iresine rhizomatosa: 6; **19**
Iris
  Blue: 172–73
  Crested: 173
  Dwarf: 173
Iris
  cristata: 173; **604**
  prismatica: 173
  tridentata: 173
  verna: 173
  virginica: 172; **603**
Ironweed: 104

Isopyrum biternatum: 15
Isotria
  medeoloides: 177
  verticillata: 177
Iva
  angustifolia: 120
  microcephala: 119–20; **432**

Jack-in-the-pulpit: 156–57
Jacob's-ladder: 75
Jacquemontia: 72
Jacquemontia tamnifolia: 72; **263**
Jatropa stimulosa: 46
Jeffersonia diphylla: 16; **56**
Jewelweed: 48
Joe-pye-weed: 106
Joint-vetch: 36
JUNCACEAE: 160
Juncus
  effusus: 160
  roemerianus: 160; **560**
  tenuis: 160; **559**
Jussiaea
  decurrens: 55
  repens: 55–56
Justicia
  americana: 96; **357**
  ovata: 96–97

King-devil: 135
Knapweed, Spotted: 132
Knotgrass: 142–43
Knotweed, Japanese: 4–5
Kosteletzkya virginica: 49–50; **183**
Krigia
  biflora: 132–33
  cespitosa: 133
  dandelion: 132
  montana: 132; **483**
  occidentalis: 132
  oppositifolia: 133
  virginica: 132; **482**
Kuhnistera pinnata: 34

Lachnanthes
  caroliana: 171; **597**
  tinctoria: 171
Lactuca
  canadensis: 134; **487**

graminifolia: 134
  serriola: 134
Ladies'-tresses, Spring: 177–78
Lady-rue: 15–16
Lady's-slipper
  Pink: 174
  Showy: 175
  Yellow: 174–75
Lady's-thumb: 4
Lamb's-quarters: 5
LAMIACEAE: 78
Lamium
  amplexicaule: 81; **296**
  purpureum: 81
Larkspur, Dwarf: 13–14
Lathyrus
  hirsutus: 38
  latifolius: 38; **141**
  pusillus: 38
Leafcup, Yellow: 118
Lepidium
  campestre: 19
  virginicum: 19; **66**
Lespedeza
  Creeping: 37
  Hairy: 36–37
Lespedeza
  capitata: 37
  hirta: 36–37; **136**
  procumbens: 37
  repens: 37; **137**
  violacea: 37
Lettuce
  White: 134–35
  Wild: 134
Leucanthemum
  leucanthemum: 129–30
  vulgare: 129–30; **471**
Liatris
  aspera: 108; **394**
  elegans: 108
  gracilis: 108
  graminifolia: 107–8
  helleri: 108; **395**
  microcephala: 108
  pilosa var. *pilosa:* 107–8; **393**
  squarrosa: 108
Ligusticum
  canadense: 59; **218**

scothicum: 59
LILIACEAE: 161
Lilium
  bulbiferum: 164
  canadense: 164
  catesbaei: 164; **574**
  grayi: 164; **575**
  lancifolium: 164
  michauxii: 163; **572**
  philadelphicum: 164
  superbum: 164; **573**
  tigrinum: 164
Lily
  Bell: 164
  Carolina: 163–64
  Clinton: 165
  Orange: 164
  Pine: 164
  Turk's-cap: 164
Lily Family: 161
Lily-of-the-valley: 166
Linaria
  canadensis: 86
  floridana: 86
  vulgaris: 86–87; **319**
Lindernia
  monticola: 90; **333**
  saxicola: 91
Liparis
  liliifolia: 180; **629**
  loeselii: 180
Lippia: 78
Lippia
  lanceolata: 78
  nodiflora: 78
Listera
  australis: 178
  smallii: 178; **623**
Lithospermum
  canescens: 78
  carolinense: 77; **284**
Liverleaf: 14
Lizard's-tail: 1
Lizard's-tail Family: 1
Lobelia, Purple: 102–3
Lobelia
  cardinalis: 103; **379**
  elongata: 102; **378**
  glandulosa: 102–3

Lobelia (cont'd.)
 inflata: 103
 puberula: 103
LOGANIACEAE: 63
Logania Family: 63
Looking-glass, Venus': 102
Loosestrife
 Fringed: 62
 Lance-leaved: 62
 Purple: 54
 Swamp: 62
 Whorled: 62
Loosestrife Family: 54
Lophiola: 171
Lophiola
 americana: 171
 aurea: 171; **598**
Lotus corniculatus: 34; **124**
Lotus-lily: 11
Lotus-lily Family: 11
Lousewort: 93
Lovage: 59
 Scotch: 59
Lovegrass, Purple: 150
Love-vine: 71
Ludwigia
 alternifolia: 56; **205**
 decurrens: 55; **203**
 linearis: 56; **204**
 linifolia: 56
 peploides: 55–56
 uruguayensis: 56
Lupine
 Lady: 31
 Sundial: 32
Lupinus
 diffusus: 31
 perennis
  ssp. *gracilis:* 32
  ssp. *perennis:* 32; **116**
 villosus: 31; **115**
Lycopus
 americanus: 83–84; **308**
 asper: 84
Lysimachia
 ciliata: 62
 lanceolata: 62; **227**
 quadrifolia: 62; **226**

 terrestris: 62; **228**
 tonsa: 62
LYTHRACEAE: 54
Lythrum salicaria: 54; **199**

Macuillamia rotundifolia: 90
Madder, Field: 99–100
Madder Family: 98
Maianthemum
 canadense: 165; **580**
 racemosum: 165; **579**
Malaxis
 spicata: 179
 unifolia: 179; **627**
Mallow
 Bristly: 48
 Seashore: 49–50
Mallow Family: 48
MALVACEAE: 48
Mandrake: 16
Manfreda
 tigrina: 172
 virginica: 172; **601**
Mappia origanoides: 83
Marshallia: 128
Marshallia
 graminifolia var. *cynanthera:*
  127–28; **464**
 obovata: 128
 ramosa: 128
 tenuifolia: 127–28
Marsh-dayflower: 159
Marsh-fleabane: 116
Marsh-hay: 148
Marsh-mallow, Smooth: 49
Marsh-marigold: 12
Marsh-pink: 63–64
Martiusia mariana: 38
Matelea
 baldwyniana: 71
 carolinensis: 70; **258**
 decipiens: 70
 flavidula: 70
 floridana: 70
 gonocarpa: 71; **259**
 obliqua: 70–71
Mayapple: 16
May-pop: 53–54

Meadow-beauty
  Pale: 55
  Smooth: 54–55
  Yellow: 55
Meadow-beauty Family: 54
Meadow-parsnip: 59
Meadow-rue, Tall: 15
Melampyrum lineare: 93; **343**
Melanthera: 124
Melanthera
  hastata: 124
  nivea: 124; **450**
MELASTOMACEAE: 54
Melilotus
  albus: 32
  indicus: 32; **117**
  officinalis: 32
Melochia
  corchorifolia: 50; **184**
  hirsuta: 50
  spicata: 50
  villosa: 50
Melothria pendula: 101; **373**
Mentha arvensis: 84; **309**
MENYANTHACEAE: 66
Menyanthes trifoliata: 66; **242**
Mercury, Three-seeded: 46
Mertensia
  maritima: 77
  paniculata: 77
  virginica: 77; **283**
Mexican-clover: 99
Mikania
  cordifolia: 107
  scandens: 107; **392**
Milfoil: 129
Milk-pea, Climbing: 39–40
Milkweed
  Aquatic: 68
  Common: 68–69
  Curly: 68
  Four-leaved: 70
  Fragrant: 69
  Green: 70
  Red: 67
  Sandhill: 70
  Swamp: 69
  White: 67–68

Whorled-leaf: 68
Milkweed Family: 67
Milkwort
  Pink: 45
  Short: 44
  Tall: 44
Milkwort Family: 43
Millet, Italian: 145
Mimosa
  microphylla: 27; **99**
  nuttallii: 27
  quadrivalvis: 27
Mimulus
  alatus: 88
  ringens: 88; **326**
Mint
  Field: 84
  Horse: 82
  Stone: 83
Mint Family: 78
Minuartia
  caroliniana: 9; **31**
  uniflora: 9–10
Mitella diphylla: 24
Miterwort: 24
Moccasin-flower: 174
Mock-strawberry: 26–27
Modiola caroliniana: 48; **178**
Mollugo verticillata: 7; **21**
Monarda
  citriodora: 82
  didyma: 82; **300**
  punctata: 82; **301**
Moneses uniflora: 60; **221**
Monkey-flower: 88
Monocotyledons: 137
Monotropa
  hypopithys: 60–61; **222**
  uniflora: 61
Monotropsis odorata: 61
Morning-glory
  Coastal: 73–74
  Common: 72
  Fiddle-leaf: 72
  Ivyleaf: 73
  Red: 73
Morning-glory Family: 71
Mouse-bloodwort: 136

Mouse-ear: 136
Muhlenbergia, Pink: 146–47
Muhlenbergia
   capillaris var. *filipes:* 146–47
   filipes: 146–47; **519**
Mullein
   Moth: 86
   Woolly: 86
Murdannia
   keisak: 159; **554**
   nudiflorum: 159
Muricanda dracontium: 157
Mustard Family: 18

Nabalus trifoliatus: 134–35
Nama
   affinis: 77
   corymbosa: 76–77
   ovata: 77
Nasturtium officinale: 20
Nelumbo, Yellow: 11
Nelumbo
   lutea: 11; **37**
   nucifera: 11
NELUMBONACEAE: 11
Nettle, Stinging: 1
Nettle Family: 1
Never-wet: 156
Nightshade, Sticky: 85
Nightshade Family: 85
Nothoscordum
   bivalve: 163; **571**
   fragrans: 163
   gracile: 163
Nuphar lutea: 12; **40**
   ssp. *orbiculata:* 12
   ssp. *sagittifolia:* 12
Nutgrass, Yellow: 151–52
Nut-rush: 155
Nuttallanthus
   canadensis: 86; **318**
   floridanus: 86
NYMPHACEAE: 11
Nymphaea
   mexicana: 12
   odorata: 12; **39**
Nymphoides
   aquatica: 66; **243**

cordata: 66
lacunosa: 66

Obolaria virginica: 65; **237**
Oclemena reticulare: 112; **408**
October-flower: 5
Oenothera
   biennis: 57; **208**
   fruticosa: 56–57
      ssp. *glauca:* 56–57; **207**
   grandiflora: 57
   humifusa: 57
   laciniata: 57
   linifolia: 57
   parviflora: 57; **209**
   speciosa: 56; **206**
   triloba: 56
ONAGRACEAE: 55
Onion, Wild: 163
Orange-grass: 51
Orchid
   Cranefly: 179
   Greenfly: 180
   Green Woodland: 175
   Purple-fringed: 176
   Ragged: 175–76
   Showy: 175
   Snowy: 176
   White-fringed: 176
   Yellow-fringed: 176
ORCHIDACEAE: 174
Orchid Family: 174
Orchis
   rotundifolia: 175
   spectabilis: 175
Ornithogalum
   nutans: 165
   thyrsoides: 165
   umbellatum: 165; **577**
Orobanche
   minor: 94
   uniflora: 94; **347**
Orontium aquaticum: 156; **545**
Orpine Family: 22
OXALIDACEAE: 42
Oxalis
   acetosella: 43
   corymbosa: 43

debilis var. *corymbosa:* 43
grandis: 42
martiana: 43
montana: 43; **157**
priceae ssp. *colorea:* 42
recurva: 42
stricta: 42; **155**
violacea: 42–43; **156**

Paint-brush
Devil's: 135
Indian: 93
Painted-leaf: 47
Panax trifolius: 58; **212**
Panicum, Seaside: 143–44
Panicum
amarum: 143–44; **508**
virgatum: 143; **507**
PAPAVERACEAE: 17
Parnassia
asarifolia: 24; **88**
caroliniana: 25
grandifolia: 24–25
Parthenium
auriculatum: 119
integrifolium: 119; **431**
Partridge-pea: 28
Small-flowered: 29
Paspalum
dilatatum: 142
distichum: 142–43
notatum: 142; **506**
urvillei: 142; **505**
Passiflora
incarnata: 53–54; **197**
lutea: 54
PASSIFLORACEAE: 53
Passion-flower: 53–54
Passion-flower Family: 53
Pea, Everlasting: 38
Pedicularis
canadensis: 93; **344**
lanceolata: 93
Peltandra
glauca: 156
sagittifolia: 156; **546**
virginica: 156
Pencil-flower: 36

Pennisetum
americanum: 145; **514**
glaucum: 145
pennisetum: 145
Pennywort: 65
Asiatic: 58
Seaside: 58
Penstemon
australis: 87; **323**
brittonorum: 88
calycosus: 88
canescens: 88; **324**
dissectus: 88
laevigatus: 88; **325**
multiflorus: 88
penstemon: 88
smallii: 88
tubiflorus: 88
Peppergrass: 19
Pepper-root: 20
Pepperwort: 19
Peramium pubescens: 178
Persicaria hydropiperoides: 4
Peruvian-daisy: 127
Petalostemon
caroliniensis: 34
gattingeri: 34
pinnatus: 34
purpureus: 34–35
Phacelia: 76
Fringed: 76
Phacelia
bipinnatifida: 76; **278**
dubia: 76; **280**
fimbriata: 76
maculata: 76
purshii: 76; **279**
ranunculacea: 76
Phaethusa virginica: 124
Phlox
Annual: 75
Blue: 74
Smooth: 75
Summer: 75
Thick-leaf: 74
Phlox
amplifolia: 75
carolina: 74; **273**

Phlox (cont'd.)
  divaricata: 74; **272**
  drummondii: 75; **275**
  glaberrima: 75
  latifolia: 74
  maculata: 74
  paniculata: 75; **274**
  stolonifera: 74
Phlox Family: 74
Phragmites
  australis: 149; **526**
  communis: 149
Phyla
  lanceolata: 78
  nodiflora: 78; **287**
Physalis
  heterophylla: 85; **312**
  viscosa: 85
Physostegia
  angustifolia: 80
  intermedia: 80
  purpurea: 80
  virginiana: 80; **294**
Phytolacca
  americana: 6
    var. *rigida:* 6
  rigida: 6; **20**
Pickerel-weed: 160
Pickerel-weed Family: 159
Pigweed: 5
Pimpernel, Water: 61
Pineapple Family: 158
Pine-sap: 60–61
Pineweed: 51
Pinguicula
  lutea: 94; **348**
  planifolia: 94
  pumila: 94; **349**
Pink
  Fire: 10
  Wild: 10
Pink Family: 8
Pipewort: 157–58
Pipewort Family: 157
Piriqueta: 53
Piriqueta
  caroliniana: 53
  cistoides ssp. *caroliniana:* 53; **196**
Pitcher-plant: 20–21

  Hooded: 21
  Parrot: 21
  Trumpet: 21
Pitcher-plant Family: 20
Pityopsis
  adenolepis: 110
  graminifolia: 109–10
    var. *tenuifolia:* 110
  pinifolia: 109; **398**
Pixie Family: 61
PLANTAGINACEAE: 97
Plantago
  aristata: 97
  lanceolata: 97; **360**
  major: 97; **358**
  maritima: 97–98
  rugelii: 97
  virginica: 97; **359**
Plantain
  Common: 97
  English: 97
  Hoary: 97
Plantain Family: 97
Platanthera
  blephariglottis var. *blephariglottis:*
    176; **617**
  ciliaris: 176; **616**
  clavellata: 175; **613**
  cristata: 176
  grandiflora: 176
  integra: 175
  lacera var. *lacera:* 175; **614**
  leucophaea: 175–76
  nivea: 176; **618**
  peramoena: 176
  psycodes: 176; **615**
Pluchea
  camphorata: 116
  foetida: 116; **421**
  odorata: 116; **420**
  purpurascens: 116
  rosea: 116
Plumegrass, Sugarcane: 140
Poa annua: 150–51; **530**
POACEAE: 138–40
Podophyllum peltatum: 16; **55**
Pogonia
  Large Whorled: 177
  Small Whorled: 177

Pogonia ophioglossoides: 177; **619**
Poinsettia, Wild: 47
Poinsettia heterophylla: 47
Pokeweed: 6
Polemonium reptans: 75; **276**
Polianthes virginica: 172
Polygala: 45
  Fringed: 45
  Large-flowered: 44
  Slender: 45
Polygala
  balduinii: 44
  brevifolia: 44–45
  cruciata: 44; **164**
  curtissii: 45; **165**
  cymosa: 44; **162**
  grandiflora: 44; **161**
  incarnata: 45; **166**
  lutea: 44; **160**
  mariana: 45
  nana: 43; **159**
  paucifolia: 45; **167**
  polygama: 44
  ramosa: 44; **163**
  ramosior: 44
  sanguinea: 45
POLYGALACEAE: 43
POLYGONACEAE: 2
Polygonatum
  biflorum: 166; **581**
  canaliculatum: 166
  pubescens: 166
Polygonella
  croomii: 5
  gracilis: 5
  polygama: 5; **15**
Polygonum
  arifolium: 4
  bicorne: 4
  cespitosum: 4
  cuspidatum: 4–5; **14**
  densiflorum: 4
  hydropiperoides: 4; **13**
  lapthifolium: 4; **11**
  longistylum: 4
  pensylvanicum: 4; **12**
  persicaria: 4
  sachalinene: 5
  sagittatum: 3; **10**

  setaceum: 4
Polymnia
  canadensis: 118
  laevigata: 118
  uvedalia: 118
POLYMONIACEAE: 74
Polypremum: 63
Polypremum procumbens: 63; **232**
Pond-lily, Yellow: 12
Pontederia
  cordata: 160; **558**
  lanceolata: 160
PONTEDERIACEAE: 159
Pony-foot: 71
Poppy Family: 17
Porteranthus
  stipulatus: 25
  trifoliatus: 25; **90**
Portulaca, Hairy: 8
Portulaca
  coronata: 8
  grandiflora: 8
  oleracea: 8; **26**
  pilosa: 8; **25**
  smallii: 8
  umbraticola ssp. *coronata:* 8
PORTULACACEAE: 7
Potentilla
  anserina: 26; **95**
  argentea: 26
  canadensis: 26; **93**
  erecta: 26; **94**
  intermedia: 26
  simplex: 26
Prenanthes trifoliata: 134–35; **489**
Primrose Family: 61
Primrose-willow: 55–56
PRIMULACEAE: 61
Prunella
  laciniata: 80
  vulgaris: 80; **293**
Pseudognaphalium
  helleri ssp. *helleri:* 118
  obtusifolium ssp. *obtusifolium:*
    117–18; **425**
Pterocaulon
  pycnostachyum: 117
  undulatum: 117
  virgatum: 117; **422**

Puccoon: 77–78
Puncture-weed: 43
Purple-tassels: 34–35
Purslane, Common: 8
Purslane Family: 7
Pussy-toes: 117
Pycnanthemum
  albescens: 83
  incanum: 83; **305**
  montanum: 83; **306**
  pycnanthemoides: 83
Pylostachya
  balduinii: 44
  cymosa: 44
  lutea: 44
  nana: 43
  ramosa: 44
PYROLACEAE: 60
Pyrrhopappus carolinianus: 134; **488**

Quaker-ladies: 98
Quamoclit
  coccinea: 73
  hederifolia: 73
  vulgaris: 73
Queen-Anne's-lace: 60
Queen's-delight: 46

Rabbit-bells: 31
Rabbit-tobacco: 117–18
Ragweed
  Common: 120
  Giant: 120
Ragwort, Southern: 131
Railroad-vine: 73
Rain-lily: 171
Ramps: 163
RANUNCULACEAE: 12
Ranunculus
  abortivus: 15; **51**
  allegheniensis: 15
  bulbosus: 14; **49**
  carolinianus: 14–15
  fascicularis: 14; **50**
  hispidus var. *nitidus:* 14–15
  micranthus: 15
Ratibida
  columnifera: 122
  pinnata: 122; **442**

Rattlesnake-master: 59, 172
Rattlesnake-plantain, Downy: 178
Rattlesnake-weed: 135
Redroot: 171
Redstem-filaree: 42
Redtop: 147
Reed
  Common: 149
  Giant: 149
Rhexia
  alifanus: 54–55; **200**
  cubensis: 55
  lutea: 55; **202**
  mariana var. *mariana:* 55
  nashii: 55; **201**
  nuttallii: 55
  petiolata: 55
Rhinanthus
  crista-galli: 93
  minor ssp. *minor:* 93; **345**
Rhynchosia, Erect: 40–41
Rhynchosia
  difformis: 40
  erecta: 40
  latifolia: 40
  michauxii: 41
  reniformis: 40–41
  tomentosa: 40; **148**
Rhynchospora
  colorata: 155
  corniculata: 154; **540**
  latifolia: 155; **541**
  macrostachya: 154–55
Richardia
  brasiliensis: 99
  scabra: 99; **364**
Robins-plantain: 115
Rock-portulaca: 7
Rorippa nasturtium-aquaticum: 20; **70**
ROSACEAE: 25
Rose Family: 25
Roseling: 159
Rose-mallow: 49
  Swamp: 49
Roseroot: 22–23
Rosin-weed, Starry: 118–19
Rotala ramosior: 54; **198**
RUBIACEAE: 98
Rubus chamaemorus: 25; **91**

Rudbeckia
   hirta: 121; **438**
   laciniata: 121–22; **440**
   maxima: 121
   mollis: 121
   nitida: 121; **439**
Rue-anemone: 15
Ruellia: 96
Ruellia
   caroliniensis: 96; **356**
   humilis: 96
Rumex
   acetosella: 3
   crispus: 3; **9**
   hastatulus: 3; **8**
   obtusifolius: 3
   pulcher: 3
Rush
   Black: 160
   Footpath: 160
Rush Family: 160
Rustweed: 63
Ryegrass, Wild: 151

Sabatia
   Rose-pink: 64
   Upland: 64
   White: 64
Sabatia
   angularis: 64; **234**
   bartramii: 64
   brachiata: 64
   calycina: 64
   capitata: 64; **235**
   difformis: 64; **236**
   dodecandra: 63; **233**
      var. *dodecandra:* 63
      var. *foliosa:* 63
   gentianoides: 64
   macrophylla
      var. *macrophylla:* 64
      var. *recurvans:* 64
   quadrangula: 64
Sabulina
   brevifolia: 9–10
   caroliniana: 9
Saccharum
   alopecuroides: 140
   brevibarbe: 140

   giganteum: 140; **499**
Sacciolepis
   indica: 144
   striata: 144; **509**
Sage
   Blue: 81
   Lyre-leaved: 81–82
Sagittaria, Narrow-leaved: 138
Sagittaria
   engelmanniana: 138
   graminea: 138; **497**
   latifolia: 138; **496**
   longirostra: 138
   montevidensis: 138
   subulata: 138
Salt-hay: 148–49
Salvia
   azurea: 81; **298**
   coccinea: 82
   lyrata: 81–82; **299**
Samolus parviflorus: 61; **225**
Sandbur
   Dune: 145–46
   Southern: 146
Sandwort: 9–10
Sanguinaria canadensis: 17; **59**
Saponaria officinalis: 11; **36**
Sarothera
   drummondii: 51
   gentianoides: 51
Sarracenia
   alata: 21; **76**
   drummondii: 20–21
   flava: 21
   leucophylla: 20–21; **73**
   minor: 21; **75**
   oreophila: 21
   psittacina: 21; **74**
   purpurea: 21
   rubra: 21
   sledgei: 21
SARRACENIACEAE: 20
SAURURACEAE: 1
Saururus cernuus: 1; **1**
Saxifraga
   aizoides: 24; **85**
   careyana: 23–24
   caroliniana: 24
   michauxii: 23; **83**

Saxifraga (*cont'd.*)
  micranthidifolia: 23
  virginiensis: 23; **84**
SAXIFRAGACEAE: 23
Saxifrage
  Early: 23–24
  Mountain: 23
  Yellow Alpine: 24
Saxifrage Family: 23
Schenoplectus americanus: 153; **537**
Schizachyrium: 140
Schizachyrium scoparium: 141–42; **503**
    var. *littoralis:* 142
Schrankia
  microphylla: 27
  nuttallii: 27
Scirpus
  americanus: 153
  cyperinus: 153; **536**
  robustus: 153; **535**
Scleria
  reticularis: 155
  triglomerata: 155; **542**
SCROPHULARIACEAE: 86
Scrophularia lanceolata: 87; **320**
Scutellaria
  elliptica: 79, 80; **290**
  incana: 79
  integrifolia: 79; **291**
    var. *hispida:* 79
    var. *integrifolia:* 79
  montana: 80; **292**
  multiglandulosa: 79
  parvula: 80
Sea-milkwort: 62–63
Sea-oats: 150
Sea-purslane: 7
Sea-rocket: 19
Sedge: 155–56
  Fraser: 156
  Whitetop: 155
Sedge Family: 151
Sedum: 22
Sedum
  glaucophyllum: 22
  nevii: 22
  pulchellum: 22; **79**
  pusillum: 22
  rosea: 22–23; **80**

  smallii: 23
  ternatum: 22; **78**
Seedbox: 56
Senecio
  anonymus: 131; **476**
  aureus: 130; **474**
  glabellus: 130
  millefolium: 131
  obovatus: 130
  smallii: 131
  tomentosus: 131; **475**
Senna
  Coffee: 28
  Wild: 28
Senna
  herbicarpa: 28
  marilandica: 28; **100**
  obtusifolia: 28; **101**
  occidentalis: 28; **102**
  tora: 28
Sensitive-brier: 27
Sericocarpus asteroides: 112; **407**
Serinia oppositifolia: 133
Sesbania: 35
Sesbania
  exaltata: 35; **130**
  macrocarpa: 35
  vesicaria: 35
Sesuvium
  maritimum: 7
  portulacastrum: 7; **22**
Setaria
  corrugata: 145
  geniculata: 145
  glauca: 145
  italica: 145
  lutescens: 145
  magna: 145; **513**
  parviflora: 145; **512**
  viridis: 144–45; **511**
Shamrock: 33
Shepherd's-needle: 125–26
Sherardia arvensis: 99–100; **367**
Shinleaf, One-flowered: 60
Shinleaf Family: 60
Shooting-star: 63
Sibaria virginica: 20
Sickle-pod: 28
Sida: 48

Sida
  acuta: 48
  carpinifolia: 48
  elliottii: 48
  rhombifolia: 48; **179**
  spinosa: 48
Silene
  baldwinii: 10
  caroliniana: 10; **34**
    var. *caroliniana:* 10
    var. *wherryi:* 10
  dichotoma: 10; **35**
  noctiflora: 10 – 11
  ovata: 10
  pensylvanica: 10
  polypetala: 10; **32**
  regia: 10
  rotundifolia: 10
  virginica: 10; **33**
Silphium
  asteriscus: 118 – 19
    var. *laevicaule:* 118 – 19;
      428
  dentatum: 118 – 19
  laevigatum: 119
  trifoliatum: 119
Silverrod: 111
Silver-weed: 26
Sisyrinchium
  albidum: 173; **606**
  angustifolium: 174
  atlanticum: 173 – 74; **607**
  brownei: 173
  exile: 173
  minus: 173
  rosulatum: 173
Skullcap
  Hairy: 79
  Large-flowered: 80
  Narrow-leaved: 79
Smallanthus uvedalius: 118; **427**
Smartweed, Dock-leaved: 4
Smilacina racemosa: 165
Snakehead: 87
Snakeroot
  Black: 13
  Small-leaved White: 107
  White: 107
Sneezeweed: 128, 129

Soapwort: 11
SOLANACEAE: 85
Solanum
  carolinense: 85; **313**
  dulcamara: 85; **315**
  rostratum: 85
  sisymbriifolium: 85; **314**
Solidago
  altissima: 110
  bicolor: 111; **403**
  caesia: 110 – 11
  canadensis: 110; **400**
  curtisii: 110 – 11; **401**
  lancifolia: 111
  odora: 112
  rigida: 112
  rugosa: 111; **402**
  sempervirens: 111; **404**
  spithamaea: 111; **405**
  stricta: 110
Solomon's-plume: 165
Solomon's-seal: 166
  False: 165
  Two-leaved: 165
Sonchus
  arvensis: 134
  asper: 133 – 34; **486**
  oleraceus: 134
Sorghastrum
  elliottii: 142
  secundum: 142; **504**
Sorrel, Wild: 3
Sow-thistle
  Common: 134
  Perennial: 134
  Spiny-leaved: 133 – 34
Spanish-moss: 158
Spanish-needles: 127
SPARGANIACEAE: 137
Sparganium
  americanum: 137; **495**
  androcladum: 137
  chlorocarpum: 137
  erectum ssp. *stoloniferum:* 137
Spartina
  alterniflora: 148
  bakeri: 149
  cynosuroides: 148; **523**
  patens: 148 – 49

Spartina (cont'd.)
  pectinata: 148
  spartinae: 148
Specularia perfoliata: 102
Speedwell
  Bird's-eye: 91
  Thyme-leaved: 91
Spider-lily: 172
Spiderwort: 159
Spiderwort Family: 158
Spigelia marilandica: 63; **231**
Spike-rush: 153–54
Spiny-pod: 70–71
Spiranthes
  cernua: 177; **621**
  longilabris: 178
  ovalis: 177
  vernalis: 177–78; **622**
Splitbeard: 141
Sporobolus
  indicus: 147; **520**
  poiretii: 147
Spreading-pogonia: 177
Spring-beauty: 7–8
Spurge: 46
  Flowering: 47
  Hairy: 47–48
  Prostrate: 47
Spurge Family: 45
Spurge-nettle: 46
Square-pod: 56
Squaw-root: 94
Squirrel-corn: 18
Stachys: 81
Stachys
  hyssopifolia: 81
  latidens: 81
  tenuifolia: 81; **297**
Star-grass, Yellow: 172
Star-of-Bethlehem: 165
Star-rush: 155
Stellaria
  corei: 8
  graminea: 9; **29**
  media: 9; **28**
  pubera: 8; **27**
Stemanthium gramineum: 161; **564**
Stenotaphrum secundatum: 146; **517**

STERCULIACEAE: 50
Sticky-cockle: 10–11
Stillingia sylvatica: 46; **172**
Stinging-nettle: 46
Stinkweed: 116
Stitchwort, Common: 9
Stomoisia
  cornuta: 95
  juncea: 95
Stonecrop
  Lime: 22
  Woods: 22
Stone-rush: 155
St. Peter's-wort, Common: 50–51
Strawberry, Wild: 25–26
Strophostyles
  helvula: 41
  leiosperma: 41
  umbellata: 41; **149**
Stylisma: 71–72
Stylisma
  aquatica: 71–72
  humistrata: 71
  patens: 71; **262**
    ssp. *angustifolia:* 71
    ssp. *patens:* 71
  villosa: 71–72
Stylophorum diphyllum: 17; **60**
Stylosanthes
  biflora: 36; **133**
  riparia: 36
Summer-farewell: 34
Sumpweed, Narrow-leaf: 119–20
Sundew: 22
Sundew Family: 22
Sundrops: 56–57
Sunflower
  Cucumber-leaved: 123
  Narrow-leaved: 122
  Pale-leaf: 123
  Purple-disc: 122–23
  Woodland: 124
Swamp-candles: 62
Swamp-pink: 161
Swertia caroliniensis: 65–66
Swordgrass: 153
Symphyotrichum
  dumosum var. *dumosum:* 113; **409**

ericoides: 114
lateriflorum: 114; **414**
novae-angliae: 113–14
novi-belgii var. *novi-belgii:* 113; **412**
patens: 114
pilosum var. *pilosum:* 113
premanthoides: 113
puniceum var. *puniceum:* 113; **410**
subulatum: 113; **411**
tenuifolium: 114; **413**
undulatum: 114–15
Syndesmon thalictroides: 15

Talinum: 7
Talinum
appalachianum: 7
calcaricum: 7
mengesii: 7
parviflorum: 7
teretifolium: 7; **23**
Tanacetum vulgare: 130; **472**
Tansy: 130
Taraxicum
erythrospermum: 133
laevigatum: 133
officinale: 133; **485**
Tear-thumb: 3–4
Tephrosia virginiana: 35; **128**
Tetragonotheca: 121
Tetragonotheca helianthoides: 121; **437**
Teucrium
canadense: 79; **289**
nashii: 79
virginicum: 79
Thalictrum
clavatum: 15–16; **54**
coriaceum: 16
dioicum: 15
macrostylum: 15
polygamum: 15
pubescens: 15; **53**
revolutum: 15
thalictroides: 15; **52**
Thermopsis: 29
Thermopsis
caroliniana: 29
fraxinifolia: 29; **106**
hugeri: 29

mollis: 29
villosa: 29; **105**
Thick-leaf Phlox: 75
Thistle
Bull: 131–32
Purple: 131
Yellow: 131
Thoroughwort: 106
Thyella tamnifolia: 72
Tiarella
cordifolia: 24; **86**
wherryi: 24
Ticklegrass: 147
Tickseed-sunflower: 126
Tickweed: 124
Tillandsia usneoides: 158; **551**
Tipularia
discolor: 179; **628**
uniflora: 179
Toadflax: 86
Toadshade
Purple: 166
Sessile: 166
Tofieldia
glabra: 161
glutinosa: 161
racemosa: 161; **561**
Toothcup: 54
Toothwort: 20
Touch-me-not: 48
Touch-me-not Family: 48
Tradescantia
hirsuticaulis: 159
ohiensis: 159
rosea: 159
virginiana: 159; **555**
Tragia urticifolia: 46; **170**
Tread-softly: 46
Trianthia
glutinosa: 161
racemosa: 161
Tribulus terrestris: 43; **158**
Trichostema
dichotomum: 79; **288**
lineare: 79
setaceum: 79
Trifolium
agaricum: 33

Trifolium (cont'd.)
  arvense: 32; **118**
  aureum: 33; **120**
  campestre: 32–33; **119**
  dubium: 33
  hybridum: 33; **122**
  incarnatum: 34
  pratense: 33; **123**
  procumbens: 32–33
  reflexum: 34
  repens: 33; **121**
  stoloniferum: 33
Trilisa
  odoratissima: 108–9
  paniculata: 109
Trillium
  Bent: 170
  Catesby's: 169
  Deceptive: 167–68
  Decumbent: 167
  Great White: 169
  Green: 168
  Least: 170–71
  Mottled: 166–67
  Narrowleaf: 167
  Nodding: 170
  Ozark Green: 168
  Painted: 168
  Pale Yellow: 168
  Persistent: 170
  Prairie: 167
  Red: 169
  Rose: 169
  Sessile: 166
  Southern Nodding: 170
  Stately: 167–68
  Stinking Benjamin: 169
  Stinking Willie: 169
  Sweet White: 170
  Twisted: 168
  Underwood: 167
  Vasey's: 169
  Yellow: 167
Trillium
  catesbaei: 169; **591**
  cernuum: 170
  cuneatum: 166; **583**
  decipiens: 167–68; **587**
  decumbens: 167; **586**

  discolor: 168; **589**
  erectum: 169; **593**
  flexipes: 170
  grandiflorum: 170; **594**
  lancefolium: 167; **584**
  luteum: 167; **585**
  maculatum: 166–67
  nivale: 168
  persistens: 170; **596**
  pusillum: 170–71
  recurvatum: 167
  reliquum: 167
  rugelii: 170; **595**
  sessile: 166
  simile: 170
  stamineum: 168; **588**
  underwoodii: 167
  undulatum: 168; **590**
  vaseyi: 169; **592**
  viride: 168
  viridescens: 168
Triodanis
  biflora: 102
  perfoliata: 102; **377**
Triosteum
  angustifolium: 101
  aurantiacum: 101
  perfoliatum: 100–101; **371**
Tripsacum dactyloides: 140; **498**
Trout-lily: 164–65
TURNERACEAE: 53
Turnera Family: 53
Turtlehead: 87
Twayblade: 180
  Kidney-leaf: 178
Twinleaf: 16
Typha
  angustifolia: 137
  domingensis: 137
  latifolia: 137; **494**
TYPHACEAE: 137

Umbrella-grass: 152
Umbrella-leaf: 16
Uniola paniculata: 150; **528**
URTICACEAE: 1
Urtica dioica: 1; **2**
Utricularia
  biflora: 96

cornuta: 95
gibba: 96
inflata: 95; **351, 352**
  var. *minor:* 95
juncea: 95; **350**
olivacea: 95
purpurea: 95; **353**
radiata: 95
subulata: 95–96; **354**
Uvularia
floridana: 162
grandiflora: 162
perfoliata: 162; **566**
puberula: 162; **567**
pudica: 162
sessilifolia: 162

Valerianella radiata: 101; **372**
Valerian Family: 101
VALERINACEAE: 101
Vanilla-plant: 108–9
Verbascum
blattaria: 86; **317**
phlomoides: 86
thapsus: 86; **316**
virgatum: 86
Verbena
  Moss: 78
  Stiff: 78
Verbena
rigida: 78; **286**
tenuisecta: 78
VERBENACEAE: 78
Verbesina
alternifolia: 124
occidentalis: 124; **451**
virginica: 124; **452**
Vernonia
acaulis: 104
altissima: 104
angustifolia: 104
flaccidifolia: 104
gigantea: 104; **380**
glauca: 104; **381**
noveboracensis: 104
pulchella: 104
Veronica
hederifolia: 91

persica: 91; **334**
serpyllifolia: 91; **335**
  var. *humifusa:* 91
  var. *serpyllifolia:* 91
Vervain Family: 78
Vesiculina purpurea: 95
Vetch
  Narrow-leaved: 37
  Smooth: 37
  Woods: 38
Vicia
acutifolia: 38
angustifolia: 37
caroliniana: 38; **140**
cracca: 37
dasycarpa: 37
hugeri: 38
sativa ssp. *nigra:* 37; **138**
villosa ssp. *nigra:* 37; **139**
Vigna: 41
Vigna luteola: 41; **150**
Viguiera: 123
Viguiera porteri: 123
Viola
affinis: 52
arvensis: 51
bicolor: 51; **188**
conspersa: 52
cucullata: 52
eriocarpa: 52
floridana: 52; **192**
glaberrima: 52
hastata: 51; **189**
lanceolata: 52; **191**
  ssp. *lanceolata:* 52
  ssp. *vittata:* 52
macloskeyi ssp. *pallens:* 53; **195**
nephrophylla: 53; **194**
pedata: 52–53; **193**
  var. *lineariloba:* 53
  var. *pedata:* 53
pensylvanica: 52
pubescens: 52
rafinesquei: 51
rostrata: 52; **190**
sororia: 52
tripartata: 52
walteri: 52
VIOLACEAE: 51

Violet
  Bird-foot: 52–53
  Florida: 52
  Green: 51
  Lance-leaved: 52
  Long-spurred: 52
  Northern Wetlands: 53
  Spear-leaved: 51–52
  Wild White: 53
Violet Family: 51
Virginia-cowslip: 77

Waldsteinia
  fragarioides
    ssp. *doniana:* 27; **97**
    ssp. *fragarioides:* 27
  lobata: 27
  parviflora: 27
Warea: 18
Warea
  amplexifolia: 18
  cuneifolia: 18; **65**
  sessilifolia: 18
Wart-cress: 19
Watercress: 20
Water-hyacinth: 159–60
Water-hyssop
  Blue: 89–90
  Smooth: 90
Waterleaf: 75–76
Water-lily: 12
Water-lily Family: 11
Water-nymph: 12
Water-pepper: 4
Water-plantain Family: 138
Water-primrose: 56
Water-shield: 11
Water-willow: 96–97
White-trumpet: 20–21
Wild-bean: 41
Wild-buckwheat: 2–3
Wild-cucumber: 101
Wild-ginger: 2
Wild-goldenglow: 126
Wild-indigo
  Blue: 30

Cream: 29–30
  Hairy: 31
  Pineland: 30
  White: 30
Wild-quinine: 119
Wildrice, Annual: 146
Windflower: 14
Wingstem: 124
Winter-cress: 19
Woodgrass
  Drooping: 142
  Elliott's: 142
Wood-lily: 165
Wood-poppy: 17
Wood-sage: 79
Wood-shamrock: 43
Wood-sorrel
  Northern: 43
  Violet: 42–43
  Yellow: 42

Xanthium
  echinellum: 120
  strumarium: 120; **435**
XYRIDACEAE: 157
Xyris ambigua: 157; **548**

Yarrow, Common: 129
Yellow-eyed-grass: 157
Yellow-eyed-grass Family: 157
Yellow-rattle: 93
Yellow-rocket: 19
Yerbadetajo: 120–21

Zephyranthes
  atamasca: 171; **599**
  candida: 171
  simpsonii: 171
Zigadenus densus: 161–62; **565**
Zizania aquatica: 146; **518**
Zizaniopsis miliacea: 146
Zizia
  aptera: 59; **216**
  aurea: 59
  trifoliata: 59
ZYGOPHYLLACEAE: 43

Photo Credits

Except as noted below, all photographs are by the authors. The authors gratefully acknowledge all who contributed photographs.

Plate 4	Bill E. Duyck	Plate 265	Carol L. Howel Gomez
Plate 7	Reed Crook	Plate 277	Hugh and Carol Nourse
Plate 11	Reed Crook	Plate 278	Carol L. Howel Gomez
Plate 40	Cheryl McCaffery	Plate 279	Christina Bird
Plate 43	Hugh and Carol Nourse	Plate 280	James Allison
Plate 44	Allan Armitage	Plate 282	Reed Crook
Plate 47	Hugh and Carol Nourse	Plate 286	Hugh and Carol Nourse
Plate 54	Hugh and Carol Nourse	Plate 290	Hugh and Carol Nourse
Plate 56	Hugh and Carol Nourse	Plate 292	Hugh and Carol Nourse
Plate 57	Bill E. Duyck	Plate 297	Hugh and Carol Nourse
Plate 58	Richard T. Ware	Plate 302	Hugh and Carol Nourse
Plate 61	Bill E. Duyck	Plate 303	James Allison
Plate 62	Alfred Schotz	Plate 338	Carol L. Howel Gomez
Plate 65	Michael Moore	Plate 343	Hugh and Carol Nourse
Plate 70	Allan Armitage	Plate 346	Carol L. Howel Gomez
Plate 77	Dennis Horn	Plate 348	Alfred Schotz
Plate 100	Carol L. Howel Gomez	Plate 355	Hugh and Carol Nourse
Plate 102	Carol L. Howel Gomez	Plate 357	Richard T. Ware
Plate 106	Richard T. Ware	Plate 395	Bill E. Duyck
Plate 119	University of Georgia Herbarium collection	Plate 398	James Allison
		Plate 405	Bill E. Duyck
Plate 126	Carol L. Howel Gomez	Plate 435	Reed Crook
Plate 127	Hugh and Carol Nourse	Plate 441	Allan Armitage
Plate 128	Alfred Schotz	Plate 453	Hugh and Carol Nourse
Plate 140	Carol L. Howel Gomez	Plate 464	Alfred Schotz
Plate 148	Richard T. Ware	Plate 466	Richard T. Ware
Plate 164	Richard T. Ware	Plate 544	Bill E. Duyck
Plate 172	Robert Wyatt	Plate 546	Hugh and Carol Nourse
Plate 173	Fred Mileshko	Plate 554	Reed Crook
Plate 212	Bill E. Duyck	Plate 562	Dennis Horn
Plate 215	Alfred Schotz	Plate 567	Carol L. Howel Gomez
Plate 222	Bill E. Duyck	Plate 575	Bill E. Duyck
Plate 235	Hugh and Carol Nourse	Plate 576	Hugh and Carol Nourse
Plate 237	Hugh and Carol Nourse	Plate 578	Hugh and Carol Nourse
Plate 238	Al Good	Plate 581	Fred Mileshko
Plate 241	Hugh and Carol Nourse	Plate 582	Hugh and Carol Nourse
Plate 248	Reed Crook	Plate 584	Dennis Horn
Plate 255	Hugh and Carol Nourse	Plate 590	Bill E. Duyck

Wilbur and Marion Duncan, the leading experts on the flora of the southeastern United States, have developed a useful new guide to identify wildflowers found east of the Mississippi. Richly illustrated with over 600 color photographs, *Wildflowers of the Eastern United States* describes more than 1,100 eastern species from Maine to northern Florida, including forbs, grasses, rushes, and sedges. More than 700 of these species also are found west of the Mississippi.

Wildflowers of the Eastern United States describes range, blooming season, and typical habitat for each species and includes a list of plants with unusual traits as a further aid to identification. The succinct descriptions together with color photographs provide the most definitive, recognizable characteristics necessary for positive identification. Few wildflower guides give such absolute diagnostic capability. General readers as well as professionals will find the book accessible, accurate, and easy to use. A glossary and line drawings define and illustrate botanical terminology, and the authors provide a brief guide to plant structure. Whether you are an amateur naturalist or a professional botanist, *Wildflowers of the Eastern United States* will be a welcome addition to your library, classroom, or backpack.

Wilbur H. Duncan and Marion B. Duncan
are coauthors of the *Smithsonian Guide to
Seaside Plants* and *Trees of the Southeastern
United States*. Wilbur H. Duncan is
coauthor, with Leonard E. Foote, of
*Wildflowers of the Southeastern United
States* and professor emeritus of botany at
the University of Georgia.